Progress in Mathematics
Volume 170

Series Editors
Hyman Bass
Joseph Oesterlé
Alan Weinstein

Alan L.T. Paterson

Groupoids, Inverse Semigroups, and their Operator Algebras

Birkhäuser
Boston • Basel • Berlin

Alan L.T. Paterson
Department of Mathematics
University of Mississippi
University, MS 38677

Library of Congress Cataloging-in-Publication Data

Paterson, Alan L.T., 1944-
    Groupoids, inverse semigroups, and their operator algebras / Alan
L.T. Paterson.
      p. cm.—(Progress in mathematics; v.170)
    Includes bibliographical references and index.
    ISBN 0-8176-4051-7 (hardcover: acid-free paper).
    1. Groupoids.   2. Inverse semigroups.   3. Operator algebras.
    I. Title.   II. Series: Progress in mathematics (Boston, Mass.);
vol. 170.
    QA181.P37   1998
    512'.2—dc21
                                     98-47617
                                         CIP

AMS Subject Classification: Primary—22A22, 46L05, 20M18, 58H05, 46L87
Secondary—20L15, 22D30, 43A07, 22A26, 43A65, 05B45, 52C22

Printed on acid-free paper.
© 1999 Birkhäuser Boston   *Birkhäuser*

ISBN 0-8176-4051-7
ISBN 3-7643-4051-7

Typeset in LaTeX by the author.
Printed and bound by Edwards Brothers, Inc., Ann Arbor, MI.
Printed in the United States of America.

9 8 7 6 5 4 3 2 1

To Christina
with love

... *als wahrhafte Unendlichkeit, in sich zurückgebogen wird deren Bild der Kreis, die sich erreicht habende Linie, die geschlossen und ganz gegenwärtig ist, ohne Anfangspunkt und Ende.*

G.W.F. Hegel, *Wissenschaft der Logik, Erster Teil*

Contents

Preface

In recent years, it has become increasingly clear that there are important connections relating three mathematical concepts which *a priori* seem to have nothing much in common. These are *groupoids, inverse semigroups* and *operator algebras*. The main (though not the only) objective of this book is to explore the connections between these three concepts. In the description below, we will, for ease of presentation, omit detailed references to the literature, leaving these in the main body of the text. We assume a basic knowledge of measure theory, functional analysis and operator algebras throughout the book.

The first chapter introduces groupoids and inverse semigroups, and motivates in a non-formal way the connections between groupoids, inverse semigroups and operator algebras developed in the text, discussing in particular what happens in the case of the Cuntz groupoid, the Cuntz inverse semigroup and the Cuntz C^*-algebra.

The notion of a *groupoid* seems to have originated with Brandt in 1927. It is most elegantly defined as *a small category with inverses*. In terms of algebra, a groupoid can intuitively be regarded as a set with a partially defined multiplication for which the usual properties of a group hold whenever they make sense. Of course, every group is a groupoid but there is a wide variety of naturally occurring groupoids that are not groups. For example, an equivalence relation R on a set X is a groupoid with the product given by $((x, y), (y, z)) \to (x, z)$ and inverse given by $(x, y)^{-1} = (y, x)$.

For any groupoid G, the *unit space* G^0 is defined (by analogy with the group case) to be the set $\{xx^{-1} : x \in G\}$. Groups are groupoids with exactly one unit. There are two natural maps $r, d : G \to G^0$ given by $r(x) = xx^{-1}, d(x) = x^{-1}x$. These maps are called respectively the *range* and *source* maps, the terminology being justified by thinking of a groupoid as a small category with inverses. The product ab of $a, b \in G$ is defined if and only if $r(b) = d(a)$.

A good way to see the need for groupoids is by observing that if we think of groups in terms of symmetries of a set, then these symmetries are defined on the *whole* of the set. But in many situations (for example, in differential geometry) the important "symmetry" is not one associated with globally defined bijections. A very simple example of this is given by the equivalence relation R above. Clearly the symmetry and the transitive properties of the relation qualify as symmetry conditions – the first is even called that name! – but there is in general no obvious group acting on X with orbit equivalence relation equal to R. As A. Weinstein pointed out ([273]), the algebraic structure characterizing symmetry in general – in particular, including *non-homogeneous* symmetry – is provided by groupoids.

The inability of the group concept to cover all important geometric symmetry – for example, in the case of the geometry of Riemannian manifolds – led to the recognition of the inadequacy of Klein's Erlangen Program. A discussion of this is given by Mark Lawson in Chapter 1 of his book [157]. An example of the inadequacy of the Klein program is illustrated by the transition from classical crystallography, which the program fits well for the most part, to the modern study of quasicrystals. This is discussed briefly in **4.2**, Example 5. The reader is referred to the article [273] of Alan Weinstein for the view of groupoids as "unifying internal and external symmetry".[1]

Many natural groupoids have been found to be essential for the study of a wide range of mathematics and mathematical physics. For example, in topology, there is the *fundamental groupoid*. A quotient of this groupoid, crucial in the study of foliations and their C^*-algebras, is the *holonomy groupoid*. (This groupoid will be treated in detail in **2.3**.) In analysis, one can mention transformation groupoids, the Cuntz groupoid, virtual groups and Toeplitz groupoids. In algebraic geometry, groupoids were used by Grothendieck to investigate moduli spaces. In physics, they have proved useful. Even in crystallography, at the level of microscopic symmetry, one has to consider *screw operators* (a rotation coupled with an admissible translation which at the macroscopic level gets suppressed and results in a pure rotation). In that context, instead of looking at the usual symmetry groups, one has to consider groupoids associated with these groups ([136, p.105ff.]). As we will see in Example 5 of Chapter 4, groupoids become even more fundamental in dealing with the noncommutative spaces arising within the contexts of quasicrystals and tilings.

Also deformation quantization is naturally interpreted in terms of Lie (smooth) groupoids, in particular, in terms of the gauge groupoid, and the tangent groupoid of A. Connes. In noncommutative geometry, groupoids are good candidates for noncommutative locally compact spaces and manifolds, and produce the C^*-algebras normally regarded as "noncommu-

---

[1] We can regard the external symmetry as determined by a *group* of automorphisms, and the internal symmetry (involving, for example, the local isotropy groups) as the additional symmetry captured by the groupoid structure. As an illustration (cf. [273]), a tiling of the bathroom floor (rather than of the whole plane) clearly does have some kind of symmetry. For example, in the middle of the floor, the pattern is very clear. But the symmetry of the tiling cannot be expressed in the group terms of external symmetry. (You have to take into account, for example, cutting round the bath and the pipes!) The groupoid techniques of internal symmetry *do* capture the symmetry of that situation.

tative" spaces. The noncommutative geometric context provides through groupoids the algebra of observables of a quantum mechanical system. In the study of Poisson manifolds, Lie groupoids and Lie algebroids play an esssential role.

We can motivate the second concept mentioned above, that of an *inverse semigroup*, by using a similar line of reasoning to that for groupoids. Since *globally* defined bijections on a set X are no longer adequate for describing symmetries, it is natural to replace them by partially defined "bijections", i.e. maps which are bijections from a subset of X onto another subset of X. It is reasonable to require a set of these, as in the group case, to have natural algebraic properties. When we do this, we obtain the second concept above, that of an *inverse semigroup*. This is a set of partial bijections on some X which is a semigroup under composition and is closed under inversion. (The composition of two maps is defined wherever it makes sense.)

Chapter 2 discusses inverse semigroups, locally compact groupoids and Lie groupoids.

In a little more detail, inverse semigroups can be defined abstractly (independently of any particular X), and as such, have been much studied in the algebraic theory of semigroups. In the topological and differential geometric context, they appear as *pseudogroups*, where the maps are taken to be homeomorphisms between open subsets of a topological space. (Extra differentiable properties are, of course, required in the differential geometric context.) There is a standard way of constructing a groupoid associated with a pseudogroup, i.e. that of taking the sheaf groupoid of *germs* of the maps in the pseudogroup. (We will look more closely at this in the study of localizations in Chapter 3. We note also that Chapter 4 investigates the construction of a canonical groupoid associated with an arbitrary inverse semigroup.) So we go from the pseudogroup, which is local in character, to the groupoid by "going to the points".

Inverse semigroups relate naturally to operator algebras through their semigroup algebras. Given an inverse semigroup S, the space $\ell^1(S)$ is, in the obvious way, a Banach *-algebra under convolution. The enveloping C^*-algebra of the semigroup algebra is the universal C^*-algebra $C^*(S)$ of S. As in the case of locally compact groups, there is also a regular representation of S on $\ell^2(S)$ which generates the *reduced C^*-algebra* $C^*_{red}(S)$ of S. The regular representation, regarded as a homomorphism on $\ell^1(S)$, is faithful. The von Neumann algebra generated by $C^*_{red}(S)$ is called $VN(S)$.

So inverse semigroups relate to operator algebras through the former's representations. The set up is parallel to representation theory for discrete groups, but the determination of the representation theory for interesting inverse semigroups is actually very difficult in general. To make progress, we have to move over to the groupoid context, where a powerful representation theory has been developed.

When we do move over from the inverse semigroup context to the group-oid context a new factor comes into play which ennables analysis techniques to be applied. *The groupoids associated with inverse semigroups have a topology.* One can see this in the case discussed earlier, that of the sheaf groupoid of germs of a pseudogroup. As a sheaf, it is more than just a groupoid – it has, of course, a topology. It is an example of a *locally compact groupoid*, one of a special class of locally compact groupoids called the *r-discrete* groupoids.

A locally compact groupoid is intuitively a groupoid which is a locally compact space for which the product and inversion maps are continuous. However extra technical conditions are required for obtaining a satisfactory theory – see below. For the immediate discussion, the reader can take the groupoid to be Hausdorff.

Inverse semigroups are most obviously present in the context of r-discrete groupoids alluded to above and which we now define. For any locally compact groupoid G, let G^{op} be the family of open subsets A of G such that the restrictions r_A, d_A of r, d to A are homeomorphisms onto open subsets of G. In the case of a group, G^{op} is just the set of singleton subsets of G and so can be identified with G. For an r-discrete groupoid G, the set G^{op} is an inverse semigroup where the product of two G^{op}-sets A, B is the set of all possible products ab with $a \in A, b \in B$, and $A^{-1} = \{a^{-1} : a \in A\}$. An *r-discrete* groupoid is a locally compact groupoid G where G^{op} is "very large" in the sense that G^{op} *is a basis for the topology of* G. The reason for the *r-discrete* terminology is that in such a groupoid, given any unit u, the set $G^u = \{x \in G : r(x) = u\}$ is a *discrete* subset of G.

While the locally compact groupoids naturally associated with inverse semigroups are the r-discrete groupoids, an important objective of the book is to give a discussion of *general* locally compact groupoids because of their great importance in fields such as analysis, differential geometry and non-commutative geometry. Indeed, the theory of locally compact groupoids is extensive, and its treatment in the present book does not claim to be comprehensive.[2]

To obtain a satisfactory theory of general locally compact groupoids, we need to address two further issues. First, just as we require a left Haar measure in order to construct the group algebra and hence the C^*-algebras of a locally compact group, so we require some corresponding measure theoretic structure on a locally compact groupoid G. This structure is that of a *left Haar system* $\{\lambda^u\}$, where each λ^u is a positive regular Borel

---

[2]The reader is referred to the book *Coordinates in Operator Algebra* ([179]) by Paul Muhly for a number of important topics and results not covered in the present work. In the differential geometric context, a regrettable omission from the present book is any discussion of symplectic groupoids and Poisson manifolds. References for these are [38, 140, 163, 267].

measure on G^u and (in the appropriate senses) the λ^u's vary continuously and invariantly.

It would be nice if the existence and uniqueness of a left Haar system for G followed from the axioms as in the locally compact group case. In fact, neither fact is true. There is a simple example of a locally compact groupoid which has no left Haar system. Trivial groupoids (equivalence relation groupoids $X \times X$ for a locally compact Hausdorff space X) in general have many different left Haar systems. (The situation is much better in the smooth case as we will see below.)

But many locally compact groupoids have natural left Haar systems. For example, in the r-discrete case, we can take each λ^u to be counting measure. Since left Haar systems are indispensable for analysis on groupoids, we incorporate the existence of a left Haar system into the definition of *locally compact groupoid*. With a left Haar system available, the space $C_c(G)$ of continuous, complex-valued functions on G with compact support becomes an involutive convolution algebra.

Secondly, many important locally compact groupoids that arise in practice are not Hausdorff. This applies to holonomy groupoids and to the r-discrete groupoids associated with inverse semigroups. Accordingly we have to allow for a locally compact groupoid to be *non-Hausdorff*. However, in all cases, there is enough of the Hausdorff property in the groupoid to make the topology manageable. The modifications needed to handle the non-Hausdorff case are not always trivial, and for this reason, the theory of locally compact groupoids is developed in the book without assuming the Hausdorff condition.

No treatment of locally compact groups could be regarded as complete without a discussion of Lie groups. The corresponding claim for groupoids is equally valid, and the last section of Chapter 2 is devoted to *Lie* groupoids. Such groupoids are of great importance in Poisson geometry and noncommutative geometry. A brief discussion of the differential geometry required in this section is given in Appendix F. A Lie groupoid is a groupoid that is a manifold for which the product and inversion maps are smooth (C^∞). For this to make sense, we require that G^0 (and the G^u's) be Hausdorff submanifolds, and that the range and source maps be submersions. After considering briefly the approach by Connes to convolution on a Lie groupoid, we turn to the appropriate notion of *left Haar system* for a Lie groupoid, that of a *smooth* left Haar system. For such a system $\{\lambda^u\}$, the λ^u's are required to be strictly positive, smooth measures and to "vary" smoothly. For groupoid purposes, smooth left Haar systems are all equivalent and so are effectively unique. Using densities and the Lie algebroid of G – the Lie algebroid of a Lie groupoid corresponds to the Lie algebra of a Lie group – we show that there always exists a smooth left Haar system on G. So at least Lie groupoids, like locally compact groups, have essentially unique (smooth) left Haar systems. In particular, every Lie groupoid is a locally compact groupoid.

We then examine three important classes of Lie groupoids. These are the holonomy groupoids of foliations, the r-discrete versions of the holonomy groupoids and the tangent groupoid of a manifold. These not only give very interesting examples of locally compact groupoids, but are of fundamental importance in noncommutative geometry, and the author hopes that readers will be encouraged to explore the use of these groupoids in the work of Connes. In the case of a foliation, inverse semigroups put in an appearance as well! In fact, the r-discrete holonomy groupoid is obtained from the germs of a certain inverse semigroup of partial diffeomorphisms acting on a submanifold of the given manifold which is transverse to the leaves. The theme of constructing an r-discrete groupoid from an inverse semigroup acting on a locally compact Hausdorff space is taken up again in the third section of Chapter 3.

In that chapter, we first consider representation theory for locally compact groupoids. This theory, described in the remarkable book of J. Renault, parallels the representation theory for a locally compact group and mediates the connection between such groupoids and operator algebras. In particular, the theory gives, for any locally compact groupoid G, the universal C^*-algebra $C^*(G)$ and the *reduced* C^*-algebra $C^*_{red}(G)$ of G.

The author believes that the technical difficulties of the theory (involving Hilbert bundles and quasi-invariant measures) is one of the reasons why groupoids are sometimes regarded as "off-putting". There is, of course, a balance to be struck, on the one hand, with being overly detailed so that the key ideas behind the main results are obscured, and, on the other hand, with stressing ideas at the expense of proof, so that the results become oracular. The author has tried to preserve this balance, and hopes that the approach adopted in the book will contribute to an appreciation of the intrinsic beauty and importance of groupoids, and help in overcoming a not uncommon psychological aversion to the concept that he himself initially experienced! The measure theoretic difficulties are ameliorated in the r-discrete case, and for this reason, that case is sometimes used to motivate and illustrate the general theory.

However, we have not given a proof of the key "disintegration" result of the representation theory of locally compact groupoids – that the representations of the convolution algebra $C_c(G)$ of a locally compact groupoid G are the integrated forms of the representations of G. This fundamental result is due to J. Renault and is a profound generalization of the theorem that for a locally compact group H, the *-representations of $L^1(H)$ are the integrated forms of the unitary representations of H. For a complete proof of the groupoid disintegration theorem, the reader is referred to the book [179] of Paul Muhly. Instead, the disintegration theorem is replaced by a related but different theorem for an r-discrete groupoid G, which is what is needed to relate the operator algebras of inverse semigroups and r-discrete groupoids. In this theorem, the groupoid representations of G are identified with a certain natural class of representations of a "large" inverse subsemigroup S of G^{op}. The proof of this theorem is inspired by an earlier proof of the disintegration theorem (for a special class of locally compact Hausdorff groupoids) by J. Renault, and relies on a result of A. Guichardet. Not surprisingly, the theorem also gives as a corollary the disintegration theorem for r-discrete, not necessarily Hausdorff, groupoids.

As indicated above, an inverse semigroup S relates to an r-discrete groupoid G when it sits as a subsemigroup in G^{op}. When this is done, the inverse semigroup naturally acts on the unit space G^0 by partial homeomorphisms. So to make further progress, we need to investigate inverse semigroup actions on a locally compact Hausdorff space X. This is done in the third section of Chapter 3. Inverse semigroup actions were already prefigured in the earlier discussion of the r-discrete holonomy groupoid. Such actions by inverse semigroups of partial homeomorphisms were studied in depth by A. Kumjian, who called them *localizations*. But we need to consider the case where an inverse semigroup acts on X without actually being given as an inverse semigroup of partial homeomorphisms. In this case, we construct an r-discrete groupoid $G(X, S)$ associated with the action. The action of S on X induces an action β of S on the C^*-algebra $C_0(X)$ in the natural way. (The elements of S act as isomorphisms on closed ideals of the C^*-algebra.) A theory of inverse semigroup covariance algebras has been developed by N. Sieben. Using this theory, we can identify the C^*-algebra of $G(X, S)$ with the covariance algebra $C_0(G^0) \times_\beta S$.

Using the theory of Chapter 3, we are then in a position to address in Chapter 4 the problem of determining the representation theory of an inverse semigroup S in terms of associated r-discrete groupoids. The r-discrete groupoids involved are of a special kind. They are *ample* groupoids. (The notion is due to J. Renault.) An ample groupoid is a locally compact groupoid for which the family G^a of *compact G^{op}-sets* is a basis for the groupoid. The set G^a is an inverse subsemigroup of G^{op}. This implies that the unit space of G is 0-dimensional. We discuss in **4.2** a number of examples motivating the connection between inverse semigroups and ample groupoids. These include the Toeplitz inverse semigroups and their associated Wiener-Hopf groupoids, as well as Cuntz and vertex inverse semigroups and groupoids.

We also discuss an example arising out of the physics of quasicrystals. Quasicrystals are modelled in terms of tilings, good examples of which are the *Penrose tilings*. J. Kellendonk has used an inverse semigroup S associated with such a tiling to construct an r-discrete groupoid G which can be regarded as describing the local structure of patterns of tiles. The

C^*-algebra of the r-discrete groupoid is the algebra of observables for discrete models of particle systems on the tiling, and Kellendonk obtains information about the gap labelling of the spectrum of the discrete version of the associated Schrödinger operator by using the K_0-group of this C^*-algebra. We obtain a closely related r-discrete groupoid as a reduction of a vertex groupoid. The unit space of each of these groupoids can be regarded as a "noncommutative space" of tilings. We also discuss a similar and motivating noncommutative space of Penrose tilings studied earlier by Connes.

An *S-groupoid* is effectively an ample groupoid G which has a homomorphic image of S sitting inside G^a which determines the topology and algebra of G. The key result of the third section of Chapter 4 is the construction of what can reasonably be called the *universal S-groupoid* $G_{\mathbf{u}}$ of S. The inverse semigroup S sits faithfully in $G_{\mathbf{u}}{}^a$, and $G_{\mathbf{u}}$ determines *every* S-groupoid in a natural way. We then discuss the motivating examples earlier in terms of the theory of *S*-groupoids, in particular, calculating explicitly the universal groupoids of the Cuntz inverse semigroups.

The relationship between the representation theories of inverse semigroups and locally compact groupoids turns out to have a very satisfactory solution in terms of the universal groupoid. In fact, we show that canonically, $C^*(S) \cong C^*(G_{\mathbf{u}})$ and $C^*_{red}(S) \cong C^*_{red}(G_{\mathbf{u}})$. This then makes available the representation theory of locally compact groupoids for determining the representation theory of inverse semigroups.

To illustrate the usefulness of these results, we prove in the final section of the book the following amenability result: *the left regular representation von Neumann algebra $VN(S)$ of an inverse semigroup S is amenable (injective) if every maximal subgroup of S is amenable.*

The book includes six appendices dealing with topics relating to those covered in the main text.

The author is grateful to the reviewers for their insightful suggestions and comments which substantially improved the text. He is also grateful to Ann Kostant of Birkhäuser, for her advice, encouragement and patience.

The great debt that the present work owes to the remarkable book and papers of Jean Renault will be apparent. The author also wishes to express his thanks to a number of mathematicians for their helpful, generous advice on some of the material presented here. These include John Duncan, Jerry Kaminker, Johannes Kellendonk, Mark Lawson, Paul Muhly, David Pask, John Quigg, Iain Raeburn, Arlan Ramsay and Alan Weinstein. He especially wishes to thank Paul Muhly for enthusiastically sharing with the author his remarkable insights into groupoids. Of course, the author takes full responsiblity for any errors in the text.

Alan L. T. Paterson
University of Mississippi

CHAPTER 1

Introduction

To save unnecessary repetition, throughout this work, unless the contrary is explicitly stated, all inverse semigroups are countable, all locally compact Hausdorff spaces have a countable basis, all Hilbert spaces are separable and all representations of *-algebras on Hilbert spaces are assumed non-degenerate.[1]

The objective of this chapter is to introduce groupoids and inverse semigroups and motivate a central theme of the book: that of the relationship between inverse semigroups, groupoids, and operator algebras. In particular, we will illustrate this relationship by discussing how the Cuntz inverse semigroup, the Cuntz groupoid and the Cuntz algebra are involved with one another. In the process, we will amplify some of the other book themes sketched in the Preface. We start with a general philosophical point.

In view of the complexity of operator algebras in general, it is clearly helpful when the algebra is derived from a simpler object, the latter determining significant properties of the former. In practice, one of the most important instances of this situation is when the operator algebra is generated by a representation of a "simpler object", that of a locally compact group G. Well-known, classical illustrations of the usefulness of knowing that the algebra is generated by a group are the following two results. The first ([72, p.320]) asserts that the left regular representation of a discrete group generates a factor if every conjugacy class of the group other than $\{e\}$ is infinite. The second ([245]), which we shall generalize later (in **4.5**)[2] asserts that the left regular representation of a discrete group generates an injective von Neumann algebra if and only if the group is amenable.

---

[1] So for us, a representation of a *-algebra A on a Hilbert space $\mathcal{H}$ is a *-homomorphism T from A into $B(\mathcal{H})$ such that the span of the vectors $T(a)(\xi)$, where $a \in A$ and $\xi \in \mathcal{H}$, is dense in $\mathcal{H}$.

[2] In the book, bold face (as in **4.5**) will be used for section references to distinguish them from equation references.

However there is large number of C^*-algebras of great interest which do not seem to be naturally associated with a group. These include the Cuntz algebras, the Cuntz-Krieger algebras, Path algebras, Wiener-Hopf algebras, the C^*-algebras arising from group partial actions, the C^*-algebras associated with inverse semigroups, the C^*-algebras of foliations, the C^*-algebras associated with quantization deformations and the C^*-algebras arising from tilings ([56, 65, 66, 94, 118, 141, 142, 143, 144, 180, 187, 192, 194, 197, 230]). For most of these algebras, it is not immediately clear what simpler object to substitute for the group G of the preceding paragraph.

It will turn out that for most of the above algebras (and many more), there are in fact *two* such objects in terms of which their structure can be elucidated. The first of these is called an *inverse semigroup*. This is a purely algebraic object (no topology or measure theory involved) which is a natural generalization of a group and has been much studied in the algebraic theory of semigroups (eg. [50, 51, 133, 202]). The second is both algebraic and topological and carries an analogue of Haar measure. It is that of a *locally compact groupoid* ([230, 179]).

For present motivational purposes, let us consider the Cuntz algebra O_n $(1 \leq n < \infty)$. Cuntz ([65, 69]) considers isometries $s_1, \ldots, s_n$ on a Hilbert space $\mathcal{H}$ such that

$$\sum_{i=1}^{n} s_i s_i^* = 1. \tag{1.1}$$

(It is easy to construct examples of such s_i.) He showed that the s_i's generate a simple, purely infinite C^*-algebra O_n independent of the choice of the generators s_i.

With the objective of relating this algebra to a simpler object, something like a group, we might propose to look at the multiplicative semigroup Σ_n generated by the s_i's and their adjoints. Unfortunately, addition is involved (as well as multiplication) in (1.1) and since addition is not available in Σ_n, it is not clear how one can usefully express (1.1) in the Σ_n context. However, Renault ([230, p.141f.]) introduced an elegant abstract semigroup closely related to Σ_n. Indeed, it easily follows from (1.1) that

$$s_i^* s_j = \delta_{ij} 1 \tag{1.2}$$

for all i, j, and the equalities of (1.2) involve only multiplication of the s_i's and their adjoints. Renault then defined the *Cuntz semigroup* to be the abstract semigroup S_n generated by a unit 1, a zero z_0 and elements s_i, t_i $(1 \leq i \leq n)$ subject to the relations

$$t_i s_j = \delta_{ij} 1. \tag{1.3}$$

(The t_j replace the adjoints s_j^* of the Hilbert space setting.) Clearly, in any product of s_j's and t_i's not equal to 1 or z_0, use of (1.3) shows that we can group all of the s_j's together followed by all of the t_i's grouped together. In fact, apart from 1, z_0, simple calculations of ([65]) give that the elements of S_n can be written uniquely in the form

$$s_\alpha t_\beta = s_{\alpha_1} \cdots s_{\alpha_r} t_{\beta_k} \cdots t_{\beta_1} \tag{1.4}$$

with $\alpha = (\alpha_1, \cdots, \alpha_r)$, $\beta = (\beta_1, \cdots, \beta_k)$ and $1 \leq \alpha_i, \beta_j \leq n$. Note the reverse order in which the t_{β_i} are written in the expression for t_β. Note also that one of the strings α, β is allowed to be empty. The product in S_n is determined by (1.4) and (1.3) – see (4.8) and (4.9) for the exact formulae. Inspired by the spatial version, one can define an involution on S_n by

$$(s_\alpha t_\beta)^* = s_\beta t_\alpha \tag{1.5}$$

and there is the natural representation of S_n taking S_n onto Σ_n, a representation which thus generates O_n.

A straightforward calculation using (1.4) and (1.3) shows that for any $s \in S_n$, there exists a unique element $t \in S$ – actually $t = s^*$ – such that both $sts = s$ and $tst = t$. *But this is precisely the definition of an* inverse semigroup!

So by definition, an *inverse semigroup* is ([133, p.129]) a semigroup S with the property that for each element $s \in S$, there exists a unique element $t \in S$ such that

$$sts = s \qquad tst = t. \tag{1.6}$$

We usually set $t = s^*$. Obviously, every group is an inverse semigroup with $s^* = s^{-1}$. It is immediate that if e is idempotent in S, then $e^* = e$ – so e is to be thought of as a "projection". The set $E(S)$ of idempotents of S is very important in inverse semigroup theory. Notice, for example, that for every $s \in S$, (1.6) gives $ss^* \in E(S)$.

For the convenience of the reader, the well-known, simple properties of inverse semigroups that we need will be proved in **2.1**. In particular, the map $s \rightarrow s^*$ is an involution on S and $E(S)$ is *commutative*. In fact, $E(S)$ is a semilattice. By definition, a semilattice is just a commutative idempotent semigroup T. The order on T is given by: $e \leq f$ if and only if $ef = e$, and for any $e, f \in E(S)$, we have $e \wedge f = ef$.

Inverse semigroups are most easily thought of in terms of partial one-to-one maps. Indeed, if Y is a set, then the set $\mathcal{I}(Y)$ of all partial one-to-one maps on Y (i.e. the set of one-to-one maps between subsets of Y) is an inverse semigroup in the natural way. The product of two such maps is just the composition of the two defined wherever it makes sense, and the involution operation is just that of inversion. Every inverse semigroup can be realized as an inverse subsemigroup of some $\mathcal{I}(Y)$. (This is the analogue of the Cayley theorem for groups and is due to Vagner and Preston.)

The Cayley theorem realizes a group as a permutation group acting on itself, and this action of the group, extended to an action on its ℓ^2-space, gives the *left regular representation* of the group. Similarly, the inverse semigroup version of Cayley's theorem also gives a *-representation of an inverse semigroup S on its ℓ^2-space. This representation (introduced by B. Barnes([11])) is fittingly called the *left regular representation* of S. Under this representation (and indeed *any* representation) of S, the elements of S are taken to *partial isometries* of the Hilbert space.

A partial isometry ([186, p.50ff.]) on a Hilbert space $\mathcal{H}$ is an operator $T \in B(\mathcal{H})$ such that TT^* is a projection. This is equivalent to T^*T being a projection so that T is a partial isometry whenever T^* is. Geometrically, T is a partial isometry on $\mathcal{H}$ if and only if it is isometric on $\ker(T)^\perp$, the latter being the range of T^*T. The range of T^*T is called the *initial subspace* of T while the range of TT^* is called the *final subspace* of T. So T is isometric from its initial subspace onto its final subspace and vanishes on the orthogonal complement of this initial subspace. On the other hand, T^* restricted to the final subspace is the inverse of the restriction of T to the initial subspace, and vanishes on the orthogonal complement of the final subspace. So if T is a partial isometry, then $TT^*T = T$ and $T^*TT^* = T^*$, exactly the inverse semigroup conditions of (1.6).

Indeed, in terms of operators on a Hilbert space, the natural analogue of a partial one-to-one map is a *partial isometry*. (Think, for example, of products of the unilateral shift and its adjoint as one-to-one maps on subsets of the canonical orthonormal basis.) Now every *-semigroup of partial isometries on a Hilbert space is, by the preceding paragraph, obviously an inverse semigroup. Conversely, every inverse semigroup can be realized as a *-semigroup of partial isometries on a Hilbert space (Proposition 2.1.4). This result, due to Duncan and Paterson ([78]), is an easy consequence of the existence of the left regular representation for S.

A result of Wordingham shows that this representation is actually faithful when extended canonically to the convolution algebra $\ell^1(S)$. In particular, this assures us that S has a good representation theory. The reduced C^*-algebra $C^*_{red}(S)$ is the C^*-algebra generated by the left regular representation of S. The universal C^*-algebra of S defined as for the group case is denoted by $C^*(S)$.

We have seen that, from an analytical point of view, an inverse semigroup S is readily thought of as a *-semigroup of partial isometries on a Hilbert space. However, to investigate the C^*-algebras associated with S, we need to have concrete information about the representation theory of S. Apart from the very accessible left regular representation, such information is difficult to obtain using inverse semigroup theory, and so despite the elegance and accessibility of this theory, we need another approach which will have built into it techniques for dealing with inverse semigroup representations. This approach is based on the notion of groupoids to which we now turn.

As discussed in the Preface, the full notion of symmetry is expressed in groupoid language. We also noted there that equivalence relations are groupoids. Given an equivalence relation, it is often important to obtain information about its set of equivalence classes. In practice, a direct approach to this set gives rise to serious problems. In such a situation, it is often more productive to remain with the equivalence relation as a groupoid and *not* take the quotient!

For example, Weinstein comments ([273]) that A. Grothendieck used groupoids extensively in his investigations into algebraic geometry, in particular using them to deal with the "difficult" equivalence relations arising from moduli spaces. In the analytical context, instead of having to deal with an equivalence relation whose family of equivalence classes may well have trivial (and hence useless) natural topological or measure theoretic structures, *we stay with the equivalence relation itself* and use its groupoid algebraic and analytical structure.

Somewhat along similar lines, in Mackey's theory of virtual groups ([170],) the virtual subgroup which replaces the stabilizer of a point in the case of orbit spaces that are not separated, is an *ergodic groupoid* for which operator algebras can be constructed using the groupoid algebra. (See, for example, the discussion in [219].) This is the source of groupoids in operator algebra theory.

In the context of noncommutative geometry, this leads to groupoids being regarded as "noncommutative" locally compact spaces (and manifolds), the intermediates which produce the other view of noncommutative spaces, viz. as C^*-algebras. Groupoids also arise naturally from physics.[3] Indeed Connes ([56, pp.33-39]) points out that Heisenberg's discovery of quantum mechanics effectively involved replacing the convolution algebra of an abelian group with a groupoid convolution algebra, viz. a matrix algebra. Further, deformation quantization is intepreted in terms of the *gauge groupoid* ([265, 63, 152, 263]). As discussed in [152], this groupoid is associated with the *tangent groupoid* of A. Connes. In his book, Connes shows, for example, how the tangent groupoid of a manifold provides a setting for deformations and the Atiyah-Singer index theorem. (We will discuss the construction of the tangent groupoid in **2.3**.)

Returning to the theme of this chapter, the connection of groupoids with the inverse semigroups discussed earlier was there right from the start. Indeed, the notion of *groupoid* was first introduced and named by H. Brandt in 1927 ([20]), and he associated with them a class of inverse semigroups

---

[3]An account of noncommutative geometry with particular relevance to physical theories is given in the book by Landi ([151]).

now called *Brandt semigroups*. (This is discussed by Clifford and Preston in [50, 3.3].) From the point of view of the following work, the quickest way to see the connection between groupoids and an inverse semigroup is through the groupoid G_S below associated with any inverse semigroup S.

Firstly, what is meant by a *groupoid*? The main intuition here is that a groupoid is *a set with a partially defined multiplication for which the usual properties of a group hold whenever they make sense*.

An elegant way to specify a groupoid is to define it as *a small category with inverses*. Indeed, let G be such a category. Since the category is "small", its objects form a set G^0, the set of *units* of G. The groupoid G is then identified with its set of morphisms, whose elements are "arrows" x from one object (the *source*[4] $d(x)$ of x) to another (the *range* $r(x)$ of x). Since the category G has inverses, every member x of G has an inverse x^{-1}. By considering identity morphisms, we see that the source and range maps d and r map G onto G^0. The set of pairs (x, y) in $G \times G$ for which xy is defined is denoted by G^2, which is called the set of *composable pairs* of the groupoid. Obviously, a product xy of elements x, y of G makes sense if and only if $d(x) = r(y)$, since the latter equality just says that the range of y is the same as the source of x so that the morphisms x, y compose.

Now every unit u can be identified with the identity morphism at u so that we can regard $G^0 \subset G$. The source and range maps, d, r are then given by:

$$d(x) = x^{-1}x \qquad r(x) = xx^{-1}. \tag{1.7}$$

Basic facts that follow immediately are: $xd(x) = x = r(x)x$, $d(x^{-1}) = r(x)$, $r(x^{-1}) = d(x)$, $x^{-1} = x$ if x is a unit, two units can be multiplied only if they coincide, $x^2 = x$ if and only if x is a unit, and if $xy = z$ in G, then $(y^{-1}, x^{-1}) \in G^2, z^{-1} = y^{-1}x^{-1}, r(x) = r(z), d(y) = d(z)$. These and similar elementary facts will be used without comment in the sequel. When confronted with such a fact, the reader is recommended to think of a groupoid in the above categorial terms, i.e. in terms of morphisms and their inverses, composition, domains and ranges. The fact should then be clear.

For each $u \in G^0$, let

$$G^u = \{x \in G : r(x) = u\} (= r^{-1}(\{u\})), \tag{1.8}$$
$$G_u = \{x \in G : d(x) = u\} (= d^{-1}(\{u\})). \tag{1.9}$$

We note that $G_u^u = G^u \cap G_u$ is closed under the product and inversion operations and satisfies the group axioms. The group G_u^u is called the *isotropy group* at u.

---

[4]Often the symbol s is used for the source map. However, like Renault ([230]), we have preferred to use d for this map, reserving s for an inverse semigroup element.

The G^u's and G_u's give (in general) different ways of fibering G over G^0, and much of analysis on groupoids consists in combining something on G^0 with something on the G^u's (or G_u's) to obtain something on G.

We also need to define groupoids from an axiomatic point of view. The reason for this is that we want to be able to think of a groupoid as being (like a locally compact group) an abstract *set* equipped with certain specified algebraic, topological, differential geometric and measure theoretic structures, analysis on groupoids having analysis on locally compact groups as a special, important, motivating case.

The axioms for the groupoid G are obtained by writing down the requirements for being a "small category with inverses" and removing redundancies. We now give the explicit list of axioms which will be used throughout this work. This list is given in the paper [115] of P. Hahn, where it is stated that it was suggested in a conversation by G. Mackey. (An equivalent list of axioms for a groupoid is given by A. Ramsay in [219, p.255].)

A groupoid is a set G together with a subset $G^2 \subset G \times G$, a product map $(a,b) \rightarrow ab$ from G^2 to G, and an inverse map $a \rightarrow a^{-1}$ (so that $(a^{-1})^{-1} = a$) from G onto G such that:

(i) if $(a,b), (b,c) \in G^2$, then $(ab,c), (a,bc) \in G^2$ and

$$(ab)c = a(bc); \qquad (1.10)$$

(ii) $(b, b^{-1}) \in G^2$ for all $b \in G$, and if (a,b) belongs to G^2, then

$$a^{-1}(ab) = b \qquad (ab)b^{-1} = a. \qquad (1.11)$$

The reader is invited to prove, using only the axioms, basic groupoid facts such as that $(x,y) \in G^2$ if and only if $d(x) = r(y)$ and that if $(x,y) \in G^2$, then $(y^{-1}, x^{-1}) \in G^2$ and $(xy)^{-1} = y^{-1}x^{-1}$. Such proofs (at least for the author) are not immediate, and this is why it is recommended that the category approach, in which the proofs *are* obvious, be used when dealing with such basic facts.

We will meet many examples of groupoids in the book, and important examples of groupoids were mentioned in the Preface. As commented in the Preface, any group is a groupoid (with the identity element as its only unit). Further, every set X is a groupoid G of units – take $G^2 = \{(x,x) : x \in X\}$ and $x^{-1} = x$. A much more interesting example of a groupoid is an equivalence relation R on a set X. Here, $R^2 = \{((x,y),(y,z)) : (x,y),(y,z) \in R\}$ and the product and inversion maps are given by $(x,y)(y,z) = (x,z)$ and $(x,y)^{-1} = (y,x)$. Groupoid equivalence relations with some additional structure are of great importance in von Neumann algebra theory. (For example, the Krieger factors are the groupoid regular von Neumann algebras associated with the orbit equivalence relations given by ergodic transformations (cf. [56, 465ff.]).)

A useful groupoid that arises in algebraic topology is the *fundamental groupoid* of a topological space X ([164, 125, 24]). For a fundamental *group*, one fixes a base point x_0 and looks at homotopy classes of paths in X beginning and ending at x_0. For the fundamental *groupoid*, paths are free to begin and end anywhere. The groupoid is then the set of homotopy classes $[\gamma]_h$ of paths γ in X, and the product and inversion for the groupoid are defined in the obvious way. So, for example, the product of two classes is defined when the terminal point of a representative of the first path is the initial point of a representative of the second. The source and range maps take a homotopy class $[\gamma]_h$ to the terminal and initial points of γ respectively. As we will see later, the holonomy groupoid of a foliated manifold, discussed in **2.3**, can be obtained as a quotient groupoid of a fundamental groupoid. It is left to the reader to check that the fundamental group $\pi(X, x_0)$ is just the isotropy group for the fundamental groupoid at the unit x_0.

Groupoids "stick together" in a very easy way. Any disjoint union $\cup_{\alpha \in A} G_\alpha$ of groupoids G_α is itself a groupoid, the inverse being the obvious operation, and the product being determined by the G_α-products on $G^2 = \cup_{\alpha \in A} G_\alpha^2$. In this respect, groupoids are much more flexible than groups. In particular, a bundle of groups is a groupoid, and so the tangent bundle of a manifold is a (Lie) groupoid. Another useful operation is that of the *reduction* G_T of a groupoid G by a subset T of its unit space G^0. Here, we define the *reduction* of a groupoid G by a subset T of G^0 as follows:

$$G_T = \{t \in G : r(t), d(t) \in T\}. \tag{1.12}$$

It is easy to check that G_T is a subgroupoid of G with unit space T.

The groupoid structure also fits in well with groupoid actions. Of particular importance is the case of the *transformation groupoid* (sometimes called the *transformation group groupoid*) G associated with a transformation group (X, H). We are given a set X on which a group acts invertibly on the right. The transformation groupoid G is defined to be $X \times H$. We take G^2 to be the set of all pairs $((x, h), (xh, k))$ and the product and inverse maps for G are given by:

$$(x, h)(xh, k) = (x, hk) \qquad (x, h)^{-1} = (xh, h^{-1}). \tag{1.13}$$

It is an easy exercise to check that G is a groupoid.

The transformation groupoid for a transformation group is a special case of a much more general construction in which the group H is replaced by a groupoid H'. Here, H' acts in a "fibered" way on the right on a set X and there is an associated groupoid $X * H'$. (A detailed discussion of this important construction is given by Muhly ([179]).) For our purposes, we won't require the details of this construction. But we will need the special case of it in which a groupoid G acts on itself, and so give the product and inversion maps explicitly for the resultant groupoid $G * G$.

Here, $G * G$ is just G^2 itself, the set of composable pairs of G ([115]). The reader is invited to prove, by checking the above groupoid axioms, that, if

$$(G^2)^2 = \{((x,y),(z,t)) \in G^2 \times G^2 : z = xy\},$$

then G^2 is a groupoid with product and inversion given by:

$$(x,y)(xy,w) = (x,yw) \quad (x,y)^{-1} = (xy,y^{-1}).$$

Since every group is a groupoid, we might wonder if inverse semigroups can be regarded as groupoids as well. The following very elementary result shows that an inverse semigroup is actually a groupoid G_S (when its multiplication is sutiably limited). In the case where $S = \mathcal{I}(X)$, the fact that G_S is a groupoid was noted by Ramsay ([219, p.256]). However the basic idea (of allowing products only when range and domain match) goes back to Ehresman in his work on pseudogroups and ordered groupoids. (See Example 2, **2.3**.)[5]

As a set, G_S is just the inverse semigroup S. The set of composable pairs G_S^2 is just $\{(s,t) \in S \times S : s^*s = tt^*\}$. The product map on G_S^2 is just the (restricted) product in S and for $s \in S$, we take $s^{-1} = s^*$. For $e \in E$, the idempotent semilattice of S, clearly $G_S^e = r^{-1}(\{e\}) = \{s \in S : ss^* = e\}$ and $(G_S)_e = d^{-1}(\{e\}) = \{s \in S : s^*s = e\}$.

Proposition 1.0.1 G_S *is a groupoid with unit space E.*

Proof. It follows directly from (1.6) that $(s^{-1})^{-1} = s$ for all $s \in S$. We check the groupoid conditions (1.10), (1.11) above for G_S. Let $(s,t),(t,v) \in G_S^2$. Then $(st)^*st = t^*s^*st = t^*tt^*t = t^*t = vv^*$ so that $(st,v) \in G_S^2$. Similarly, $(s,tv) \in G_S^2$. Since S is associative, $(st)v = s(tv)$ and this gives (1.10).

Let $s \in S$. Then $s^*s = s^*s^{**}$ and so $(s,s^*) \in G_S^2$. If $(s,t) \in G_S^2$, then $s^*(st) = s^*st = tt^*t = t$. Similarly, if $(w,s) \in G_S^2$, then $(ws)s^* = w$. This gives (1.11). Obviously, $G_S^0 = d(G_S) = \{s^*s : s \in S\} = E$.   □

The groupoids with which we will be concerned in this work are *locally compact*. These groupoids, which are discussed in **2.2**, are, in particular, *topological groupoids* in the sense that they are equipped with a topology for which multiplication and inversion are continuous. A technical complication is that such groupoids are not always Hausdorff, but there are enough

---

[5]A discussion of Ehresmann's work on ordered groupoids is given by Mark Lawson in his book [157].

locally compact Hausdorff sets in the groupoid to make the necessary analysis go through. For example, the unit space G^0 of such a groupoid is a locally compact Hausdorff subset, as is $G^u = \{a \in G : r(a) = u\}$ for every unit u of G. Using an idea of Connes, for such a groupoid, we can define a good version of $C_c(G)$ which in the Hausdorff case is just the usual space of continuous complex-valued functions on G with compact support. We also need the notion of a left Haar system which corresponds to left Haar measure in the locally compact group case.

To motivate this, note that left multiplication by $a \in G$ gives a bijection from the set $\{b \in G : r(b) = d(a)\} = G^{d(a)}$ onto the set $\{c \in G : r(c) = r(a)\} = G^{r(a)}$. What a left Haar system does is to assign a positive regular Borel measure λ^u supported by the locally compact Hausdorff space G^u, u ranging over the unit space of G, in such a way that left multiplication by any $a \in G$ takes $\lambda^{d(a)}$ onto $\lambda^{r(a)}$. This is the groupoid version of the invariance of left Haar measure in the group case. Regrettably, the topology for a locally compact groupoid, unlike that for a locally compact group, does not automatically determine an (effectively unique) left Haar system, and so such a system has to be presupposed. However, many locally compact groupoids do have (effectively unique) left Haar systems. This applies to r-discrete and Lie groupoids, both of which are discussed in the text. (The definition of *r-discrete groupoid* is given below.) In the r-discrete case, the canonical left Haar system is particularly simple, for each unit u, each λ^u in that case being just counting measure on G^u. In the Lie case, the λ^u's are smooth measures (i.e. locally equivalent to Lebesgue measure) and vary smoothly over the groupoid.

Given a left Haar system on a locally compact groupoid G, we can, as in the group case, form a convolution product and involution for the functions in $C_c(G)$ so that the latter becomes a *-algebra. A representation of G is defined with respect to a Hilbert bundle $\{H_u\}$ ($u \in G^0$) with an appropriately invariant probability measure on G^0, the representation taking each $a \in G$ to a unitary operator from $H_{d(a)}$ onto $H_{r(a)}$ and products in G to composition products of these unitary operators. (The measure determines the necessary notions of measurability and null sets in the precise definition (**3.1**).) J. Renault showed that, as in the group case, integrating up such a representation with respect to the left Haar system gives a *-representation of $C_c(G)$, the Hilbert space of the representation being the L^2-space of sections of the Hilbert bundle. Renault also showed that all representations of $C_c(G)$ are obtained in this way. (See Theorem 3.1.1 for the precise statement.)

Every locally compact groupoid G admits many "left regular representations", one for each probability measure on the unit space. When expressed in integrated form, these determine the *reduced C^*-algebra* of the groupoid, $C^*_{red}(G)$. This C^*-algebra is the groupoid version of the reduced C^*-algebra of a locally compact group. There is also a universal C^*-algebra $C^*(G)$ of the groupoid defined by completing $C_c(G)$ under its largest C^*-norm.

These C^*-algebras associated with a locally compact groupoid are of great practical importance. For in many situations where mathematicians are looking for a C^*-algebra, there is a natural locally compact groupoid present and the desired C^*-algebras come automatically from the groupoid by the representation theory discussed above. This occurs in the theory of deformation quantization, where we look for a C^*-algebra in which the deformation can take place. For example, for a manifold M, in the case where one wants to deform $C_0(T^*M)$ into the algebra $\mathcal{K}$ of compact operators, the tangent groupoid (discussed in **2.3**) gives the C^*-algebra. Another example is in the work of J.Kellendonk on the modelling of quasicrystals.

As noted above, the class of *r-discrete* groupoids plays an important role in the book. The precise definition of this class is, perhaps, a little strange at first sight, and it will be helpful for motivation to consider first the transformation groupoid $X \times H$ associated with a transformation group (X, H). Recall that H is a group with a right action on a set X, and the product and inversion maps are given by (1.13).

However, in addition, we now take X to be a locally compact Hausdorff space and H to be a discrete (countable) group with a continuous right action on X. Then $G = X \times H$ is a locally compact groupoid when G is given the product topology. The unit space G^0 is identified with X via the map $(x, e) \to x$ where e is the identity of H, and is an open subset of G. Each G^x is countable. Indeed if $x \in X$, then $G^x = \{(v, h) \in G : (v, h)(vh, h^{-1}) = r((v, h)) = (x, e)\} = \{x\} \times H$ and this is a discrete countable subset of G. We take the measure λ^x for the left Haar system to be counting measure. The C^*-algebra $C^*(G)$ of this groupoid is isomorphic to the crossed-product C^*-algebra $C_0(X) \times_\beta H$, where β is the left action of H on $C_0(X)$ determined by its corresponding right action on X.

Each element $h \in H$ can be realized as a "horizontal slice" of $G = X \times H$ by associating it with $A_h = X \times \{h\}$. These rather special subsets of G can be characterized in groupoid language using the range and source maps r and d. Indeed, using (1.13), we find that $r((x, h)) = x$ and $d((x, h)) = xh$. In particular, in the terminology of Renault ([230, p.10]), A_h is a *G-set*, where a subset of a groupoid is called a *G-set* when both r and d are one-to-one on the subset. Not only that, the restrictions of r and d to A_h are homeomorphisms from A_h onto open subsets (X in both cases) of X. By replacing X in the definition of A_h by any open subset of X, we obtain a basis for G such that the restriction of each of r, d to any basis member is a homeomorphism onto open subsets of X. This motivates the definition of an r-discrete groupoid.

For a (general) locally compact groupoid G, define G^{op} to be the family of open Hausdorff subsets U of G such that both $r \mid_U, d \mid_U$ are homeomorphisms onto open subsets of G. (In particular, the elements of G^{op} are G-sets.) The groupoid G is called *r-discrete* if G^{op} is a basis for the topology of G. (So in the r-discrete case, the family G^{op} is *large*.) In particular, the transformation groupoid $X \times H$ is r-discrete. As we shall see, the class of r-discrete groupoids is abundant. If G is r-discrete, then G^0 is open and every G^u is discrete. As commented earlier, a left Haar system is given by the counting measures on the G^u's.

The relevance of inverse semigroups to r-discrete groupoids can be motivated by the transformation groupoid $X \times H$, since any $U \times \{h\}$ can, via the agency of r and d, be regarded as the restriction of h to U, and this is a *homeomorphism between open subsets of* X, a partial one-to-one map, an element of the inverse semigroup $\mathcal{I}(X)$! Of course, general r-discrete groupoids (such as the holonomy groupoids considered in **2.3**) are more complex than transformation groupoids, but it turns out that the inverse semigroup aspect of G^{op} is very transparent if we just use the groupoid structure: for an r-discrete groupoid G, G^{op} *is an inverse semigroup under set multiplication in the groupoid, with involution given by set inversion.*

Multiplying two subsets of a discrete group will, of course, in general give a set much bigger than the ones that we started with, since every member of one set is allowed to multiply every member of the other set. But in the case when $A, B \in G^{op}$ the number of groupoid products that we can form are very strictly controlled. Indeed, both A, B are G-sets, so that given $a \in A$, there is at most one $b \in B$, namely the b (if there is one) whose range is the domain of a, for which ab is defined. (The only G-sets in a discrete group are the singletons or the empty set!) As we will see, the relationship between inverse semigroups and r-discrete groupoids is implemented by representing the semigroup as a subsemigroup of some G^{op}.

We saw above how a transformation group (X, H) gave rise to an r-discrete groupoid, its transformation groupoid. Since every r-discrete groupoid G is naturally associated with an *inverse semigroup* (viz. G^{op}), we might expect there to be an inverse semigroup version (X, S) of a transformation group which gives rise to an r-discrete groupoid. This, in fact, is the case, and as we shall see, there are many other naturally occurring examples of such pairs (X, S). Of particular interest are the pairs arising in foliation theory. In that case, S is the *holonomy pseudogroup* and X is a suitable transversal for the foliation. (This is discussed in **2.3**.)

Leading into the inverse semigroup version of (X, H), we first discuss what is meant by a *right action* of an inverse semigroup S on a set X. We then move on to the role played by the set $E(S)$ of idempotents of S in constructing such X's, and illustrate such a construction in the case $S = S_n$, the Cuntz inverse semigroup. This will motivate the S-actions that are central to the present work, that of *localizations*.

We start with an inverse semigroup S and we want to associate r-discrete groupoids with S. The first task is to replace the group H above by the inverse semigroup S acting on the right on some X by partial one-to-one maps, and then try to construct a groupoid inspired by the transformation groupoid. So we would have an inverse semigroup antihomomorphism $s \to \alpha_s$ from S into $\mathcal{I}(X)$ where each α_s is a one-to-one map from a subset D_s of X onto another subset $R_s = D_{s^*}$ of X. Such a map $s \to \alpha_s$ is called a *right action* on X. But what X should we choose? On reflection, it is clear that X should somehow be related to the set $E(S)$ of idempotents of S. For if $e \in E(S)$, then $e^2 = e$ and $e^* = e$, so that α_e can be identified with the identity map on D_e, *a subset of* X. In other words, $E(S)$ is "picking out" subsets of X. Of course, these subsets do not tell us how the inverse semigroup elements act on the *elements* of the subsets, and this information will need to be included at some stage.

However, this does focus attention on $E(S)$ and it is natural to look first for an action of S on $E(S)$. In the trivial case where S is a group H, the only element of $E(S)$ is the identity element e and we do have an (uninspiring!) right action of H on $\{e\}$ given by conjugation: $\alpha_g(e) = g^{-1}eg = e$.

Much more interesting is the case of a general inverse semigroup S. As we will see (3.50), S also admits a conjugation right action $s \to \beta_s$ on $E(S)$. Here $D_s = \{e : e \leq ss^*\}$, where $e \leq ss^*$ means that $e(ss^*) = e$. (Recall that $E(S)$ is a semilattice and that $ss^* \in E(S)$.) For $e \in D_s$, $\beta_s(e) = s^*es$ and β_s is a one-to-one map from D_s onto D_{s^*}. (This is easy to see if we think of S in terms of partial one-to-one maps or in terms of partial isometries.)

A systematic investigation into the groupoids associated with S for general S will be undertaken in Chapter 4. However, for the present, let us see what happens with this S-action on $E(S)$ in the case where S is the Cuntz inverse semigroup S_n discussed earlier. (To avoid notational clash, we will write $s \to T_s$ in place of the $s \to \beta_s$ of the preceding paragraph.)

We first calculate the idempotents of S_n. Of course, $1, z_0$ are idempotents. Taking $s = s_\alpha t_\beta$, we obtain the other idempotents $ss^* = s_\alpha t_\alpha$ which we can identify with the string α. (This is obtained by multiplying out $(s_\alpha t_\beta)(s_\beta t_\alpha)$ using (1.4) and (1.2).) So $E(S_n) = \{z_0\} \cup Y \cup \{1\}$ where Y is the set of finite strings α. The semilattice ordering on $E(S_n)$, where $e \leq f$ if and only if $ef = e$, is given by $z_0 \leq \alpha \leq 1$ for all $\alpha \in Y$, and for all $\alpha, \delta \in Y$, $\delta \leq \alpha$ if for some γ, we have $\delta = \alpha\gamma$, i.e. if δ is α followed by another string γ (possibly empty). The domain D_s of T_s is then D_α, the set of all strings $\alpha\gamma$, and $T_s : D_\alpha \to D_\beta$ is, using (1.4), given by: $T_s(\alpha\gamma) = (s_\alpha t_\beta)^* s_\alpha\gamma t_\alpha\gamma s_\alpha t_\beta = \beta\gamma$. So Y is invariant under the right S_n-action, and since Y is where the real interest lies, we will concentrate on it below, discarding 1 and z_0.

For the purposes of representation theory, the right action of S_n on Y is unsatisfactory since Y is discrete and one would expect a *compact* space in its place. For example, to take an easier case, if S was unital and idempotent (i.e. a semilattice), then by Gelfand theory, its irreducible representations form a compact Hausdorff space X: this is the space of non-zero semicharacters on S with the pointwise topology, and includes $S = E(S)$ as a dense subset in a natural way. In that case, if we stayed with S (which corresponds to Y above), we would miss all of the representation theory of S sitting in X outside Y.

So we want to replace Y by a compact space Z. The constructive procedure of **4.3** for general S takes the "guesswork" out of finding a space such as Z. However in the present case where $S = S_n$, there is an obvious way to do it. By Tychonoff's theorem, $Z = \{1, 2, \ldots n\}^{\mathbf{P}}$ is compact in the product topology and there is a natural right action of S_n on Z. This action is formally exactly the same as its action on Y: the elements of Z are still strings, only infinite in length. So the set D_α of the preceding paragraph gets replaced by the compact open set D'_α of infinite words $\alpha\gamma$ where $\gamma \in Z$. Each T_s is then a homeomorphism from D'_α onto D'_β.

So we now have a right S_n-action on Z, the domain of each T_s being an open (compact) subset of Z. A more detailed analysis of S_n in its action on Z is given in **4.3**, Example 3. The pair (Z, S_n) is an example of what we will call in **3.3** a *localization*. A localization is a pair (X, S) where an inverse semigroup S acts on the right by partial homeomorphisms on a locally compact Hausdorff space X, the family of the domains of these partial homeomorphisms being a basis of open subsets for the topology of X. The term *localization* was introduced by Kumjian and the theory of localizations studied in Kumjian's fundamental paper ([148]). One of the main themes of Chapter 3 of the present work is that localizations are effectively the same as r-discrete groupoids.

Our definition of the term slightly extends Kumjian's notion. In his definition, the inverse semigroup is actually given as an inverse semigroup of partial homeomorphisms on X. A good example of such a localization arises in foliation theory, where S is a pseudogroup.

Let us consider for the present the case of a general localization (X, S). We will return to the specific localization (Z, S_n) later. We saw above that in the case of a group H acting on a locally compact space X, the covariance C^*-algebra $C_0(X) \times_\beta H$ was isomorphic to the groupoid C^*-algebra $C^*(G)$ where G is the transformation groupoid associated with the action of H on X. This leads naturally to the following question.

What should be the versions of $C_0(X) \times_\beta H$ and G in the case of the action of the inverse semigroup S on X, and will the version of the first also be isomorphic to $C^*(G)$?

At this stage, the reader may wonder what the point of r-discrete groupoids is, since we have their C^*-algebras identified with covariance algebras which, one might think, could be studied by adapting group covariance C^*-algebra techniques to their inverse semigroup counterparts. Here are four reasons justifying the groupoid viewpoint.

First, the covariance approach without the groupoid viewpoint is rather like studying arbitrary abelian C^*-algebra without knowing that they are $C_0(X)$'s: the groupoid in that case is just X regarded as a groupoid of units. In the general case, the role of X for the study of $A \times_\beta S$ where A is abelian is played by the groupoid, as it were the "space" associated with this non-commutative C^*-algebra. Second, we know that the representations of $A \times_\beta S$ are given by covariant pairs – but how do you determine them? In the groupoid case, this question is easier to handle since, as briefly referred to earlier, there is a well-developed representation theory for groupoids in terms of (what are called) quasi-invariant measures (**3.1**) (which can be obtained ([230, p.24ff.]) from *any* probability measure on the unit space by saturation), induced representations and Hilbert bundles. Third, in the foliation context, there are important differential geometric concepts such as holonomy which do not appear at the level of a covariance algebra $A \times_\beta S$. However in the groupoid context, holonomy appears very naturally (**2.3**). (Indeed, in that context, the groupoid is called the *holonomy groupoid*.) Last, amenability of a C^*-algebra or von Neumann algebra is a very useful property for the algebra to possess. There are elegant amenability results for covariance algebras involving group actions (see, for example, [52] and the discussion in [196, (2.35)]), but amenability for inverse semigroups is much more complex than amenability for groups. But if we translate this problem over to the groupoid context, there is a deep theory developed by Renault ([230]) which can be used to study this problem. (See, for example, **4.5** for the use of groupoids in the corresponding problem for the von Neumann algebra $VN(S)$ for inverse semigroups.)

The relationship between the representation theories of inverse semigroups and groupoids is investigated in detail in the fourth chapter. In Chapter 3, the primary situation under discussion was that of a localization, the semigroup S acting on a space. In Chapter 4, we look at the representations of S as such, no space X being initially involved. However, as illustrated by the localization (Z, S_n) above, the spatial side will enter using the idempotent semilattice of S. It was noted that Z is totally disconnected, and this has the consequence that the Cuntz groupoid G_n is more than just r-discrete – it has a basis G^a of G-sets that are *compact* open (Hausdorff) subsets of the groupoid. Such a groupoid will be called *ample*, and the groupoids with which we will be concerned in Chapter 4 will all be ample. Ample groupoids were effectively introduced by Renault in [230, p.20]. The unit space of an ample groupoid, like Z in the Cuntz case, is totally disconnected. (As discussed in the Preface, a number of examples illustrating the connection between inverse semigroups and ample groupoids are given in **4.2**.)

In **4.3**, we obtain a canonical ample groupoid $G_{\mathbf{u}}$, called the *universal groupoid* for S which contains all of the groupoid information needed for the representation theory of S. The theory developed in Chapter 3 proves helpful in obtaining $G_{\mathbf{u}}$. Just as we did for S_n above, we consider the natural action of S on its idempotent semilattice $E(S)$. In the S_n case, finite strings from $\{1, 2, \ldots, n\}$ suggested, using Tychonoff's theorem, the compact space Z of infinite strings from $\{1, 2, \ldots, n\}$ on which S_n acted as well, giving a localization (Z, S_n) which in turn gave the Cuntz groupoid G_n.

For S in general, $E(S)$ is not concretely represented by strings. What we do instead, however, is look at the space $X = X(S)$ of *non-zero semicharacters* on $E(S)$ which, either in the pointwise topology or equivalently, as the space of non-zero multiplicative linear functionals on the commutative Banach algebra $\ell^1(E(S))$, is a locally compact Hausdorff space. (The space Z in the S_n-case can be regarded as a closed subset of the corresponding X.) There is an obvious right action of S on X and we naturally want to apply the fundamental construction of Theorem 3.3.2 to obtain an r-discrete groupoid. Unfortunately, the pair (X, S) is not usually a localization since the domains of the partial homeomorphisms associated with S need not form a basis for X. However, we can find in the semigroup algebra a larger inverse semigroup S', closely related to S, to which the action of S on X naturally extends, and the pair (X, S') *is* a localization. The fundamental construction then applies to (X, S') and yields an ample groupoid. This groupoid is the *universal groupoid* $G_{\mathbf{u}}$ for S. The construction of the universal groupoid is followed by examples of universal groupoids, in particular, the determination of the universal groupoid for the Cuntz inverse semigroups S_n ($1 \le n \le \infty$). (The universal groupoid of S_n is not the same as the Cuntz groupoid.)

The results of **4.4** show that as far as C^*-algebras are concerned, the connection between S and its universal groupoid $G_{\mathbf{u}}$ is as perfect as could possibly have been hoped for. In particular the universal C^*-algebra of S is isomorphic to the universal C^*-algebra of $G_{\mathbf{u}}$ (Theorem 4.4.1) and the reduced C^*-algebra of S is isomorphic to the reduced C^*-algebra of $G_{\mathbf{u}}$ (Theorem 4.4.2). So groupoid techniques become available for studying $C^*(S)$ and $C^*_{red}(S)$, and indeed we give simple examples of apparently new representations of S obtained from this groupoid perspective.

An illustration of the use of groupoid techniques to solve a problem in the analysis of inverse semigroups is given in **4.5**. A classical result of Schwartz [245] referred to at the start of this chapter gives that a discrete group G is amenable if and only if its reduced von Neumann algebra $VN(G)$ is amenable (injective). What about the corresponding problem for an inverse semigroup S? We show (Theorem 4.5.2) that if every maximal subgroup of S is amenable then the reduced von Neumann algebra $VN(S)$ of S is amenable. This gives a very simple, purely algebraic criterion for determining the amenability of $VN(S)$. The author does not know if the converse is true. While it seems plausible that there is an "inverse semigroup" proof of this result on the amenability of $VN(S)$, the author has been unable to discover it.

To prove Theorem 4.5.2, the problem is effectively transferred over to the universal groupoid $G_{\mathbf{u}}$. Recall earlier that S can be regarded as the groupoid G_S of Proposition 1.0.1. Now it is shown in Proposition 4.4.6 that this groupoid sits densely in $G_{\mathbf{u}}$. The inverse semigroup S relates to the groupoid $G_{\mathbf{u}}$ through the mediation of G_S. In particular, we can use G_S to relate the reduced C^*-algebra of S to the reduced C^*-algebra of $G_{\mathbf{u}}$, and applying a result of Renault on amenable groupoids yields the above criterion for the amenability of $VN(S)$.

CHAPTER 2

Inverse Semigroups and Locally Compact Groupoids

2.1 Inverse semigroups

In this section we give a brief and largely self-contained account of the results on inverse semigroups that will be required in the sequel. The results we need from the algebraic theory of semigroups are well-known, and are contained in the standard textbooks (such as [50, 51, 133, 202]). However, for the convenience of the reader and for the purpose of establishing notation, an account concentrating on what we will need later from that theory is desirable.

Recall (Chapter 1) that a semigroup S is called an *inverse semigroup* if for each $s \in S$, there is a unique element $t \in S$ such that

$$sts = s \quad tst = t. \tag{2.1}$$

We write this element t as s^*. The map $s \rightarrow s^*$ will be called the *involution* on S. (The justification for this terminology is given by (ii) of Proposition 2.1.1.) It is clear that there is at most one involution giving an inverse semigroup structure on a semigroup.

In the algebraic theory of semigroups, s^{-1} is usually used in place of s^*. The former fits in well with the interpretation of inverse semigroup elements as partial one-to-one maps (see below) as well as with groupoid terminology, while the latter is natural in the context of the Banach *-algebras associated with S. We will feel free to use either s^* or s^{-1} depending on the context.

Recall also that a very important example of an inverse semigroup is given by $S = \mathcal{I}(Y)$, the set of partial one-to-one maps on a set Y. So each element of $\mathcal{I}(Y)$ is a bijection from a subset A of Y onto another subset B of

Y. The set $\mathcal{I}(Y)$ is a semigroup where the product of two such bijections is just composition, defined wherever it makes sense. For example, if $T_1, T_2 \in \mathcal{I}(Y)$ with $T_1 : A \to B$ and $T_2 : C \to D$, then $T_1 T_2 : T_2^{-1}(D \cap A) \to T_1(D \cap A)$ is given by : $T_1 T_2(c) = T_1(T_2(c))$. The element T_1^* is taken to be T_1^{-1}. It is easily checked that $\mathcal{I}(Y)$ is an inverse semigroup. Every inverse semigroup can be realised as a *-subsemigroup of some $\mathcal{I}(Y)$ as we shall see below.

The set $E(S)$ of idempotents in S plays an important role in the theory. It is clear from (2.1) that for any $s \in S$, we have $(ss^*)^2 = (ss^*s)s^* = ss^*$ so that every ss^* belongs to $E(S)$. The next result gives simple, useful facts about the *-operation and about $E(S)$.

Proposition 2.1.1   *Let S be an inverse semigroup with idempotent set $E(S)$. Then :*

(i) *$e^* = e$ for all $e \in E(S)$ and the elements of $E(S)$ commute, so that $E(S)$ is a commutative idempotent semigroup (i.e. is a semilattice); further, if $s \in S$ and $e \in E(S)$, then $ses^* \in E(S)$ and $ses^* \le ss^*$;*

(ii) *for all $s, t \in S$, we have $s^{**} = s$ and $(st)^* = t^* s^*$;*

(iii) *if T is a semigroup and $\phi : S \to T$ is a surjective homomorphism, then the map $\phi(s) \to \phi(s^*)$ is well-defined and is the involution for an inverse semigroup structure on T;*

(iv) *A homomorphism between two inverse semigroups preserves the involution.*

Proof.   (i) Since for $e \in E(S)$, $eee = e = eee$, the definition of *inverse semigroup* gives $e^* = e$.

Now let $e, f \in E(S)$. Then ([133, p.130]) one readily checks that $g = f(ef)^* e \in E(S)$ and that

$$g(ef)g = g \quad (ef)g(ef) = ef.$$

Hence by (2.1) and (i), we have $(ef)^* = g = g^*$ and so $ef = g \in E(S)$. Similarly, $fe \in E(S)$, and since (2.1) is satisfied with $s = fe, t = ef$, we have $fe = (ef)^* = ef$. So $E(S)$ is a semilattice, a commutative idempotent semigroup.

Using the commutativity of $E(S)$, it is easy to check that $ses^* \in E(S)$, and since $(ses^*)ss^* = se(s^*ss^*) = ses^*$, we obtain that $ses^* \le ss^*$.

(ii) This is directly checked using (2.1) and (i).

(iii) We claim that the idempotents of T commute. Indeed, suppose that $k = \phi(p)$ is an idempotent in T. Let $e = p(p^2)^* p \in S$. Using (2.1), the element e is an idempotent. Since $\phi(p^2) = k^2 = k = \phi(p)$, we obtain, using (2.1), that $\phi(e) = \phi(p^2)\phi((p^2)^*)\phi(p^2) = \phi(p^2(p^2)^*p^2) = \phi(p^2) = \phi(p) =$

k. The commuting property for idempotents in T then follows from the corresponding property for idempotents in S ((i)).

Let $v \in T$. There is an element $t \in T$ such that $vtv = v, tvt = t$. For if $v = \phi(s)$, we can take $t = \phi(s^*)$. Suppose that there is another element $u \in T$ with $vuv = v, uvu = u$. Since the idempotents vt, vu commute, we have $vt = (vuv)t = (vt)(vu) = (vtv)u = vu$. Similarly, $tv = uv$. So

$$t = tvt = uvt = uvu = u$$

and uniqueness follows.

(iv) follows from (iii). □

In group theory, a quotient group is the set of coset equivalence classes for a normal subgroup. In the semigroup context, such a coset equivalence relation is replaced by a *congruence*.

Here, an equivalence relation $\sim$ on a semigroup T is called a *congruence* ([133, p.21]) if $as \sim at$ and $sa \sim ta$ whenever $s \sim t$ in T and $a \in T$. For any congruence $\sim$ on T, the set of equivalence classes $T/ \sim$ is also a semigroup in the natural way, and the map ϕ which sends an element of T to its equivalence class is a surjective homomorphism. Conversely, if W is a semigroup and ψ is a homomorphism from T onto W, then ψ defines a congruence $\sim$ on T, where two elements of T are $\sim$-equivalent if and only if they have the same ψ image. The factor inverse semigroup $T/ \sim$ is identifiable with W and under this identification, ψ is just the map that sends an element of T to its equivalence class. By reversing multiplication in $T/ \sim$, it obviously follows that congruences also correspond to surjective antihomomorphisms. Every relation on S generates a congruence on S, the smallest congruence on S containing the relation.

It follows from the involution preserving property of homomorphisms on inverse semigroups ((iv) of Proposition 2.1.1) that if T is an inverse semigroup and $\sim$ is a congruence on T, then $s \sim t$ if and only if $s^* \sim t^*$. In particular, $T/ \sim$ is an inverse semigroup in the natural way.

A particularly simple kind of congruence on a semigroup T is that of a *Rees congruence*. Let I be an ideal of T. Then ([50, p.17]) the *Rees congruence modulo I* is the congruence ρ_I on T defined as follows : for $a, b \in T$, $a\rho_I b$ means that either $a = b$ or both $a, b \in I$. It is obvious that ρ_I is a congruence on T. The quotient T/ρ_I is called the *Rees factor semigroup* and is usually written T/I. It is obvious that T/I is just T with the ideal I collapsed to zero.

An inverse semigroup S relates naturally ([133, p.140]) to groups through a certain congruence σ_S. For $s, t \in S$, we say that $s\sigma_S t$ whenever there exists $e \in E = E(S)$ such that $es = et$. The next result shows that σ_S is a congruence on S, for which S/σ_S is a group $G(S)$. This group is the "largest" group which can be obtained from S as a homomorphic image. It can be roughly thought of in terms of S as a "concertina" which is pushed "flat" by collapsing all of the idempotents together into the identity of the group.

Proposition 2.1.2 *The relation σ_S is a congruence on S and $G(S) = S/\sigma_S$ is a group. Further, $G(S)$ is the maximal group homomorphic image of S in the sense that if $\psi : S \to G$ is a surjective homomorphism and G is a group, then ψ factors through $G(S)$.*

Proof. Trivially σ_S is reflexive. If $s, t, u \in S$ and $e, f \in E(S)$ are such that $es = et$, $ft = fu$ then $(ef)s = (ef)u$ so that σ_S is transitive. Now suppose that $s\sigma_S t$ and that $u \in S$ and let e be as above. Then using (i) of Proposition 2.1.1, $(ueu^*)us = (ueu^*)ut$ and $esu = etu$ giving that $us\sigma_S ut$ and $su\sigma_S tu$. So σ_S is a congruence on S. Clearly all idempotents are σ_S-equivalent. From the proof of (iii) of Proposition 2.1.1, every idempotent in $G(S)$ is the image of an idempotent. So $G(S)$ has one idempotent and it easily follows from the definition of *inverse semigroup* that $G(S)$ is a group.

Finally, if $\psi : S \to G$ is a homomorphism with G a group and $es = et$ as above, then $\psi(e)$ is the identity of G and so $\psi(s) = \psi(t)$. So ψ factors through $G(S)$. □

The inverse semigroup S is called *E-unitary* if E is a σ_S-equivalence class. The theory of E-unitary semigroups shows that this class of inverse semigroups is large. Many examples of such inverse semigroups are given by the following construction reminiscent of that of covariance algebras in operator algebras. Let X be a semilattice and Γ a group acting on X (on the left) as a group of automorphisms. Then ([133, p.182]) $S = X \times \Gamma$ is E-unitary with product and involution given by: $(x, g)(y, h) = (x(g(y)), gh)$, $(x, g)^* = (g^{-1}(x), g^{-1})$. It is easy to check that $S/\sigma_S = \Gamma$. A generalization of this construction due to D. B. McAlister yields all E-unitary inverse semigroups. The reader is referred [202, Ch.6,7] for a full discussion of E-unitary inverse semigroups.

The main relevance of E-unitary inverse semigroups for the present work is that, as we shall see in Corollary 4.3.2, these inverse semigroups have the desirable property that their universal groupoids are *Hausdorff*.

The important *Vagner-Preston theorem*, the inverse semigroup analogue of the Cayley theorem for groups, effectively identifies an inverse semigroup S with a subsemigroup of $\mathcal{I}(S)$. It is the key to the most accessible Hilbert space representation of S, the left regular representation.

Proposition 2.1.3 *For each $s \in S$ let $D(s) = \{t \in S : tt^* \leq s^*s\}$ and $R(s) = \{w \in S : ww^* \leq ss^*\}$. For each $s \in S$, define $\gamma_s : D(s) \to R(s)$ by: $\gamma(s)(t) = st$. Then $\gamma(s)$ is a bijection from $D(s)$ onto $R(s)$ and the map γ is an inverse semigroup isomorphism from S into $\mathcal{I}(S)$.*

Proof. Let $s \in S$ and $t, t' \in D(s)$ be such that $st = st'$. Then since $tt^*, t't'^* \leq s^*s$, we have

$$t = (s^*s)(tt^*)t = s^*st = s^*st' = (s^*s)(t't'^*)t' = t'$$

so that $\gamma(s)$ is one-to-one. Since for $t \in D(s)$, $(st)(st)^* \leq ss^*$ (Proposition 2.1.1, (i)), we have $st \in R(s)$. Similarly, if $w \in R(s) = D(s^*)$, then $s^*w \in D(s)$ and $s(s^*w) = w$. So $\gamma(s)$ is a bijection from $D(s)$ onto $R(s)$.

Let $s, t \in S$ and a belong to the domain of $\gamma(t) \circ \gamma(s)$. Then $aa^* \leq s^*s$ and $(sa)(sa)^* = s(aa^*)s^* \leq t^*t$, giving $aa^* = s^*(s(aa^*)s^*)s \leq s^*t^*ts = (ts)^*(ts)$. So $a \in D(ts)$. Conversely, if $w \in D(ts)$, then $ww^* \leq s^*t^*ts \leq s^*s$, so that $w \in D(s)$. Further, $(sw)(sw)^* = sww^*s^* \leq s(s^*t^*ts)s^* \leq t^*t$, so that $sw \in D(ts)$. So the domain of $\gamma(t) \circ \gamma(s)$ equals that of $\gamma(ts)$ and these two maps coincide. It now follows that γ is a homomorphism. Finally, suppose that $\gamma(s) = \gamma(t)$. Then $D(s) = D(t)$, and since $s^* \in D(s)$, we have $ss^* = \gamma(s)(s^*) = \gamma(t)(s^*) = ts^*$. By symmetry, $tt^* = st^*$. It follows that $t = (tt^*)t = st^*t = s(s^*s)(t^*t) = (ss^*)s(t^*t) = (ts^*)s(t^*t) = t(t^*t)(s^*s) = (ts^*)s = ss^*s = s$. So γ is an isomorphism. $\square$

Representation theory for inverse semigroups (*via* the mediation of r-discrete groupoids) is (in one form or another) a central theme of the present work. That there are interesting representations is shown by the fact that the map γ above can be regarded as a representation of S on $\ell^2(S)$, and in this guise, is the *left regular representation* of S. Before going on to this, we will discuss some basic facts about the representations of an inverse semigroup S.

Parallel to the group case, the representations of S will be identifiable with the *-representations of the Banach *-algebra $\ell^1(S)$. A typical element of $\ell^1(S)$ is, of course, of the form $f = \sum_{s \in S} a_s s$ where each $a_s \in \mathbf{C}$ and $\|f\|_1 = \sum_{s \in S} |a_s| < \infty$. Then $\ell^1(S)$ is a Banach *-algebra under the natural convolution product

$$\left(\sum_{s \in S} a_s s, \sum_{t \in S} b_t t\right) \to \sum_{s,t \in S} a_s b_t st$$

and involution given by $(\sum_{s \in S} a_s s)^* = \sum_{s \in S} \overline{a_s} s^*$. Unfortunately, $\ell^1(S)$ usually does not have a bounded approximate identity. (The Clifford semigroup considered in Appendix C gives an example of this.) A useful dense *-subalgebra of $\ell^1(S)$ is $\mathbf{C}(S)$, the (algebraic) semigroup algebra over the field $\mathbf{C}$, the subalgebra of *finite* linear combinations of the elements of S.

We now define exactly what we mean by a *representation* of the inverse semigroup S. (Notice that non-degeneracy is built into the definition.) A *representation* of S on a (separable) Hilbert space $\mathcal{H}$ is a *-homomorphism from S into $B(\mathcal{H})$ such that the span of the vectors $\pi(s)(\xi)$ for $s \in S$ and

$\xi \in \mathcal{H}$ is dense in $\mathcal{H}$. (The latter requirement is equivalent to the condition: if $\xi \in \mathcal{H}$ and $\pi(S)\xi = \{0\}$, then $\xi = 0$.) Given such a representation π, the operator $\pi(e)$ is a projection if $e \in E(S)$. Note that if $e \leq f$ in $E(S)$, then $ef = e$, and applying π to this equality gives $\pi(e) \leq \pi(f)$, so that π is *increasing* on $E(S)$.

Further, for each $s \in S$, the operator $\pi(s)$ is a partial isometry on $\ell^2(S)$ since $\pi(s)\pi(s^*) = \pi(ss^*)$ is a projection. It follows that π takes S into an inverse semigroup of *partial isometries* on a Hilbert space (or equivalently, into a *-semigroup of partial isometries on a Hilbert space). As we will see below (Proposition 2.1.4), it is a consequence of the Vagner-Preston theorem that *every* inverse semigroup can be realized as an inverse semigroup of partial isometries on a Hilbert space. This is probably the most helpful way for an analyst to think of inverse semigroups.

As in the group case, a representation π of S gives rise canonically (by taking "linear combinations") to a non-degenerate *-representation, also denoted by π, of $\mathbf{C}(S)$ on $\mathcal{H}$. Since every $\pi(s)$ is of norm ≤ 1 (since it is a partial isometry), it follows that π is norm decreasing on $\mathbf{C}(S)$ and so extends to a (non-degenerate) representation of $\ell^1(S)$ of norm ≤ 1. Conversely, every (non-degenerate) *-representation of the Banach *-algebra $\ell^1(S)$ is of norm ≤ 1 and is given by a representation of S.

The universal C^*-algebra $C^*(S)$ of S is the completion of $\ell^1(S)$ (or $\mathbf{C}(S)$) under the norm $\|.\|$, where for $f \in \ell^1(S)$,

$$\|f\| = \sup_{\pi} \|\pi(f)\|,$$

the sup being taken over the set of representations π of S. Of course, strictly we only have at this stage that $\|.\|$ is a C^*-seminorm on $\ell^1(S)$, but, as we will see later, through the agency of the reduced C^*-norm on $\ell^1(S)$, $\|.\|$ is actually a norm.

We now discuss how the Vagner-Preston theorem, translated into functional analytic terms, gives a representation of S on $\ell^2(S)$. This important representation is called the *left regular representation* of S and this interpretation of the Vagner-Preston theorem was first observed by Barnes ([11]). (See also [195, 78, 281].) We can regard S as an orthonormal basis for $\ell^2(S)$ and the idea is to extend each map $\gamma(s)$ of Proposition 2.1.3 to the closed span H_s of $D(s)$ in $\ell^2(S)$ in the obvious way and make it zero on $H_s^\perp$, the closed span of the complement of $D(s)$. This gives a map π_2 from S into $B(\ell^2(S))$. Now S is an orthonormal basis for $\ell^2(S)$ and $\pi_2(s)(t) = \gamma(s)(t)$ for all $s, t \in S$. Since γ is a homomorphism, so also is π_2. Simple checking shows that for $t_1, t_2 \in S$, we have $\langle \pi_2(s)(t_1), t_2 \rangle = \langle t_1, \pi_2(s^*)(t_2) \rangle$ so that π_2 is a *-homomorphism. The map π_2 is non-degenerate since the closed span of the elements $\pi_2(s)(s^*s) = s$ is the whole of $\ell^2(S)$. So π_2 is a representation of S.

For future reference, the formula for the left regular representation for S is given by:

$$\pi_2(s)(\sum_{t \in S} a_t t) = \sum_{tt^* \le s^* s} a_t st. \tag{2.2}$$

(Recall that $D(s) = \{t \in S : tt^* \le s^* s\}$, so that the right-hand side of (2.2) effectively says that $\pi_2(s)(t) = st$ if $t \in D(s)$ and is 0 otherwise. This is how π_2 was formulated in the preceding paragraph.) The next characterization of inverse semigroups was given by Duncan and Paterson in [78].

Proposition 2.1.4   *A semigroup is an inverse semigroup if and only if it is isomorphic to a *-semigroup of partial isometries on a Hilbert space.*

Proof.   If S is an inverse semigroup, then by the above, π_2 is a *-homomorphism from S onto a *-semigroup of partial isometries on $\ell^2(S)$. Since for all $s, t \in S$, $\pi_2(s)(t) = \gamma(s)(t)$, Proposition 2.1.3 gives that π_2 is an isomorphism. Conversely, if S is a *-semigroup of partial isometries on a Hilbert space, then, for each $T \in S$, the operator T^* is the unique element U of S for which $TUT = T, UTU = U$, and so S is an inverse semigroup by definition.   □

As discussed above, the representation π_2 can be regarded as a *-representation of the Banach *-algebra $\ell^1(S)$ on $\ell^2(S)$. Obviously, $\|\pi_2\| = 1$. The C^*-algebra generated by $\pi_2(S)$, which is the norm closure of $\pi_2(\ell^1(S))$, is the *reduced C^*-algebra* of S, and is denoted by $C^*_{red}(S)$.

As an instructive example of an inverse semigroup reduced C^*-algebra, let us consider the case where S is the Cuntz inverse semigroup S_n with n finite (Chapter 1). To calculate what this C^*-algebra is, let π_2 be the left regular representation of S_n on $\ell^2(S_n)$. Recall that S_n has a unit 1, a zero z_0, the other elements being the $s_\alpha t_\beta$'s of (1.4). It is easily checked (using (1.5), (4.8) and (4.9)) that $D(1) = S_n$, $D(z_0) = \{z_0\}$ and $D(s_\alpha t_\beta) = \{z_0\} \cup \{s_{\beta\gamma} t_{\beta'} : \gamma, \beta'$ are strings $\}$. Write $\ell^2(S_n) = \ell^2(T_n) \oplus \mathbf{C}z_0$ where $T_n = S_n \sim \{z_0\}$. Each of the preceding Hilbert space summands is an invariant subspace of $\ell^2(S_n)$ for $\pi_2(S_n)$, and the map $s \rightarrow (\pi_2(s) - \pi_2(z_0), \pi_2(z_0))$ implements an isomorphism $C^*_{red}(S_n) \cong A \times \mathbf{C}$. To determine A, let $V_i = \pi_2(s_i) - \pi_2(z_0)$. Then V_i is an isometry and $P = I - \sum_{i=1}^n V_i V_i^*$ is the orthogonal projection onto $\mathbf{C}1$. By [65, Proposition 3.1] – see also [69, V.5] – A is isomorphic to the *Cuntz-Toeplitz algebra*, an extension of the Cuntz algebra O_n by the algebra $\mathcal{K}$ of compact operators on a separable Hilbert space. (A copy of $\mathcal{K}$ is given by the closed ideal of A generated by P.)

These fundamental C^*-algebras associated with S, $C^*(S)$ and $C^*_{red}(S)$, are quite easy to define, as we have seen. However, to study these algebras in terms of S alone poses difficulties, and we will see later (**4.3**) that the groupoid approach to these algebras gives powerful tools for their study. One non-trivial result about $C^*_{red}(S)$, obtained by the combinatorial techniques of semigroup theory, is the elegant result Theorem 2.1.1 due to Wordingham ([281]), whose proof is now given. His result gives that π_2 is faithful on $\ell^1(S)$. The groupoid significance of the space of semicharacters of a semilattice, used in Wordingham's proof, will become apparent in **4.3**.

Let E be a semilattice and X be the set of non-zero semicharacters $\alpha : E \to \{0,1\}$. (So X is the set of non-zero homomorphisms from E into the two-point multiplicative semigroup $\{0,1\}$.)

For $e \in E$ define $\bar{e} : E \to \{0,1\}$ by : $\bar{e}(f) = 1$ if $f \geq e$ and is 0 otherwise. It is easily checked that $\bar{e} \in X$. Let $Z = \{\bar{e} : e \in E\}$. In the following lemma, whose proof uses a modification of a result of Hewitt and Zuckerman ([122]), the set X is regarded as a subset of $\ell^\infty(E) = (\ell^1(E))^*$. (Under this identification, X is the Gelfand space of the commutative Banach algebra $\ell^1(E)$.)

Lemma 2.1.1   *The set $\overline{E}$ separates the elements of $\ell^1(E)$.*

Proof.   Suppose that the preceding statement is false. Then there exists non-zero $g \in \ell^1(E)$ such that $\bar{e}(g) = 0$ for all $e \in E$. We will derive a contradiction. We can suppose that g is of the form $\sum_{r=1}^\infty a_r e_r$ where the e_r's are distinct, $\{a_r\} \in \ell^1$ and $a_1 \neq 0$.

We first show[1] that $\alpha(g) = 0$ for all $\alpha \in X$. Indeed, let $\alpha \in X$ and $F = \{e \in E : \alpha(e) = 1\}$. Then F is a subsemilattice of E which is a net under the semilattice ordering. By taking products of e's in F, we can obtain an element $\bar{e}$ with $e \in F$ coinciding with α on any preassigned finite number of elements of E. So $\alpha = \lim_{f \in F} \bar{f}$ in the pointwise topology on E. Since the norm of every element of X in $\ell^\infty(E)$ is 1, the set X is uniformly bounded and it follows that the convergence in the above limit is weak* in $\ell^\infty(E)$ and hence that

$$\alpha(g) = 0. \tag{2.3}$$

Now the e_i's (involved in the expansion of g) are distinct, and if $e, f \in E$ are such that $e \leq f, f \leq e$ then $e = f$. It follows that there exists, for each $r \geq 2$, an element $f_r \in \{e_1, e_r\}$ such that $\overline{f_r}(e_1) \neq \overline{f_r}(e_r)$. In particular,

$$\mid \overline{f_r}(e_1) - \overline{f_r}(e_r) \mid = 1. \tag{2.4}$$

For each $n \geq 2$ let $\phi_n \in \ell^\infty(E)$ be the "Lagrangian interpolation function":

$$\phi_n(e) = \prod_{r=2}^n \frac{\overline{f_r}(e) - \overline{f_r}(e_r)}{\overline{f_r}(e_1) - \overline{f_r}(e_r)}.$$

From (2.4) and the fact that every $\overline{f_r}(e)$ is either 1 or 0, it follows that $\|\phi_n\|_\infty = 1$. Further, $\phi_n(e_1) = 1$ and $\phi_n(e_r) = 0$ for $2 \leq r \leq n$. Multiplying

---

[1] This also follows directly from Proposition 4.3.1 later.

out the formula for ϕ_n gives it as a finite sum of multiples of elements of X (since any non-zero product of two semicharacters is itself a semicharacter). Hence from (2.3) we have

$$0 = \phi_n(g) = a_1 + \sum_{n+1}^{\infty} a_r \phi_n(e_r). \tag{2.5}$$

Using the uniform boundedness of the ϕ_n and letting $n \to \infty$ in (2.5), we obtain that $a_1 = 0$ and a contradiction. $\square$

Corollary 2.1.1 *Let E be a semilattice. Then the Gelfand transform $g \to \hat{g}$ from $\ell^1(E)$ into $C_0(X)$ is faithful.*

We now give Wordingham's theorem.

Theorem 2.1.1 *Let S be an inverse semigroup. Then the left regular representation $\pi_2 : \ell^1(S) \to B(\ell^2(S))$ is faithful.*

Proof. Let $g \in \ell^1(S)$ be such that $\pi_2(g) = 0$. Write $g = \sum_{s \in S} a_s s$ in $\ell^1(S)$ with each $a_s \in \mathbf{C}$. For each $e \in E(S)$, let

$$g_e = \sum_{s^*s=e} a_s s. \tag{2.6}$$

Then $g_e \in \ell^1(S)$ and $g = \sum_{e \in E(S)} g_e$. It is sufficient to show that for $f \in E(S)$, we have $g_f = 0$. (For this will imply that $g = 0$.)
    Now

$$0 = \pi_2(g) = \sum_{e \in E(S)} \pi_2(g_e). \tag{2.7}$$

Further, by the definition of π_2, for any $e, f \in E(S)$ and $s \in S$ such that $s^*s = e$, we have that $\pi_2(s)(f) = sf$ if $f \le e$ and is zero otherwise. Fix $f \in E(S)$. So $\pi_2(g_e)(f) = g_e f$ if $f \le e$ and is zero otherwise. It follows using (2.7), that

$$\sum_{f \le e} g_e f = 0. \tag{2.8}$$

Now let $u \in E(S)$ with $f \ge u$. Using (2.8),

$$\sum_{f \le e} g_e u = \left(\sum_{f \le e} g_e f \right) u = 0. \tag{2.9}$$

Let $t \in S$. Expanding the left-hand side of (2.9) and equating the t-coefficient to 0 gives

$$\sum \{ a_s : f \le s^*s, su = t \} = 0. \tag{2.10}$$

Let $F_f = \{v \in E(S) : f \leq v\}$. Then F_f is a subsemilattice of $E(S)$. For each $v \in F_f$, let $\beta_v^t = \sum\{a_s : v = s^*s, su = t\}$ and $\beta^t = \sum_{v \in F_f} \beta_v^t v \in \ell^1(F_f)$. Then for $e \in F_f$,

$$\bar{e}(\beta^t) = \sum_{v \geq e} \beta_v^t = \sum\{a_s : e \leq s^*s, su = t\} = 0$$

using (2.10) with f replaced by e. By Lemma 2.1.1, $\beta^t = 0$. So every $\beta_v^t = 0$. In particular, since $f \in F$, the coefficient $\sum\{a_s : s^*s = f, sf = t\}$ of f in β^s is 0. Noting that if $s^*s = f$ then $s = s(s^*s) = sf$, we obtain

$$g_f = \sum_{s^*s=f} a_s s = \sum_{s^*s=f} a_s sf = \sum_{t \in S}(\sum\{a_s : s^*s = f, sf = t\})t = 0$$

as required. $\square$

2.2 Locally compact and r-discrete groupoids

Recall (Chapter 1) that a groupoid G is a set with a partially defined multiplication on a subset G^2 of $G \times G$ and an inverse map $a \to a^{-1}$ satisfying (1.10) and (1.11). Also, the unit space G^0 of G is the subset of elements xx^{-1} where x ranges over G. The range map $r : G \to G^0$ and source map $d : G \to G^0$ are defined by (1.7). For $x, y \in G$, the pair (x, y) belongs to the set G^2 of composable pairs if and only if $d(x) = r(y)$. For each $u \in G^0$, the subsets G^u, G_u are given by ((1.8), (1.9)): $G^u = r^{-1}(\{u\}), G_u = d^{-1}(\{u\})$. Note that $(G^u)^{-1} = G_u$ since $d(x^{-1}) = r(x)$.

The groupoids with which we will be concerned are topologized in a way analogous to that of a locally compact group. The groupoid G is called a *topological groupoid* if it has a topology for which product and inversion are continuous. (The topology on the domain G^2 of the product map is the relative topology inherited from $G \times G$.) Let G be a topological groupoid. Then the maps r, d are continuous: for example, r is continuous since it is a composition $x \to (x, x^{-1}) \to xx^{-1}$ of continuous maps. Now suppose in addition that G^0 is Hausdorff in the relative topology. (This is true by definition for any locally compact groupoid.) Then $\Delta = \{(u, u) : u \in G^0\}$ is closed in $G^0 \times G^0$. For if $(u_\delta, u_\delta) \to (u, v)$ in $G^0 \times G^0$, then $u_\delta \to u$ and $u_\delta \to v$, and since G^0 is Hausdorff, we have $u = v$. Since $d \times r : G \times G \to G^0 \times G^0$ is continuous and Δ is closed in $G^0 \times G^0$, it follows that

$$G^2 = (d \times r)^{-1}(\Delta) \tag{2.11}$$

is closed in $G \times G$.

The topological groupoids with which we will be concerned are the *locally compact* groupoids. An (initially irksome) problem that we are confronted with is that the locally compact groupoids that arise in practice are often not Hausdorff. A simple example of a non-Hausdorff locally compact groupoid associated with an inverse semigroup is given in Appendix C. The same is true in other contexts. We will see that it applies, for example, to the case of the holonomy groupoids associated with foliations. We will discuss holonomy groupoids in detail in **2.3**, the section that studies Lie groupoids. We will meet many examples of locally compact groupoids throughout the text. In particular, all Lie groupoids are locally compact groupoids.

Fortunately, there is enough of the Hausdorff property even in the non-Hausdorff situation to ennable the arguments that work for locally compact Hausdorff groupoids to be adapted to the general case. Nonetheless, the precise definition of a locally compact groupoid requires care and will now be given.

Definition 2.2.1 A *locally compact groupoid* is a topological groupoid G which satisfies the following axioms:

(i) G^0 is locally compact Hausdorff in the relative topology inherited from G;

(ii) there is a countable family $\mathcal{C}$ of compact Hausdorff subsets of G such that the family $\{C^o : C \in \mathcal{C}\}$ of interiors of members of $\mathcal{C}$ is a basis for the topology of G;

(iii) every G^u is locally compact Hausdorff in the relative topology inherited from G;

(iv) G admits a left Haar system $\{\lambda^u\}$. (This is defined below.)

Observe that by (i), every singleton subset of G^0 is closed in G^0, so that by the continuity of r, d, each G^u, G_u is closed in G. Note that from (ii), every point has an arbitrarily small compact Hausdorff neighborhood and every open subset of G is locally compact in the usual sense of point set topology ([145, p.146]). Singletons in G are also closed, and the topology of G has a countable basis. Also G^0 and every G^u, G_u has a countable basis.

Usually, (iv) is not required for the definition of a locally compact groupoid, but we have, for convenience, included it in the definition since analysis on groupoids essentially requires a left Haar system. See the discussion of left Haar systems below.

Since many of the locally compact groupoids that arise in practice are not Hausdorff, the usual definition of $C_c(G)$, i.e. as the space of continuous complex-valued functions on G with compact support, is no longer adequate (as it can be a very small space). Instead, following the idea of Connes in [54], one defines $C_c(G)$ for such a groupoid G as the span of those complex-valued functions f that are continuous with compact support on an open

Hausdorff subset, each of the functions f being defined to be zero outside that open Hausdorff set. Property (ii) above ensures that there is a basis for the topology of G consisting of open (locally compact) Hausdorff sets, so that $C_c(G)$ is a "large" space. In general, the functions in $C_c(G)$ are not continuous on G but $C_c(G)$ has its usual meaning when G is Hausdorff, each f above itself being, in that case, continuous on G with compact support.

Also, in general, if $f \in C_c(G)$, then by (iii) above, the restriction of f to any G^u belongs to $C_c(G^u)$. Conversely, if $g \in C_c(G^u)$, then there exists $f \in C_c(G)$ which restricts to g. To see this, cover the support of g in G^u by a finite number of open (locally compact) Hausdorff sets $U_1, \ldots, U_n$ in G. Using a partition of unity argument in the locally compact Hausdorff space G^u, we can suppose that $g = \sum_{i=1}^n g_i$ where $g_i \in C_c(G^u \cap U_i)$. Now extend each g_i to a function $f_i \in C_c(U_i)$ and set $f = \sum_{i=1}^n f_i \in C_c(G)$.

For $f : G \to \mathbf{C}$ define $\check{f} : G \to \mathbf{C}$ by:

$$\check{f}(x) = f(x^{-1}). \tag{2.12}$$

Then $\check{f} \in C_c(G)$ if $f \in C_c(G)$.

We note that the fact that G^2 is closed in $G \times G$ ensures that every point of G^2 has a compact, Hausdorff neighborhood in G^2. The corresponding space $C_c(G^2)$ is defined as for $C_c(G)$. If $U, V \subset G$ we define

$$U \dot{\times} V = (U \times V) \cap G^2. \tag{2.13}$$

In particular, the family of sets $U \dot{\times} V$, where U, V are open subsets of G, forms a basis of locally compact Hausdorff sets for the topology of G^2.

Although the above definition of $C_c(G)$ works well in the non-Hausdorff groupoid context, certain properties of its Hausdorff version do not hold in general. For example, as pointed out by Renault ([233]), $C_c(G)$ is not usually closed under pointwise products. It can also be shown that in general, $f \in C_c(G)$ does not imply that $\mid f \mid \in C_c(G)$. (A counterexample to both is provided by the example of Appendix C.) However, it is obvious that every $C_c(G)$-function is Borel measurable. The definition of a left Haar system for G, referred to in (iii) above, is formulated as for the Hausdorff case ([230, pp.16-17]).

Definition 2.2.2. A *left Haar system* for a locally compact groupoid G is a family $\{\lambda^u\}$ $(u \in G^0)$, where each λ^u is a positive regular Borel measure on the locally compact Hausdorff space G^u, such that the following three axioms are satisfied:

(i) the support of each λ^u is the whole of G^u;

(ii) for any $g \in C_c(G)$, the function g^0, where

$$g^0(u) = \int_{G^u} g \, d\lambda^u$$

belongs to $C_c(G^0)$;

(iii) for any $x \in G$ and $f \in C_c(G)$,

$$\int_{G^{d(x)}} f(xz)\,d\lambda^{d(x)}(z) = \int_{G^{r(x)}} f(y)\,d\lambda^{r(x)}(y). \qquad (2.14)$$

Note that (ii) makes sense since the restriction of g to G^u belongs to $C_c(G^u)$. Further, the integrand in the first integral of (2.14) also makes sense since if $z \in G^{d(x)}$, then $r(z) = d(x)$ so that the product xz is defined. This integrand belongs to $C_c(G^{d(x)})$ since it is the composition of the homeomorphism $z \to xz$ from $G^{d(x)}$ onto $G^{r(x)}$ and the function $f_{|G^{r(x)}}$. In the locally compact group case, (2.14) just expresses the invariance of left Haar measure; in that case, there is only one λ^u, viz. that for which u is the identity e, since $r(x) = e = d(x)$ for all x in the group. We can think of (2.14) as expressing the fact that the λ^u's are invariant under multiplication by groupoid elements as informally discussed in Chapter 1.

Since $(G^u)^{-1} = G_u$ and the map $x \to x^{-1}$ is a homeomorphism on G, we can associate with λ^u the measure $\lambda_u = (\lambda^u)^{-1}$ on G_u: so

$$\lambda_u(E) = \lambda^u(E^{-1})$$

for any Borel subset E of G_u. (The measures λ^u, λ_u can also be regarded as measures on the σ-algebra of Borel subsets of G in the obvious way.) Obviously, for $f \in C_c(G)$, we have

$$\int f\,d\lambda_u = \int \check{f}\,d\lambda^u. \qquad (2.15)$$

It follows from (ii) that the map $u \to \int_{G_u} g\,d\lambda_u$ also belongs to $C_c(G)$.

Of course in the statement of (2.14), we do not really need to put in explicitly the regions of integration $G^{d(x)}, G^{r(x)}$ involved, for they are determined by the measures $\lambda^{d(x)}, \lambda^{r(x)}$. But sometimes it is helpful to put in the regions explicitly as useful reminders, especially when the integral formulas become complicated. Alternative forms of (2.14) are

$$\int_{G^{r(x)}} f(x^{-1}z)\,d\lambda^{r(x)}(z) = \int_{G^{d(x)}} f(y)\,d\lambda^{d(x)}(y) \qquad (2.16)$$

and

$$\int_{G_{r(x)}} f(zx)\,d\lambda_{r(x)}(z) = \int_{G_{d(x)}} f(y)\,d\lambda_{d(x)}(y). \qquad (2.17)$$

The first of these follows from (2.14) by replacing x by x^{-1}, while the second is obtained by making the substitution $y \to y^{-1}$ in (2.16), replacing f by $\check{f}$, and using (2.12).

The equality (2.16) gives rise, for any $x \in G$, to an isometry from $L^1(G^{d(x)}, \lambda^{d(x)})$ onto $L^1(G^{r(x)}, \lambda^{r(x)})$ given by $f \to x * f$ where

$$(x * f)(y) = f(x^{-1}y). \tag{2.18}$$

Now we noted earlier that for $u \in G^0$ and $f \in C_c(G^u)$, the function f is the restriction to G^u of some $F \in C_c(G)$. So (2.16) holds for $f \in C_c(G^{d(x)})$, and since the map $y \to x^{-1}y$ is a homeomorphism from $G^{r(x)}$ onto $G^{d(x)}$, the density of $C_c(G^u)$ in $L^1(G^u, \lambda^u)$ gives that (2.16) holds for all $f \in L^1(G^{d(x)}, \lambda^{d(x)})$ and that the map $f \to x * f$ is an isometry as claimed. We will discuss the maps $f \to x * f$ further in **4.5**. The same isometry result holds with L^p in place of L^1 for any $p \geq 1$.

Unlike the locally compact group case, the existence of a left Haar system does not follow from properties (i), (ii) and (iii) of the definition of *locally compact groupoid*. (A counterexample, using Proposition 2.2.1, was given by A. K. Seda ([252]).) Further, simple examples exist to show that a left Haar system can be far from unique. We now look at one of these which is of special importance and which will illustrate the notions of a *locally compact groupoid* and of a *left Haar system*.

Let G be the largest equivalence relation on a locally compact, second countable, Hausdorff space X. So $G = X \times X$ and $G^2 = \{((x,y),(y,z)) : x,y,z \in X\}$. We noted in Chapter 1 that any equivalence relation is a groupoid. In particular, G is a groupoid with product and inversion maps given by $((x,y),(y,z)) \to (x,z)$ and $(x,y) \to (y,x)$ respectively. Then $G^0 = \{(x,x) : x \in X\}$ which we identify in the obvious way with X. Further, $r((x,y)) = x, d((x,y)) = y$ and $G^x = \{x\} \times X, G_x = X \times \{x\}$. (In terms of the category approach to G (Chapter 1), the object set is just X and whenever $x, y \in X$, there is exactly one arrow with domain y and range x.)

With the product topology, it is easily checked that G satisfies the first three properties of a locally compact groupoid. (It is, in addition, Hausdorff.) To get a left Haar system on G, let μ be *any* positive regular Borel measure on X whose support is G^0. For any $u \in X$, let λ^u be the measure $\delta_u \times \mu$ on $G^u = \{u\} \times X$. (This just amounts to specifying $\lambda^u(\{u\} \times A) = \mu(A)$ for any Borel subset A of X.) In "no nonsense" terms, one just identifies G^u in the obvious way with X and takes $\lambda^u = \mu$. The family $\{\lambda^u\}$ given in this way will be called *the left Haar system associated with μ*. It remains to check the three conditions above for a left Haar system. It is obvious that (i) holds. Condition (ii) just asserts the elementary fact that if $f \in C_c(X \times X)$, then the map $u \to \int_X f(u,y) \, d\mu(y)$ is continuous, while (iii) amounts to asserting the trivial identity that if $x_0, y_0 \in X$, then

$$\int_X f((x_0, y_0)(y_0, z)) \, d\mu(z) = \int_X f(x_0, z) \, d\mu(z)!$$

So G, with a choice of a measure μ on X, is a locally compact groupoid. It is called a *trivial groupoid*.[2]

Since there are, in general, many different such measures μ on a locally compact space X, it follows that the left Haar system on a trivial groupoid G is very far from being unique. (There are, in general, pairs of such measures μ that are even mutually singular!)

Let us look at another equivalence relation locally compact groupoid R_p which, as we will see in **4.3**, Example 5, arises naturally in the interpretation of the set of Penrose tilings as a noncommutative space. Let X_p be the set of sequences $\{x_n\}$ where x_n is either 0 or 1 and any 1 that occurs in the sequence is followed by a 0. This is a closed subspace of the Cantor set $\{0,1\}^{\mathbf{P}}$ and so is a compact Hausdorff space. Indeed, X_p is homeomorphic to $\{0,1\}^P$ – just regard an element of X_p as a sequence of 0's and 10's. However, for the purposes of the equivalence relation R_p on X_p that arises in the context of the Penrose tilings, it is important that we stay with X_p rather than $\{0,1\}^P$. We emphasize that in X_p, a 1 is always followed by a 0. The discussion below is an exegesis of [56, p.90].

Let R_p be the equivalence relation on X_p given by: $\{x_n\}R_p\{y_n\}$ if there exists an m such that $x_n = y_n$ for all $n \geq m$. (So two sequences are equivalent if they coincide eventually.) It is initially tempting to try to make R_p into a locally compact groupoid by giving it the relative topology inherited from $X_p \times X_p$. But R_p is not locally compact Hausdorff in this topology. For if $x = \{x_n\} \in X_p$, then any neighborhood of (x,x) in the relative product topology of R_p is going to contain a set of the form $\{(y,z) \in R_p : y_i = z_i = x_i$ for $1 \leq i \leq N\}$ for some $N \in \mathbf{P}$. But this set has a cluster point in $X_p \times X_p$ which is not in R_p.

This is a typical situation for locally compact groupoids – we will see it again with, for example, the Cuntz groupoid (**4.2**) and the Kronecker foliation equivalence relation. The relative product topology is often *not* the correct one for the groupoid (though it is in the case of the trivial groupoid). Instead, what is needed is a topology of "inductive limit" type. We now describe this in the case of R_p. For $n \in \mathbf{P}$, let

$$R_p^{(n)} = \{(x,y) \in R_p : x_i = y_i \text{ for } i \geq n\}. \qquad (2.19)$$

Each $R_p^{(n)}$ is a compact Hausdorff space since it is a closed subspace of $X_p \times X_p$. Also R_p is the increasing union of the $R_p^{(n)}$'s and so has an inductive limit topology $\mathcal{T}$. Thus a subset W of R_p is $\mathcal{T}$-open if and only if $W \cap R_p^{(n)}$ is open in $R_p^{(n)}$ for all n. It is left to the reader to check that each $R_p^{(n)}$ with the relative $\mathcal{T}$-topology is a compact open subset of R_p, and indeed that that topology coincides on $R_p^{(n)}$ with the relative prod-

---

[2]This is the terminology used by Muhly in [179]. When X is an arbitrary set, Vaisman ([263, p.138]) calls the trivial groupoid $X \times X$ the *banal groupoid*. Weinstein([273]) comments that it is also called the *coarse* groupoid.

uct topology inherited from $X_p \times X_p$. Further, $(R_p, \mathcal{T})$ is second countable and locally compact Hausdorff. In terms of convergent sequences,
$(x^r, y^r) \to (x, y)$ in R_p if and only if for some N, $(x^r, y^r) \in R_p^{(N)}$ eventually,
and $(x^r, y^r) \to (x, y)$ in $R_p^{(N)}$.

It is easy to check that $(R_p, \mathcal{T})$ is a locally compact groupoid. For
example, let us check (iii) of Definition 2.2.1. If x is in the unit space
$R_p^0 = X_p$, then R_p^x is the set of pairs (x, y) where, for some n, $y_i = x_i$ for
all $i \geq n$. This is a countable set, and by the above characterization of
convergent sequences, is a discrete subset of R_p and so is locally compact
Hausdorff.

The locally compact groupoid R_p has an interesting property – *the unit
space R_p^0 is an open subset of R_p*. For $R_p^0 = R_p^{(1)}$, which, by the above,
is open. The locally compact groupoid R_p is an example of an r-discrete
groupoid. (The class of these groupoids is defined later in the section.) In
particular, the counting measures on the R_p^x's give a left Haar system for
R_p.

For a general locally compact groupoid, the existence of a left Haar
system does impose a topological constraint on the groupoid because of the
continuity condition of (ii) above.

Proposition 2.2.1   ([230, p.17], [252]) *Let G be a locally compact groupoid. Then each of r, d is an open map from G onto G^0.*

Proof.   Let U be an open subset of G. We will show that $r(U)$ is an open
subset of G^0. Now by (ii) of Definition 2.2.1, the set U is a union of open
Hausdorff subsets V of G. So we can suppose that U is Hausdorff. Let
$u \in U$ and $f \in C_c(U)$ with $f \geq 0$ and $f(u) > 0$. By (ii) of Definition 2.2.2,
$f^0 \in C_c(G^0)$ and using (i) of Definition 2.2.2, we have $f^0(r(u)) > 0$. Since
f^0 vanishes outside $r(U)$, it follows that the open subset $\{v \in G^0 : f^0(v) >
0\}$ of G^0 is contained in $r(U)$ and contains $r(u)$. So $r(U)$ is the union of a
family of open subsets of G^0 and so is open in G^0. Since $d(U) = r(U^{-1})$,
the set $d(U)$ is also open in G^0.   □

For some purposes, it will be useful to consider Borel, rather than continuous functions on a locally compact groupoid. To this end, we now
establish some notation.

Firstly, in the usual notation, if X is a *locally compact Hausdorff* space,
then $C(X)$ is the space of continuous, bounded, complex-valued functions
on X, $C_0(X)$ is the space of functions $f \in C(X)$ that vanish at ∞, while
$C_c(X)$ is the space of functions $f \in C_0(X)$ with compact support.

Now let X be a topological space (not necessarily locally compact Hausdorff). The family of compact subsets of X is denoted by $\mathcal{C}(X)$. The Borel
algebra of X is denoted by $\mathcal{B}(X)$, the σ-algebra generated by the open
subsets of X. We denote by $B(X)$ the space of bounded, complex-valued,

Borel functions on X, and by $B_c(X)$ the subspace of those $f \in B(X)$ which vanish outside a compact subset of X.

Now let G be a locally compact groupoid. It is straight-forward to show that $B_c(G)$ can be defined in a similar way to that of $C_c(G)$: it is the span of those bounded, complex-valued, Borel functions on G that vanish outside a compact subset of an open Hausdorff subset of G. Unlike $C_c(G)$ in general, the spaces $B(G), B_c(G)$ are closed under the formation of pointwise products and the taking of absolute values. Also, for any $g \in B_c(G)$ and any $u \in G^0$, the restriction $g_{|G^u} \in B_c(G^u)$, so that $g_{|G^u} \in L^1(G^u, \lambda^u)$. It follows that the functions in (ii) and (iii) of Definition 2.2.2 are defined in the $B_c(G)$ case as well as in the $C_c(G)$ case. In other words, the function g^0 on G^0, where $g^0(u) = \int_{G^u} g \, d\lambda^u$, and the functions $x \to \int_{G^{d(x)}} f(xz) \, d\lambda^{d(x)}(z)$ and $x \to \int_{G^{r(x)}} f(y) \, d\lambda^{r(x)}(y)$ on G are defined for $g \in B_c(G)$.

We want to show (cf. [230, p.61]) that for such g and f, the function $g^0 \in B_c(G^0)$, and the equality (2.14) holds. In other words, we want to show that the $B_c(G)$-version of the axioms in Definition 2.2.2 holds, as well as the earlier $C_c(G)$-version. To this end, the following lemma, suggested to the author by a referee, is useful. (The author is grateful to that referee for pointing out an error in the author's original approach.)

Let X be a locally compact Hausdorff space. A sequence $\{f_n\}$ in $B_c(X)$ is said to converge pointwise to $f \in B_c(X)$ in a *compactly bounded way* if $f_n \to f$ pointwise, and there exists $C \in \mathcal{C}(X)$ and a positive real number k such that for all n,

$$| f_n | \leq k\chi_C$$

Lemma 2.2.1   *Let X be a second countable locally compact Hausdorff space and let A be the smallest subspace of $B_c(X)$ containing $C_c(X)$ and closed under pointwise convergence in a compactly bounded way. Then $A = B_c(X)$.*

Proof.   Of course, $B_c(X)$ contains $C_c(X)$ and is closed under pointwise convergence in a compactly bounded way. So A, the intersection of all subspaces of $B_c(X)$ which contain $C_c(X)$ and are closed under pointwise convergence in a compactly bounded way, exists.

Let $f \in B_c(X)$ be ≥ 0 and $C \in \mathcal{C}(X)$ be such that f vanishes outside C. Let $n \in \mathbf{P}$ and for each $i \in \mathbf{N}$, let $B_i = f^{-1}([\frac{i}{n}, \frac{i+1}{n})) \cap C \in \mathcal{B}(C)$. Let J be the (finite) set of i's such that $B_i \neq \emptyset$ and let $f_n = \sum_{i \in J} f(c_i)\chi_{B_i}$ where $c_i \in B_i$. Then f_n vanishes outside C and $\|f_n - f\|_\infty \leq 1/n$. It follows that to show that $A = B_c(X)$, we just need to prove that $\chi_B \in A$ for any Borel subset B of C. Let

$$\mathcal{M} = \{B \in \mathcal{B}(C) : \chi_B \in A\}.$$

We have to show that $\mathcal{M} = \mathcal{B}(C)$. If $\{B_n\}$ is an increasing sequence in $\mathcal{M}$ and $B = \cup_n B_n$, then $\chi_{B_n} \to \chi_B$ in a compactly bounded way, so that

$B \in \mathcal{M}$. The corresponding result holds if $\{B_n\}$ is a decreasing sequence in $\mathcal{M}$. So $\mathcal{M}$ is a monotone class.

Next, let E be a closed subset of C. We show that $E \in \mathcal{M}$. To this end, let D be any compact subset of X such that $C \subset D^o$. Since X is second countable and E is compact, there exists a decreasing sequence $\{U_n\}$ of open subsets of X with $E = \cap_n U_n$. Since $E \subset D^o$, we can suppose that every $U_n \subset D^o$. By Urysohn's lemma, there exists for each n a function $f_n \in C_c(X)$ such that $0 \leq f_n \leq 1$, $f_n(x) = 1$ for $x \in E$ and $f_n(x) = 0$ for $x \in X \sim U_n$. It follows that the bounded sequence $\{f_n\}$ of $C_c(X)$-functions converges pointwise to χ_E in a compactly bounded way. From the definition of A, it follows that $\chi_E \in A$ so that $E \in \mathcal{M}$. Now the σ-algebra $\mathcal{B}(C)$ is generated by the closed subsets of C, and since $\mathcal{M}$ is a monotone class, we have $\mathcal{M} = \mathcal{B}(C)$ as required. $\square$

Corollary 2.2.1   *Let G be a locally compact groupoid and $g \in B_c(G)$. Then $g^0 \in B_c(G^0)$ and the equality (2.14) holds.*

Proof.   We can suppose that $g \in B_c(U)$ where U is an open Hausdorff subset of G (since $B_c(G)$ is the span of those bounded, complex-valued, Borel functions on G that vanish outside a compact subset of an open Hausdorff subset of G). Let B be the subspace of functions $h \in B_c(U)$ for which the conclusion of the corollary holds (with h in place of g). Then $C_c(U) \subset B$, since $C_c(U) \subset C_c(G)$ and Definition 2.2.2 applies. Let $h_n \to h$ pointwise in $B_c(U)$ in a compactly bounded way with each $h_n \in B$. By the dominated convergence theorem, we have $h_n^0 \to h^0$ pointwise on G^0 so that $h^0 \in B_c(G^0)$. The same theorem gives that (2.14) holds for h. Hence $h \in B$. So B contains $C_c(U)$ and is closed under pointwise convergence in a compactly bounded way. By Lemma 2.2.1, $B = B_c(U)$. So $g \in B$ and the corollary is proved. $\square$

We now discuss the normed convolution *-algebra structure of $C_c(G)$ ([233, 230]). The convolution product on $C_c(G)$ is given by:

$$f * g(x) \;=\; \int_{G^{r(x)}} f(y)g(y^{-1}x)\, d\lambda^{r(x)}(y) \tag{2.20}$$

$$=\; \int_{G^{d(x)}} f(xt)g(t^{-1})\, d\lambda^{d(x)}(t), \tag{2.21}$$

the equality of the last two expressions following from (iii) of Definition 2.2.2.[3] For a function $f : G \to \mathbf{C}$, define $f^* : G \to \mathbf{C}$ by:

$$f^*(x) = \overline{f(x^{-1})} \;\; (= \overline{\check{f}(x)}). \tag{2.22}$$

---

[3]More precisely, we have $f(xt)g(t^{-1}) = h(xt)$ where $h(t) = f(t)g(t^{-1}x)$. Note that the pointwise product h may not belong to $C_c(G)$ but it will belong to $B_c(G)$, and we can use the $B_c(G)$-version of (iii) of Definition 2.2.2 with h in place of f.

(The map $f \to f^*$ will be the involution on $C_c(G)$.)

Unlike the case of a locally compact group, there is no modular function involved in the involution. In the locally compact group case, the modular function arises when we compare a left Haar measure λ with λ^{-1} (a right Haar measure). However, in the locally compact groupoid case, there are many λ^u's (in place of the essentially unique left Haar measure) and $\lambda^u, (\lambda^u)^{-1}$ live on *different* subsets of the groupoid. Under these circumstances, defining the involution as in (2.22) seems to be the only reasonable choice, but this will entail modifying the formulae involved in integrating groupoid representations later in **3.1**. (See, for example, (3.24) and (3.25).) As we will see in that section, modular functions will reappear in groupoid representation theory through quasi-invariant measures.

A useful norm on $C_c(G)$ – closely related to, though in general different from, the L^1-norm in the locally compact group case – is the *I-norm* $\|.\|_I$ due to P. Hahn ([114]). This is associated with two other norms $\|.\|_{I,r}, \|.\|_{I,d}$ where:

$$\|f\|_{I,r} = \sup_{u \in G^0} \int_{G^u} |f(t)| \, d\lambda^u(t), \qquad (2.23)$$

$$\|f\|_{I,d} = \sup_{u \in G^0} \int_{G_u} |f(t)| \, d\lambda_u(t). \qquad (2.24)$$

Then

$$\|f\|_I = \max\{\|f\|_{I,r}, \|f\|_{I,d}\}. \qquad (2.25)$$

(As we will see later, to ensure an isometric involution on $C_c(G)$, we need to take the maximum of the (I,r) and (I,d) norms, i.e. the I-norm, on $C_c(G)$.)

We need to check that the $(I,r), (I,d)$-norms are indeed norms. We deal with the (I,r)-case. Firstly, the right-hand side of (2.23) is finite. For given $f \in C_c(G)$, we have $|f| \in B_c(G)$, and since the function $u \to \int_{G^u} |f| \, d\lambda^u$ belongs to $B_c(G^0)$, we obtain that $\|f\|_{I,r} < \infty$. If $f \neq 0$, then for some $u \in G^0$, the restriction of $|f|$ to G^u is non-zero, and hence its λ^u-integral is non-zero by (i) of Definition 2.2.2. The other required norm properties for $\|.\|_{I,r}$ are obviously true.

Similarly, $\|.\|_{I,d}$ is a norm, and it immediately follows that $\|.\|_I$ is also a norm. The following useful result relates uniform convergence of sequences in $C_c(G)$ with "controlled supports" in $C_c(G)$ to their I-norm convergence. (For the Hausdorff case, see [230, p.51].)

Proposition 2.2.2 *Let C be a compact subset of G and $\{f_n\}$ be a sequence in $C_c(G)$ such that every f_n vanishes outside C. Suppose that $f_n \to f$ uniformly in $C_c(G)$. Then $f_n \to f$ in the I-norm of $C_c(G)$.*

Proof. Using (ii) of Definition 2.2.1 and the compactness of C, there exist in G open Hausdorff sets $U_1, \ldots, U_n$ covering C and open Hausdorff sets

$V_1, \ldots, V_n$ such that the closure of each U_i in V_i is compact. Let $F_i \in C_c(V_i)$ be such that $F_i \geq \chi_{U_i}$. In particular, F_i is positive. Let $F = \sum_{i=1}^n F_i$. Then $F \in C_c(G)$ and $F \geq \chi_C$. Hence $| f_n - f | \leq | f_n - f | F$, and we have

$$
\begin{aligned}
\| f_n - f \|_{I,r} &= \sup_{u \in G^0} \int_{G^u} | f_n(t) - f(t) | \, d\lambda^u(t) \\
&\leq \sup_{u \in G^0} \int_{G^u} | f_n(t) - f(t) | \, F(t) \, d\lambda^u(t) \\
&\leq \| f_n - f \|_\infty \| F^0 \|_\infty \\
&\to 0
\end{aligned}
$$

as $n \to \infty$. Similarly, $\| f_n - f \|_{I,d} \to 0$ and so the same conclusion holds for the I-norm. $\square$

We now prove that $C_c(G)$ is a normed *-algebra under the I-norm with convolution multiplication. For the reader who is content with the r-discrete case, an easy proof can be given for that case using (2.30).

Theorem 2.2.1   ([230, Ch.2]) *Let G be a locally compact groupoid. Then $C_c(G)$ is a separable, normed *-algebra under convolution multiplication and the I-norm, and the involution is isometric.*

Proof.   We first prove separability for $C_c(G)$. By (ii) of Definition 2.2.1, there exists a countable basis $\{U_i\}$ of open locally compact Hausdorff subsets for the topology of G. Each U_i is itself therefore separable. Write U_i as an increasing union of a sequence $\{D_i^n\}$ ($n \geq 1$) of compact subsets of U_i with D_i^n contained in the interior $(D_i^{n+1})^0$ of D_i^{n+1}. Now both $C_0((D_i^n)^0), C_c((D_i^n)^0)$ are separable in the sup-norm topology. Let $\mathcal{A}_i^n$ be a countable dense subset of $C_c((D_i^n)^0)$ and $\mathcal{A}_i$ be the (countable) union over n of the sets $\mathcal{A}_i^n$. Next let $\mathcal{A}$ be the (countable) set of functions which are finite sums of elements of $\cup_i \mathcal{A}_i$.

We claim that $\mathcal{A}$ is I-norm dense in $C_c(G)$. Indeed, if $f \in C_c(G)$, we can write $f = \sum_{j=1}^n f_j$ where $f_j \in C_c(W_j)$ for some open Hausdorff subset W_j of G. Covering the (compact) support of f_j in W_j by a finite number of U_i's each entirely inside W_j and using a partition of unity argument, we can express f_j as a finite sum of functions each in some $C_c(U_i)$. So we can assume that $f_j = h \in C_c(U_i)$ for some i. By construction, the support of h in U_i is contained in some $(D_i^n)^0$, and there is a sequence $\{g_k\}$ in $C_c((D_i^n)^0) \cap \mathcal{A}$ with $\| g_k - h \|_\infty \to 0$. Now the functions g_k, h vanish outside the compact set D_i^n, and so by Proposition 2.2.2, we have $\| g_k - h \|_I \to 0$. The I-norm density of $\mathcal{A}$ in $C_c(G)$ now follows, and hence also the separability of $(C_c(G), \|.\|_I)$.

The fact that $C_c(G)$ is closed under convolution for locally compact *Hausdorff* groupoids follows from Lemma C.0.2. This proof can be adapted to general locally compact groupoids at the price of a little more complication. The proof of the associative property, which will now be given, looks messy, but really reduces to the groupoid properties that $(xz)y = x(zy)$ whenever one or other makes sense and that $(zy)^{-1}z = y^{-1}$ ($(z,y) \in G^2$).

So let $f, g, h \in C_c(G)$ and $x \in G$. Then using the convolution formula (2.21) four times, the facts that $d(y^{-1}) = r(y)$ and $r(y) = d(x) = r(z)$ for the x, y, z in the proof, (2.14) and Fubini's theorem, we obtain that $f * (g * h)(x)$ equals

$$
\int_{G^{d(x)}} f(xy) g * h(y^{-1}) \, d\lambda^{d(x)}(y)
$$

$$
= \int_{G^{d(x)}} f(xy) \, d\lambda^{d(x)}(y) \int_{G^{d(y^{-1})}} g(y^{-1}z) h(z^{-1}) \, d\lambda^{d(y^{-1})}(z)
$$

$$
= \int_{G^{d(x)}} f(xy) \, d\lambda^{d(x)}(y) \int_{G^{d(x)}} g(y^{-1}z) h(z^{-1}) \, d\lambda^{d(x)}(z)
$$

$$
= \int_{G^{d(x)}} h(z^{-1}) \, d\lambda^{d(x)}(z) \int_{G^{d(x)}} f(xy) g(y^{-1}z) \, d\lambda^{d(x)}(y)
$$

$$
= \int_{G^{d(x)}} h(z^{-1}) \, d\lambda^{d(x)}(z) \int_{G^{d(z)}} f(x(zy)) g((zy)^{-1}z) \, d\lambda^{d(z)}(y)
$$

$$
= \int_{G^{d(x)}} h(z^{-1}) \, d\lambda^{d(x)}(z) \int_{G^{d(xz)}} f((xz)y) g(y^{-1}) \, d\lambda^{d(xz)}(y)
$$

$$
= \int_{G^{d(x)}} f * g(xz) h(z^{-1}) \, d\lambda^{d(x)}(z)
$$

$$
= (f * g) * h(x).
$$

We now prove that for all $f, g \in C_c(G)$, we have $g^* * f^* = (f * g)^*$. Indeed, using (2.20), (2.16) and the fact that $d(x) = r(x^{-1})$ for all $x \in G$, we have

$$
\begin{aligned}
g^* * f^*(x) &= \int_{G^{r(x)}} g^*(y) f^*(y^{-1}x) \, d\lambda^{r(x)}(y) \\
&= \int_{G^{r(x)}} g^*(x(x^{-1}y)) \overline{f(x^{-1}y)} \, d\lambda^{r(x)}(y) \\
&= \int_{G^{d(x)}} g^*(xy) \overline{f(y)} \, d\lambda^{d(x)}(y) \\
&= \overline{\int_{G^{r(x^{-1})}} f(y) g(y^{-1}x^{-1}) \, d\lambda^{r(x^{-1})}(y)} \\
&= (f * g)^*(x).
\end{aligned}
$$

So the map $f \to f^*$ is an involution on $C_c(G)$.

Next we show that the involution is isometric for the I-norm. Indeed, by (2.12), we have

$$
\|f\|_{I,r} = \|f^*\|_{I,d} \tag{2.26}
$$

so that $\|f\|_I = \|f^*\|_I$.

To obtain that $C_c(G)$ is a normed algebra under the I-norm, we have to show that for all $f, g \in C_c(G)$,

$$\|f * g\|_I \leq \|f\|_I \|g\|_I. \qquad (2.27)$$

To this end, using (2.26) and the equality $(f*g)^* = g^* * f^*$, it is sufficient to show that

$$\|f * g\|_{I,r} \leq \|f\|_{I,r} \|g\|_{I,r}. \qquad (2.28)$$

For $u \in G^0$, using (2.20), Fubini's theorem and (2.16), we have

$$\int_{G^u} |\, f * g(x)\,|\; d\lambda^u(x)$$

$$= \int_{G^u} d\lambda^u(x) \int_{G^u} |\, f(t)\,| \, |\, g(t^{-1}x)\,| \; d\lambda^u(t)$$

$$= \int_{G^u} |\, f(t)\,| \; d\lambda^u(t) \int_{G^u} |\, g(t^{-1}x)\,| \; d\lambda^u(x)$$

$$= \int_{G^u} |\, f(t)\,| \; d\lambda^u(t) \int_{G^{d(t)}} |\, g(x)\,| \; d\lambda^{d(t)}(x)$$

$$\leq \|g\|_{I,r} \int_{G^u} |\, f(t)\,| \; d\lambda^u(t)$$

$$\leq \|f\|_{I,r} \|g\|_{I,r}.$$

The inequality (2.28) now follows.   □

The space $B_c(G)$ is also a *-algebra using the same formulae as (2.20) and (2.22). This can be proved using Lemma 2.2.1 a number of times. For example (in the Hausdorff case), to prove that $f * g \in B_c(G)$ for $f, g \in B_c(G)$, one uses the lemma first to show that for $h \in C_c(G)$, the space $\{p \in B_c(G) : p * h \in B_c(G)\}$ equals $B_c(G)$. So $f * h \in B_c(G)$. Next one uses the lemma again to show that $\{q \in B_c(G) : f * q \in B_c(G)\}$ equals $B_c(G)$, from which it follows that $f * g \in B_c(G)$. One can also define $\|f\|_{I,r}$, $\|f\|_{I,r}$ and $\|f\|_I$ for $f \in B_c(G)$, and show, exactly as in the proof of Theorem 2.2.1, that $(B_c(G), \|.\|_I)$ satisfies all of the conditions for a normed *-algebra with isometric involution *except* that $\|f\|_I$ can be 0 without f being 0. (Consider, for example, a non-zero function that is zero almost everywhere on a non-discrete locally compact group.) So $\|.\|_I$ is, in general, a seminorm, not a norm. It is easy to see – and will be discussed later – that if G is r-discrete, then $B_c(G)$ *is* a normed *-algebra.

It seems to be unknown if, in general, $C_c(G)$ always has a bounded approximate identity for $\|.\|_I$. It follows from the work of Renault ([230, p.56]) and Muhly, Renault and Williams ([184]) that $C_c(G)$ with the inductive limit topology does always have an approximate identity. The latter topology is however not usually metrizable and we will not have occasion to use it in the present work, the I-norm and pointwise convergence in a compactly bounded way sufficing for our purposes. Again, if G is r-discrete, then $C_c(G)$ *does* have a bounded approximate identity for the I-norm as we will see later. **All representations of the $*$-algebra $C_c(G)$ on a Hilbert space, considered in the book, will be assumed to be I-norm continuous.**[4]

We now turn to a fundamental issue for the present work, viz. that of how inverse semigroups and groupoids relate to one another. In [230, p.10], Renault defines a *G-set* in a groupoid G to be a subset A of G such that the restrictions r_A, d_A of the range and source maps to A are one-to-one. A brief discussion of G-sets was given in Chapter 1. The next proposition ([230, p.10]) shows that the family Σ of G-sets in G is in fact an inverse semigroup, the product being given by $(A, B) \to AB$ $(= \{ab : a \in A, b \in B, (a, b) \in G^2\})$ and the inverse map by $A \to A^{-1}$ $(= \{a^{-1} : a \in A\})$. The inverse semigroup Σ is usually uncountable.

Proposition 2.2.3 *The family Σ of G-sets in a groupoid G is an inverse semigroup under set multiplication and with set inversion as involution. The set G^0 is the unit 1 and $\emptyset$ the zero 0 for Σ.*

Proof. We first show that Σ is a semigroup. Recall that a product xy of elements $x, y \in G$ is defined only if $d(x) = r(y)$. Let $A, B \in \Sigma$ and $C = AB$. Suppose that $c_1, c_2 \in C$ and that $r(c_1) = r(c_2)$. Write $c_i = a_i b_i$ with $a_i \in A, b_i \in B$. Then $r(a_1) = r(c_1) = r(c_2) = r(a_2)$. Since r is one-to-one on A, we have $a_1 = a_2$. But then $r(b_1) = d(a_1) = d(a_2) = r(b_2)$ and since r is one-to-one on B, we have $b_1 = b_2$. So $c_1 = c_2$ and r is one-to-one on C. Similarly d is one-to-one on C, and $C \in \Sigma$. For $a \in G$, we have $r(a) = d(a^{-1}), d(a) = r(a^{-1})$, and it follows that $A^{-1} \in \Sigma$.

Next, using the G-set property for A and A^{-1}, we have

$$AA^{-1}A = \{aa^{-1}a : a \in A\} = A$$

using (1.11). Interchanging A and A^{-1} gives $A^{-1}AA^{-1} = A^{-1}$. From the definition of *inverse semigroup* it remains to show that the only $B \in \Sigma$ for which $ABA = A, BAB = B$ is $B = A^{-1}$. To show this, using the G-set property, for $a \in A$, there exists a unique $b \in B$ such that $aba = a$, and using (1.11), $b = a^{-1}aa^{-1} = a^{-1} \in A^{-1}$. So $A^{-1} \subset B$. Interchanging the roles of B, A and using $BAB = B$ gives $B \subset A^{-1}$, so that $B = A^{-1}$. So Σ is an inverse semigroup.

---

[4] Renault ([230, p.50]) requires representations of $C_c(G)$ to be continuous in the inductive limit topology on $C_c(G)$ rather than the I-norm but the fact that we can use the I-norm here is a simple folk-lore result that uses the appropriate version of [230, Ch.2, Proposition 1.4].

Next r, d are the identity maps on G^0 so that $G^0 \in \Sigma$. Also, if $a \in G$ and $u \in G^0$, then $(a, u) \in G^2$ if and only if $d(a) = r(u) = u$. So $AG^0 = \{a(d(a)) : a \in A\} = A$ and similarly, $G^0 A = A$. So G^0 is the unit for Σ. Trivially, $\emptyset$ is the zero for Σ.   □

With Σ as in the preceding proposition, the semilattice $E(\Sigma)$ is just the family of subsets of G^0. To prove this, let $A \in E(\Sigma)$. Then $A^2 = A$. Let $x \in A$. Then for some $y, z \in A$ with $(y, z) \in G^2$, we have $x = yz$. Then $r(y) = r(yz) = r(x)$, and since A is a G-set, we have $y = x$. Similarly, since $d(z) = d(x)$, we have $z = x$. So $x^2 = x$ and $x \in G^0$. So $A \subset G^0$. Conversely, it is trivial that any subset of G^0 is in $E(\Sigma)$.

Certain inverse semigroup properties become transparent in the G-set context. For example, there is ([133, p.137]) a natural partial ordering on an inverse semigroup S given by: $s \leq t$ if and only if there exists $e \in E(S)$ such that $s = te$. If S is an inverse subsemigroup of Σ in the groupoid G, the ordering on S is the natural one: $A \leq B$ if and only if $A \subset B$. The easy proof (which uses the above fact that the idempotents of S are subsets of G^0) is left to the reader.

We are interested in a class of locally compact groupoids which have a large set of topologically nice G-sets. These are the r-discrete groupoids discussed in Chapter 1. These groupoids appear in the work of Feldman and Moore ([98]), and they are discussed in detail by Renault in his book ([230, pp.18ff.]). Sometimes, r-discrete groupoids are referred to as *étale* groupoids.

Let G be a locally compact groupoid and for $A \subset G$, let r_A, d_A respectively be the restrictions of r and d to A. Recall from Chapter 1 that the family of open, Hausdorff subsets A of G such that r_A, d_A are homeomorphisms onto open subsets of G is denoted by G^{op}. Note that if $U \in G^{op}$, then $r(U) \subset G^0$ is open *in* G (as well as in G^0). Note also that G^{op} makes sense for topological groupoids.

Definition 2.2.3. A locally compact groupoid G is called *r-discrete* if G^{op} is a basis for the topology of G.

Every $A \in G^{op}$ is a G-set. The family G^{op} is usually not countable. Since open subsets of G^{op}-sets are obviously also in G^{op}, it follows that G is r-discrete if and only if $G = \cup G^{op}$. To justify the "r-discrete" terminology, we will see below that if G is r-discrete, then every r-fiber G^u is (in fact) discrete. If G is r-discrete, then the unit space $G^0 = \cup_{A \in G^{op}} r(A)$ is *open* in G. Renault ([230, pp.18-20]) actually defines G to be r-discrete if G^0 is open in G, and shows that r-discreteness in the sense of Definition 2.2.3 is equivalent to the unit space G^0 being open in G and G having a left Haar system.[5]

---

[5]Renault's definition of *locally compact groupoid* does not require the existence of a

In order to check that a topological groupoid G is r-discrete, we need only show that G^0 is locally compact Hausdorff in the relative topology, and that there is a countable family $\mathcal{C}$ satisfying (ii) of Definition 2.2.1 such that $C^o \in G^{op}$ for all $C \in \mathcal{C}$. Indeed, if this can be shown, then (i) and (ii) of that Definition trivially hold, while (iii) follows from the discreteness of each G^u (below), and (iv) from the proof of Proposition 2.2.5. So G is a locally compact groupoid, and it is r-discrete since G^{op} contains the basis $\{C^o : C \in \mathcal{C}\}$ for G.

A useful property possessed by an r-discrete groupoid G is that $A \subset G$ belongs to G^{op} if and only if A is an open, Hausdorff G-set. One implication is trivial. For the other, suppose that A is an open, Hausdorff G-set. By Proposition 2.2.1 and the openness of G^0 in G, it follows that the $r_A : A \to r(A)$ is a continuous, open map onto an open subset of G. Since r_A is one-to-one, it is a homeomorphism. Similarly, d_A is a homeomorphism onto $d(A)$, and so $A \in G^{op}$.

The class of r-discrete groupoids is very large and we will meet many such groupoids in the course of the book. Of course, any countable discrete group is an r-discrete groupoid, and more generally, any transformation groupoid for which the group acting is countable and discrete is also r-discrete. (We looked briefly at this class of r-discrete groupoids in Chapter 1.) The r-discrete groupoids naturally associated with an inverse semigroup are the *ample* groupoids defined later in this section.

The locally compact groupoid R_p discussed earlier in this section is an r-discrete groupoid. Indeed, let $(x, y) \in R_p$. Then for some N, $(x, y) \in R_p^N$. Then the set A of elements $(x_1 x_2 \ldots x_{N-1} z, y_1 y_2 \ldots y_{N-1} z)$ in R_p is easily checked to be in G^{op}, and varying N and (x, y) gives a basis for R_p. So R_p is indeed r-discrete.

The next result shows that when G is r-discrete, then G^{op} is an inverse subsemigroup of the semigroup Σ of G-sets in G.

Proposition 2.2.4 *Let G be an r-discrete groupoid. Then G^{op} is an inverse subsemigroup of Σ. The set G^0 is the unit 1 and $\emptyset$ the zero 0 for G^{op}.*

Proof. Since inversion is a homeomorphism on G which interchanges r and d, it is obvious that $A^{-1} \in G^{op}$ if $A \in G^{op}$. It remains to show that for $A, B \in G^{op}$, we have $AB \in G^{op}$. Since G^{op} is the set of open, Hausdorff G-sets and a product of two G-sets is a G-set, we just have to show that AB is open and Hausdorff. (Trivially, G^0 is the unit and $\emptyset$ the zero for G^{op}.)

Let $(u, v) \in A \dot\times B$ ((2.13)). Then $uv \in AB$. Since G is r-discrete, there exists $W \in G^{op}$ such that $uv \in W$. So r_W is a homeomorphism onto an

---

left Haar system. In Definition 2.2.3, we also do not need to assume *a priori* the existence of a left Haar system, such systems coming automatically from Proposition 2.2.5 below.

open neighborhood Z of $r(uv)$. By continuity of the product on G, there exist open neighborhoods U of u and V of v such that $UV \subset W$. We can suppose that $U \subset A$, $V \subset B$ (so that $U, V \in G^{op}$). We can also suppose that $U \subset d^{-1}(r(V))$. Then $r(UV) = r(U)$ is an open neighborhood of $r(uv)$ contained in Z. Since $W \in G^{op}$, we have that $UV = r^{-1}(r(UV)) \cap W$ is an open neighborhood of uv in AB. So AB is open.

Next, let $(a, b), (a_1, b_1) \in A \dot\times B$ and $ab \neq a_1 b_1$. Without loss of generality, we can suppose that $a \neq a_1$. Since A is Hausdorff, there exist disjoint, open neighborhoods U, U_1 of a and a_1 in A. Then $UB, U_1 B$ are open neighborhoods of ab, $a_1 b_1$ in AB. Further, by applying r, we have $UB \cap U_1 B = \emptyset$. So AB is Hausdorff.   □

We note that for r-discrete G, every G^u is discrete (thus justifying the terminology). To see this, observe that if $x \in G^u$, then there exists $U \in G^{op}$ such that $x \in U$, and since r is one-to-one on U, the singleton set $\{x\} = G^u \cap U$ is open in G^u. Since G^u has a countable basis, it follows that G^u is also countable.

We noted earlier (by way of example of the trivial groupoids) that locally compact groupoids in general do not have unique left Haar systems. However, as we shall see below, every r-*discrete* groupoid comes equipped with a canonical left Haar system, that of counting measure on the sets G^u. In fact, *all* left Haar systems on such a groupoid are equivalent in the sense of measure theory. As we shall see in Theorem 2.3.1, under reasonable smooth conditions on the measures of the left Haar system, this is also true for Lie groupoids. In fact, the following proposition, which determines all of the left Haar systems on an r-discrete groupoid G, can be regarded as a 0-dimensional version of that theorem.

Let $P_+(G)$ be the set of continuous functions $\alpha : G^0 \to (0, \infty)$. Obviously, $P_+(G)$ is a cone, i.e. closed under addition and multiplication by positive scalars. The family $\Lambda(G)$ of left Haar systems $\{\lambda^u\}$ on G is also a cone in the natural way: $\{\lambda^u\} + \{\mu^u\} = \{\lambda^u + \mu^u\}$ and for $c > 0$, $c\{\lambda^u\} = \{c\lambda^u\}$.

Proposition 2.2.5   *Let G be an r-discrete groupoid. Every $\alpha \in P_+(G)$ defines a left Haar system $\{\Gamma^u(\alpha)\}$ where for each u,*

$$(\Gamma^u(\alpha)) = \sum_{x \in G^u} \alpha(d(x)) \delta_x. \tag{2.29}$$

Conversely, every left Haar system $\{\lambda^u\}$ is of the form $\{\Gamma^u(\alpha)\}$ for some $\alpha \in P_+(G)$, and $\Gamma : P_+(G) \to \Lambda(G)$ is an isomorphism of cones.

Proof.   Let $\alpha \in P_+(G)$ and $\lambda^u = \Gamma^u(\alpha)$. We check the properties (i),(ii) and (iii) of Definition 2.2.2. Firstly, since α is strictly positive, it follows that the support of λ^u is G^u.

Next let $g \in C_c(G)$. We can suppose that $g \in C_c(U)$ where U is an open Hausdorff subset of G. (The function g is, of course, extended to be 0 outside U.) Since G^{op} is a basis for the topology of G, the (compact) support of g in U is covered by a finite number of sets $A \in G^{op}$, $A \subset U$. We can also suppose that the closure of each A in U is compact. Using a partition of unity argument for this cover, we can assume that $g \in C_c(A)$. Then

$$g^0(u) = \int_{G^u} g \, d\lambda^u = g \circ (r_U)^{-1}(u)\alpha(d_U \circ (r_U)^{-1}(u))$$

so that $g^0 \in C_c(r(U)) \subset C_c(G^0)$.

Lastly, let $x \in G$ and $f \in C_c(G)$. Then putting $y = xz$, we have

$$\int_{G^{d(x)}} f(xz) \, d\lambda^{d(x)}(z) = \sum_{r(z)=d(x)} f(xz)\alpha(d(z))$$

$$= \sum_{r(y)=r(x)} f(y)\alpha(d(y)),$$

while

$$\int_{G^{r(x)}} f(y) \, d\lambda^{r(x)}(y) = \sum_{r(y)=r(x)} f(y)\alpha(d(y)).$$

This gives (2.14).   □

Taking α to be identically 1 on G^0, we obtain the left Haar system $\{\lambda^u\}$ where λ^u is counting measure. **We will always take this to be the canonical left Haar system for an r-discrete groupoid.** (Of course, since the G^u's are discrete, all left Haar systems are trivially equivalent to the counting measure one.)

The argument of the proof of Proposition 2.2.5 gives that (in the r-discrete case) each $f \in C_c(G)$ is a linear combination of functions $g \in C_c(A)$ for some $A \in G^{op}$. This makes proofs of the basic results on locally compact groupoids often easier in the r-discrete case. For example, to show that (above) $C_c(G)$ is closed under convolution in the r-discrete case, we can assume $f \in C_c(A), g \in C_c(B)$ where $A, B \in G^{op}$. From Proposition 2.2.4, $AB \in G^{op}$. But then, recalling that the left Haar system of G consists of counting measures and using the fact that A, B are G-sets, we obtain from (2.20) that $f * g \in C_c(AB) \subset C_c(G)$ is given by the very simple formula:

$$f * g(x) = f(a)g(b) \tag{2.30}$$

where $x = ab$ for $a \in A, b \in B$. The reader is invited to give a very quick proof for the r-discrete case that $C_c(G)$ is a *-algebra under convolution.

The normed *-algebra $C_c(G)$ has a bounded approximate identity $\{f_n\}$ when G is r-discrete. Indeed, in that case G^0 is open, and by Urysohn's lemma, there exists a sequence $\{f_n\}$ in $C_c(G^0)$ such that $0 \leq f_n \leq 1$ for every n and G^0 is the increasing union of the interiors U_n of the sets $\{u \in G^0 : f_n(u) = 1\}$. Recalling that the λ^u's are counting measures – in particular, $\lambda^u(\{u\}) = 1$ – it is easily checked that

$$\|f_n\|_I = \sup_{u \in G^0} | f_n(u) | = 1$$

so that the sequence $\{f_n\}$ is a bounded for the I-norm. The fact that $\{f_n\}$ is an approximate identity for $C_c(G)$ follows since $f * f_n = f = f_n * f$ eventually for any $f \in C_c(G)$. Let us explicitly prove the $f * f_n = f$ part of this in order to illustrate the use of the convolution formula (2.21), leaving the rest of the proof to the reader. We can suppose that f vanishes outside a compact subset C of G. Since $f_n \in C_c(G^0)$, only units t $(= t^{-1})$ have to be considered in (2.21). Now for such a t and any $x \in G$, xt is defined only if $t = r(t) = d(x)$, and since $\lambda^{d(x)}$ is counting measure, we have

$$f * f_n(x) = f(xd(x))f_n(d(x)) = f(x)f_n(d(x)).$$

If n is large enough so that $d(C) \subset U_n$, we have $f * f_n = f$ as claimed.

As noted earlier, the "I-norm" is defined on $B_c(G)$ just as for $C_c(G)$ for a locally compact groupoid G, but unfortunately, it is in general a seminorm, not a norm. However, in the r-discrete case, because the λ^u's are counting measure, it *is* a norm so that *if G is r-discrete, then $(B_c(G), \|.\|_I)$ is a normed *-algebra with isometric involution.* This normed *-algebra also has a bounded approximate identity – we can take this to be the bounded approximate identity $\{f_n\}$ for $C_c(G)$ above. Even easier, the sequence $\{\chi_{U_n}\}$ is a bounded approximate identity for $B_c(G)$ where the U_n's were the subsets of G^0 used in the construction of $\{f_n\}$.

As we shall see in Chapter 4, the locally compact groupoids naturally associated with an abstract inverse semigroup S form a special class of r-discrete groupoids. An example of such a groupoid is the Cuntz groupoid G_n (Chapter 1 and **4.2**, Example 3). Such groupoids will be called *ample*. To define this, for any locally compact groupoid G, let

$$G^a = \{A \in G^{op} : A \text{ is compact}\}. \tag{2.31}$$

So G^a is the family of *compact*, Hausdorff, open G-sets A in G such that both r_A, d_A are homeomorphisms onto open subsets of G.

Definition 2.2.4 The locally compact groupoid G is called *ample* if G^a is a basis for the topology of G.

Since $G^a \subset G^{op}$, it follows that every ample groupoid is r-discrete. The adjective *ample* comes from the terminology of Renault ([230, p.20]) who, following Krieger, called G^a, in the case of a (Hausdorff) r-discrete groupoid, the *ample semigroup* of G. It is left as an exercise (cf. Proposition 2.2.4) to the reader to show that for an ample groupoid, G^a is closed under products and inversion, so that *if G is ample, then G^a is an inverse subsemigroup of G^{op}.*

If G is an ample groupoid, then since G^0 is open in G and G^a is a basis for the topology of G, it follows that G^0 is totally disconnected. Also, since G has a countable basis of open sets, and every compact, open subset of G is the union of a finite number of such open sets, it follows that the semigroup G^a is *countable*. The remainder of this section is devoted to proving three results on ample groupoids which will prove useful later.

First, the study of G^a is facilitated by the fact that characteristic functions of sets in G^a are in $C_c(G)$, and the map $A \to \chi_A$ identifies G^a with an inverse subsemigroup of $C_c(G)$.

Proposition 2.2.6 *Let G be an ample groupoid and $A, B \in G^a$. Then $\chi_A \in C_c(G)$ and*

$$\chi_A * \chi_B = \chi_{AB}, \tag{2.32}$$

$$(\chi_A)^* = \chi_{A^{-1}}. \tag{2.33}$$

Proof. Since A is an open Hausdorff subset of G and χ_A is (trivially) continuous with compact support on A, we have $\chi_A \in C_c(G)$. Now for $x \in G$, we have from (2.20) that

$$\chi_A * \chi_B(x) = \int_{G^{r(x)}} \chi_A(y)\chi_B(y^{-1}x)\, d\lambda^{r(x)}(y) = \lambda^{r(x)}(A \cap xB^{-1}).$$

Since $\lambda^{r(x)}$ is counting measure and the sets A, xB^{-1} are G-sets, we have $\chi_A * \chi_B(x)$ is 1 when $A \cap xB^{-1}$ is non-empty and is zero otherwise. But $A \cap xB^{-1}$ is non-empty if and only if $x \in AB$, and this gives (2.32). The proof of the equality (2.33) is left to the reader. □

The second result shows in particular that the span of the χ_A's above in the ample case is *I*-norm dense in $C_c(G)$.

Proposition 2.2.7 *Let G be an ample groupoid and S be an inverse subsemigroup of G^a which is a basis for the topology of G. Then the span W of the characteristic functions χ_A for $A \in S$ is I-norm dense in $C_c(G)$.*

Proof. It is sufficient to show that if $\epsilon > 0$ and $f \in C_c(U)$, where U is a Hausdorff open set in G, then there exists $h \in W$ such that $\|f - h\|_I < \epsilon$. Since G^a is a basis for G and the support of f in U is compact, the latter support is covered by a finite number of sets $A_1, \ldots, A_n$ where $A_i \in G^a$, $A_i \subset U$. Since the A_i's are compact and open in the Hausdorff space U, intersections and differences of the A_i's are also in G^a. So we can suppose

that the A_i's are disjoint. Then $f = \sum_{i=1}^n f\chi_{A_i}$ and each $f\chi_{A_i} \in C(A_i)$
$(= C_c(A_i))$. So we can suppose that U is one of the A_i's. So $U \in G^a$. Let
W_U be the subspace of W spanned by the functions χ_A where $A \in S, A \subset U$.
Now W_U is a *-subalgebra of $C(U)$ (pointwise product). The *-part is
obvious. To prove the subalgebra part, we need only show that for $A, B \in S$
with $A, B \subset U$, we have $A \cap B \in S$, for then $\chi_A\chi_B = \chi_{A\cap B} \in W_U$. This
follows since $A \cap B = (AA^{-1})B \in S$. (The latter equality follows since
$A \cup B$ is contained in a G-set. Note that in general, G^a is *not* closed under
intersections.[6]) Next, since S is a basis for the topology of G and U is
Hausdorff, it follows that the subalgebra W_U separates the points of U
and does not vanish at any point of U. By the Stone-Weierstrass theorem,
W_U is uniformly dense in $C(U)$. Let $\epsilon > 0$ and $h \in W_U$ be such that
$\|f - h\|_\infty < \epsilon$. Then since U is a G-set, each of the sets $U \cap G^u, U \cap G_u$
is either a singleton or empty, and since both λ^u, λ_u are counting measures
for every $u \in G^0$, it follows by calculating from (2.23) and (2.24) that
$\|f - h\|_I \leq \|f - h\|_\infty < \epsilon$ as required.   □

The third result shows that (Borel) measurability is preserved under the
processes of multiplying and inverting subsets in an ample groupoid. (The
ample requirement in this proposition is probably unnecessary.)

Proposition 2.2.8   *The product of two measurable subsets of an ample
groupoid G is itself measurable and the inverse of every measurable set is
measurable.*

Proof.   Let B_1, B_2 be measurable subsets of G. Since G is a countable
union of G^a-sets, it follows that each B_i is a countable union of subsets
A where A is a measurable subset of some $C \in G^a$. We can therefore
suppose that each $B_i \subset C_i$ where $C_i \in G^a$. We show first that B_1C_2
is measurable. Let $\mathcal{C} = \{B \in \mathcal{B}(C_1) : BC_2 \in \mathcal{B}(G)\}$. The union of an
increasing sequence in $\mathcal{C}$ is trivially in $\mathcal{C}$. Let $\{D_n\}$ be a decreasing sequence
in $\mathcal{C}$. Let $z \in \cap_n D_n C_2$. Then for some $d_n \in D_n, c_n \in C_2$ we have $z = d_n c_n$.
Since C_1 is a G-set and $r(d_n) = r(z)$, we have that for some $d \in C_1, d_n = d$
for all n. Then $c_n = d^{-1}z$ for all n so that $z \in (\cap_n D_n)C_2 = \cap_n(D_n C_2)$.
So $\mathcal{C}$ is a monotone class and contains the subsets of C_1 which are in G^a.
Since G^a is a basis for G and C_1 is open in G, it follows that $\mathcal{C} = \mathcal{B}(C_1)$.
So B_1C_2 is measurable. The same result holds of course with C_2 replaced
by any of its G^a-subsets. A similar argument now applies with $\mathcal{C}$ replaced
by the family $\{B \in \mathcal{B}(C_2) : B_1B \in \mathcal{B}(G)\}$ to give B_1B_2 measurable.
      The last part of the proposition follows since the inversion map is a
homeomorphism.   □

---

[6]The (useful) example of Appendix C gives a counterexample.

2.3 Lie groupoids

In this section, we will require some of the basic concepts of differential geometry, in particular, those of manifolds, submanifolds, submersions, the tangent bundle, differential forms, s-densities and foliations. A brief survey of the necessary concepts is given in Appendix F to which the reader is invited to refer as required.

Of special importance in the category of locally compact groups are the *Lie groups*. These constitute the nexus where geometry, algebra and analysis meet, as well as giving the general theory of locally compact groups motivation and richness. The class of locally compact groupoids which corresponds in the groupoid category to the class of Lie groups in the group category is that of *Lie groupoids*, the topic of this section.

Lie groupoids are of fundamental importance in noncommutative geometry and in the study of Poisson geometry. In the latter, the Lie groupoids involved are the *Poisson groupoids* (the class of which includes the *symplectic groupoids* ([38, 140, 163, 267])). In noncommutative geometry, the Lie groupoid is the noncommutative version of a (smooth) manifold in differential geometry.

It is not possible in the present work to do justice to the scope of Lie groupoids (even if the present writer were competent to do so). In particular, there will (sadly) be no discussion of Poisson groupoids – an excellent reference for that subject is the recent survey [274] by Alan Weinstein. See also [45, 153].

The present section relates to the rest of the book in a number of ways. Firstly, it provides some very important, additional examples of locally compact groupoids, as well as exhibiting a large class of groupoids with (under reasonable smoothness conditions) essentially unique left Haar systems. These groupoids also shed light on the non-Hausdorffness allowed in the definition of *locally compact groupoid*. Secondly, much of the rest of the book will be concerned with r-discrete groupoids, and r-discrete holonomy groupoids give a very important class of Lie groupoids that are r-discrete. Thirdly, another central theme of the book, that of the interaction of inverse semigroups with groupoids, emerges in the context of the r-discrete holonomy groupoid in the form of *pseudogroups*. Lastly, these groupoids, when regarded as sheaf groupoids, prefigure the localization theory of **3.3**.

The author also hopes that the discussion of Lie groupoids given in the present section will encourage the reader to explore further their roles in the fascinating worlds of noncommutative geometry and Poisson geometry. We now briefly survey the content of the section.

Lie groupoids, like Lie groups, should be manifolds. But since locally compact groupoids need not be Hausdorff, we cannot expect a Lie groupoid G to be necessarily Hausdorff. It is reasonable to define a Lie groupoid as a groupoid that is a (not necessarily Hausdorff) manifold for which the product and the inversion maps are smooth. With the definition of a locally compact groupoid in mind, it is also reasonable to require G^0 and every G^u to be Hausdorff submanifolds[7] of G. The slightly unexpected extra requirement for a Lie groupoid – that the range and source maps be *submersions* – ensures that G^2 is a submanifold of $G \times G$ so that we can talk about smoothness for the product map.

The basic properties of Lie groupoids are then developed, the major examples being postponed until later in the section. It is shown that these groupoids are *locally compact groupoids* in the sense of **2.2**. So the analysis on locally compact groupoids discussed in Chapter 3, in particular, the integration theory on which $C^*(G)$ depends, applies to these groupoids. This provides a basis for the use of the C^*-algebras of Lie groupoids in noncommutative geometry.

The problem that has to be dealt with here is the existence of a left Haar system of a Lie groupoid G. Connes deals with this problem ([53, 56]) by defining convolution not on $C_c(G)$ directly but rather on $C_c^\infty(G, \Omega)$ where Ω is a (trivial) line bundle over G involving $1/2$-densities on the G_u's and G^u's. This gives convolution defined independently of *any* left Haar system. However, analysis on locally compact groupoids (see **3.1**, **3.2**) intrinsically uses a left Haar system, and most people would probably prefer a convolution formula involving integrating a measure rather than integrating densities. There are also situations (e.g. [198]) where having an explicit left Haar system is helpful for producing, for example, asymptotic morphisms. For these reasons, we have preferred to develop convolution in terms of a left Haar system. We require an additional condition on a left Haar system for a Lie groupoid, viz. that it be *smooth*. This means that in terms of appropriate local coordinates, the Radon-Nikodym derivatives of the λ^u's are strictly positive and smooth. It turns out that *smooth* left Haar systems are unique up to equivalence, so that in this respect, Lie groupoids gratifyingly behave like locally compact groups. The connection between densities and smooth left Haar systems is clarified in Theorem 2.3.1 where such systems are identified with the strictly positive sections of the 1-density line bundle $\Omega^1(A(G)^*)$. (Here, $A(G)$ is the vector bundle over G^0 of tangent vectors along the G^u's at the units u. This is the *Lie algebroid* of G, the counterpart to the Lie algebra of a Lie group.) Since such sections obviously exist (using a partition of unity on G^0) it follows that there are smooth left Haar systems on G.

Having proved that Lie groupoids are indeed locally compact groupoids, we then turn to important examples in noncommutative geometry that

---

[7]In fact we only need to require that G^0 is a Hausdorff submanifold and that every G^u is Hausdorff.

help to justify the theory. The first of these is the *holonomy groupoid* or *graph* of a foliated manifold. (A brief introduction to foliations is given in Appendix F.) In a number of cases (no holonomy), this groupoid is just the leaf equivalence relation with a special topology on it. In general, we have to factor in *holonomy*, a concept due to Ehresmann, Haefliger and Reeb ([84, 110, 228]). Holonomy measures the behavior of leaves close to each other, and is obtained by following a path in a given leaf along adjacent leaves. An example of the use of the *holonomy groupoid G* in noncommutative geometry is in the longitudinal index theorem of Connes and Skandalis, where the analytic index of a pseudodifferential operator on the manifold, elliptic along the leaves, lies in the K-theory of $C^*(G)$ (Connes and Skandalis [61]).

A major theme of this book is that of the interconnectedness of inverse semigroups and groupoids, and so far, in our discussion of Lie groupoids, there has been no sign of inverse semigroups! However, the construction of the holonomy groupoid involves taking germs of certain local diffeomorphisms between transverse sections of the foliation. Now a local diffeomorphism is, of course, a partial one-to-one map, and such maps generate (**2.1**) inverse semigroups. The inverse semigroups associated in this way with foliations are examples of what are called *pseudogroups*. The notion goes back as far as Lie in the 1880's, and Cartan in the 1920's did work on the classification of "Lie pseudogroups".

We saw in **2.1** that the topology of an r-discrete groupoid G is determined by the "large" inverse semigroup G^{op}. Indeed, as we will see in Chapters 3 and 4, inverse semigroups determine r-discrete groupoids and conversely, and so it is natural to ask if we can obtain an r-discrete version of the holonomy groupoid using an appropriate pseudogroup. Such a groupoid was constructed by Haefliger ([113]) and by Hilsum and Skandalis ([130]), and is a *reduction* of the original holonomy groupoid. In the construction, each leaf gets replaced by a countable subset. The r-discrete holonomy groupoid is also a Lie groupoid. It is not uniquely defined (depending on a choice of a family of transverse sections) but it captures much of the information of the holonomy groupoid. For example, a result of Hilsum and Skandalis ([130]) gives that the reduced C^*-algebras of the two versions of the holonomy groupoid are Morita equivalent.

The last class of Lie groupoids examined in the section is that of the *tangent groupoids*. The concept has its algebraic-geometric origins in the work of Gerstenhaber ([102]) and was used by Baum, Fulton and Macpherson ([12]). It parallels the "blowing up" of a subvariety in algebraic geometry. It is a special case of the *normal groupoid* construction which seems to appear first in the work of Hilsum and Skandalis ([131]) and is also discussed by Weinstein ([269]).

Given a manifold M, the tangent groupoid G_M provides the framework for "deforming" the trivial equivalence relation $M \times M$ into the tangent bundle TM. At the C^*-algebra level, this corresponds to the "Heisenberg quantization" in which the algebra K of compact operators on $L^2(M)$ is "deformed" into $C_0(T^*M)$.

One starts off with two Lie groupoids: $M \times M$ and TM. Note that TM is a groupoid, with algebraic structure that of the bundle of the additive groups $T_x M$ as x ranges over M. It is obvious that $M \times M$ is a Lie groupoid, and (like every vector bundle), the groupoid TM with its bundle topology is a Lie groupoid. The tangent groupoid is defined to be

$$G_M = [(M \times M) \times \mathbf{R}^*] \cup (TM \times \{0\})$$

where $\mathbf{R}^* = \mathbf{R} \setminus \{0\}$. This is a disjoint union of groupoids, and, as observed in Chapter 1, any such union is a groupoid in the obvious way. The topology on $(M \times M) \times \mathbf{R}^*$ and TM are the usual ones, so that the really interesting feature of the topology is how to specify when a sequence in $(M \times M) \times \mathbf{R}^*$ converges to an element of $TM \times \{0\}$.

We first deal with the case where U is an open subset of $\mathbf{R}^n$. The Lie groupoid structure of G_U is made more transparent by identifying it with a certain Lie groupoid with a straightforward product sitting as an open subset in $\mathbf{R}^{2n+1}$. The tangent groupoid G_M is then dealt with using charts U for M (identified with open subsets U of $\mathbf{R}^n$). The family of sets of the form G_U give an atlas for the manifold structure of G_M. We conclude the section by using the earlier construction of smooth left Haar systems on Lie groupoids to calculate such a system for G_U. The author hopes that the reader will wish to explore further the use of the tangent groupoid in noncommutative geometry. (See, for example, the paragraph following the proof of Theorem 3.1.2.) Having surveyed the content of the section, we start now on the detailed development.

As discussed above, Lie groupoids will, in general, be manifolds that are not Hausdorff. In this section, then, the term *manifold* unqualified will not be assumed to be Hausdorff. When we require the usual notion of manifold, we will explicitly refer to it as a *Hausdorff* manifold.

A manifold then has the same definition as that given in Appendix F except that the Hausdorff condition is not assumed. A useful alternative definition of a *manifold* M is as follows. We assume that M is a set and that there is given a family $\{U_\alpha : \alpha \in A\}$ covering M and, for each α, a bijection $\phi_\alpha : U_\alpha \to \mathbf{R}^n$ mapping U_α onto an open subset of $\mathbf{R}^n$ such that every $\phi_\alpha(U_\alpha \cap U_\beta)$ is open in $\mathbf{R}^n$. We also require the transition function condition (F.1). Then it is an easy exercise in point set topology to show that M can be given a topology, with a basis consisting of sets of the form $\phi_\alpha^{-1}(W)$ where W is an open subset of $\mathbf{R}^n$. We require in addition that the topology be second countable. Then M becomes a manifold with charts (U_α, ϕ_α).

Manifolds (without the Hausdorff requirement) are not unfamiliar in differential topology. In fact, Lang, in Chapters 2 and 3 of his book ([154]), does not assume the Hausdorff condition for manifolds. Camacho and Neto ([44, p.11]) comment that such manifolds *occur naturally in certain discussions*. (One such natural occurrence is that of *Lie groupoids*.) Notions that are locally defined (such as submanifolds or vector bundles) are treated in the non-Hausdorff case just as in the Hausdorff case. We run into problems in the non-Hausdorff case when we need *partitions of unity* and this has repercussions for constructs which use such a partition. For example, the construction of Riemannian metrics fails in general, and there are problems with differential forms and the de Rham map.[8] As the following definition shows and as is suggested by the locally compact groupoid definition, additional Hausdorffness for a Lie groupoid is located in G^0 and the G^u's. We will give the definition first and then discuss it.[9]

Definition 2.3.1 A *Lie groupoid* is a groupoid G which is a manifold such that G^0 is a Hausdorff submanifold of G, every G^u is Hausdorff in the relative topology, the product and inversion maps are smooth and the range and source maps $d, r : G \to G^0$ are submersions.

We note that there is some redundancy in the above definition. As in the case of a Lie group, the smoothness of the inversion map can be shown to be a consequence of the other axioms ([164, p.85]).

An obvious question with the above definition is that it is not immediately clear how G^2 is a manifold. (This is required in order to make sense of smoothness for the product map.) Let G be a Lie groupoid and $p = \dim G^0$. Since the singleton $\{u\}$ is a submanifold (of dimension zero) and r is a submersion, it follows (Appendix F) that $G^u = r^{-1}(\{u\})$ is a (Hausdorff) submanifold of G. The same, of course, applies to G_u.

Next, we discuss the required manifold structure for G^2. To this end,

---

[8]This is observed by Brylinkski and Nistor in [40, p.342].

[9]The terminology *smooth groupoid* is sometimes used in place of *Lie groupoids*. (See, for example, the book [56, p.101] of Connes.) The terminology *Lie groupoids* was used in the important paper of Coste, Dazord and Weinstein ([63]), and seems to becoming established – see, for example, the paper of Weinstein ([274, p.15]) and the book of Vaisman ([263, p.140]). Indeed, it is natural to use the terminology *Lie groupoids* to refer to the groupoid version of *Lie groups*. References for the definition of a Lie groupoid in the Hausdorff case are [84, 85, 86, 208, 209, 210, 211, 164]. Kirill Mackenzie in his book ([168, p.84]) refers to these groupoids as *differentiable groupoids*. He uses the expression *Lie groupoids* to refer to a special class of what is called in the present book *Lie groupoids*.

In [63, 263], the authors do not assume a Lie groupoid to be Hausdorff and, as in Definition 2.3.1 below, require G^0 to be Hausdorff. They do not require the G^u's to be Hausdorff. We do require that condition as well in order to ensure that every Lie groupoid is a locally compact groupoid (Corollary 2.3.1), thus making available for Lie groupoids the representation theory for locally compact groupoids.

we argue as in [164, p.84]. The map

$$d\times r : G \times G \rightarrow G^0 \times G^0$$

is a surjective submersion (since both d and r are such). Now the diagonal $\Delta = \{(u,u) : u \in G^0\}$ is obviously a submanifold of $G^0 \times G^0$. Again by Appendix F,

$$G^2 = \{(x,y) \in G \times G : d(x) = r(y)\} = (d \times r)^{-1}(\Delta) \qquad (2.34)$$

is a submanifold of $G \times G$. This gives the manifold topology on G^2 required for the smoothness of the product map. Note that a similar argument gives that

$$G^r = \{(x,y) \in G \times G : r(x) = r(y)\}$$

is also a submanifold of $G \times G$.

We can now give some very obvious classes of Lie groupoids. Firstly, if M is a manifold, then M is of course a Lie groupoid of units. Next, the trivial groupoid $M \times M$ is a Lie groupoid. This is easy to check. For example (cf. **2.2**), G^2 is the subset $\{(x,y,y,z) : x,y,z \in M\}$ of $G \times G = M \times M \times M \times M$. The submanifold G^2 is thus identified with $M \times M \times M$ (ignoring the repeated y in (x,y,y,z)) and the product map which sends (x,y,z) to (x,z) is trivially smooth. A third example is provided by any Lie group. Finally (for the present), any vector bundle E (Appendix F) over a Hausdorff manifold M is a Lie groupoid. (This applies, for example, to the tangent bundle TM.) Indeed, E is a bundle of the vector spaces E_x each of which is, of course, an abelian additive group. So E is a bundle of groups and hence (Chapter 1) is a groupoid. The unit space of TM is $M \times \{0\}$ which is identified with M. The range and source maps are both just the map $e \rightarrow \pi(e)$, where π is the map from E to M defining the vector bundle, and each $G^x = E_x$. It is easy to check, using trivializations for E, that E with its manifold topology is a Lie groupoid.

Using (F.3), we can calculate the manifold dimensions of each G^u and G^2. Indeed, with $n = \dim G, p = \dim G^0$, we have

$$n - \dim G^u = p - \dim\{u\} = p$$

so that $\dim G^u = n - p$ (independent of u). (We will write $k = n - p$.)

Also

$$2n - \dim G^2 = \dim G \times G - \dim G^2 = \dim G^0 \times G^0 - \dim \Delta = 2p - p$$

so that $\dim G^2 = 2n - p$.

In the study of Lie groupoids, a certain type of open subset proves useful. (This kind of subset is related to the notion of a *foliation chart* for foliated manifolds (as in Appendix F).) Let V be an open subset of the Lie groupoid G. Since r is a submersion, it is an open map, so that $r(V)$ is open in G^0. The pair (V, ψ) is called an *r-fiberwise product* (cf. [191, p.7]) if there exists an open subset W of $\mathbf{R}^k$ containing 0, and ψ is a diffeomorphism from V onto $r(V) \times W$ which preserves r-fibers in the sense that $p_1(\psi(x)) = r(x)$ for all $x \in V$, where p_1 is the projection onto the first coordinate. So ψ "straightens out" each fiber $G^u \cap V$ ($u \in r(V)$) into W. We write $\psi_u = \psi_{|G^u \cap V}$. So

$$\psi_u : G^u \cap V \to \mathbf{R}^k \tag{2.35}$$

(identifying $\mathbf{R}^k$ with $\{u\} \times \mathbf{R}^k$). For such a product V, we will often leave the ψ implicit and write $V \cong r(V) \times W$. We will regard $r(V) \subset r(V) \times W$ by identifying $r(V)$ with $r(V) \times \{0\}$ (even though $r(V)$ may not be contained in V). Note that V is Hausdorff since $r(V) \times W$ is Hausdorff.

When $x \in G^0$, then there exists an r-fiberwise product pair (V, ψ) with $x \in V$, $r(V) \subset V$ and $\psi(u) = (u, 0)$ for all $u \in r(V)$.

There is a basis for the topology of G consisting of r-fiberwise product open sets. Indeed, since r is a submersion, it is locally equivalent to a projection (as discussed in Appendix F), and the result is obvious locally when r is a projection.

Next we need to define $C_c^\infty(G)$, the non-Hausdorff version of the space of C^∞-functions with compact support.[10] This is defined by Connes (([54])) effectively in the same way as was $C_c(G)$ in (**2.2**). So $C_c^\infty(G)$ is the space of complex-valued functions $f : G \to \mathbf{C}$, where f can be written as a finite sum $\sum_{i=1}^n f_i$ with each $f_i \in C_c^\infty(U_i)$ for a chart U_i of G. (A partition of unity argument shows that $C_c^\infty(G)$ has its usual meaning when G is Hausdorff.)

Clearly, $C_c^\infty(G) \subset C_c(G)$. If $\{\lambda^u\}$ is a left Haar system for G, then for any chart U of G and $f \in C_c(U)$, there exists (by the Stone-Weierstrass theorem) a sequence $\{f_n\}$ in $C_c^\infty(G)$ with supports in a fixed compact subset of U such that $f_n \to f$ uniformly on G. It follows from Proposition 2.2.2 that $C_c^\infty(G)$ *is I-norm dense in* $C_c(G)$. (The I-norm on $C_c(G)$ was defined in (2.25).)

Our next principal objective is to show that every Lie groupoid G is a locally compact groupoid. To this end, we have to check properties (i)-(iv) in the definition of *locally compact groupoid* (Definition 2.2.1). Clearly, properties (i) and (iii) hold since the spaces G^0, G^u are Hausdorff submanifolds of G (and every Hausdorff manifold is locally compact Hausdorff). We now check property (ii). This follows since G is a manifold. Indeed, for any manifold M, there is a countable basis $\mathcal{U}$ for M. Since the charts for M cover M, we can take each $U \in \mathcal{U}$ to be an open subset of a chart. Then U is a countable union of compact sets C_n with $C_n \subset C_{n+1}^o \subset C_{n+1}$

---

[10]Of course, this definition also applies for a general (non-Hausdorff) manifold.

and the family of these compact sets as U ranges over $\mathcal{U}$ satisfies property (ii) (for $M = G$).

It remains to show that G satisfies (iv) of Definition 2.2.1, i.e. that G admits a left Haar system. This is not so obvious. We will see later how to produce essentially unique, natural left Haar systems on G. But before that, we need to discuss an approach to convolution on $C_c^\infty(G)$, developed by A. Connes, which does not use a left Haar system at all. The approach uses a 1/2-density bundle on G, to which we now turn. (A discussion of s-densities on a vector bundle is given in Appendix F.)

Firstly we introduce the very important vector bundle $A(G)$, the *Lie algebroid* of G ([209, 164, 191]). As a set, $A(G) = \cup_{u \in G^0} T_u(G^u)$, so that the fiber over u is the tangent space at the point u along the submanifold G^u. We now discuss the vector bundle structure of $A(G)$. Of course, the bundle map $\pi : A(G) \to G^0$ just sends $X \in T_u G^u$ to u.

Cover G^0 by charts (U_α, ψ_α) in G that are r-fiberwise products $\cong r(U_\alpha) \times W_\alpha$. The sets $\{r(U_\alpha)\}$ form an open cover for G^0, each W_α is an open subset of $\mathbf{R}^k$ containing 0, and for each $u \in r(U_\alpha)$, we can take $\psi_\alpha(u) = (u, 0)$. We now construct maps $\tau_{\alpha\beta} : r(U_\alpha) \cap r(U_\beta) \to GL(\mathbf{R}^k)$. To this end, $r(U_\alpha) \cap r(U_\beta) \subset U_\alpha \cap U_\beta$, and so for each $u \in r(U_\alpha) \cap r(U_\beta)$, there exist open subsets W_α^u, W_β^u respectively of W_α, W_β, both containing 0, such that the restriction $h_{\alpha\beta}^u$ of the map $\psi_\alpha \psi_\beta^{-1}$ to $\{u\} \times W_\beta^u$ has range in $\{u\} \times W_\alpha^u$. We then take (cf. the tangent bundle case in Appendix F)

$$\tau_{\alpha\beta}(u) = D(h_{\alpha\beta}^u)((u, 0)).$$

The $\tau_{\alpha\beta}$'s are easily checked to be well-defined and to satisfy the cocycle condition, and are taken to be the transition maps for the *Lie algebroid* $A(G)$.

Motivation for this terminology is provided by the case where G is a Lie group. In that case, $G^0 = \{e\}$, $G^e = G$ and $A(G)$ is the tangent space to the identity e of G, i.e. it is the *Lie algebra* of G. Of course, the argument of the preceding paragraph only gives $A(G)$ as a vector bundle over G^0 without any algebraic structure. We comment in passing that the Lie algebra structure on $A(G)$ is on its *sections*, these corresponding to right invariant vector fields, not on its elements. (Of course in the Lie group case, the sections of $A(G)$ coincide with its elements.) One also needs to involve the derivative of the range map r which maps $A(G)$ into TG^0. This is called the *anchor* of $A(G)$. A detailed account of Lie algebroids in general and of $A(G)$ in particular is given by Mackenzie ([164, p.100f.]). We will not have occasion to use the Lie algebroid structure of the vector bundle $A(G)$[11] in this book.

---

[11] Another natural example of a Lie algebroid is that of the cotangent bundle T^*M of a Poisson manifold. For a discussion of the Lie algebroids arising in the study of Poisson

As described in Appendix F, for any $s \in \mathbf{R}$, the trivial complex line bundle $\Omega^s E$ is defined for any real vector bundle E over a Hausdorff manifold M, in particular, when $E = A(G)^*$ and $M = G^0$. As for the case where $E = T^*M$, in terms of local coordinates coming from an r-fiberwise product open set $V \cong r(V) \times W$, the s-densities on $A(G)^*$ are of the form $h(u) \mid k(u) \mid^s \mid dw \mid^s = h(u) \mid \omega \mid^s$ where $\omega = k(u) \, dw_1 \wedge \cdots \wedge dw_k$ and h is a complex-valued smooth function on an open subset of G^0. (Here, of course, $dw_1 \wedge \cdots \wedge dw_k$ is evaluated at $u \in G^0$.) In the case of a positive s-density, the $h(x)$ is omitted.)

Let $\Omega^s_d(G), \Omega^s_r(G)$ be the pull-back bundles

$$d^{-1}(\Omega^s(A(G)^*)), \quad r^{-1}(\Omega^s(A(G)^*))$$

respectively. (Note that although G may not be Hausdorff, vector bundles and their pull-backs still make sense.) These are complex line bundles over G. We are particularly interested in the case where $s = 1/2$. Tensoring gives another line bundle $\Omega(G)$, where

$$\Omega(G) = \Omega^{1/2}_d(G) \otimes \Omega^{1/2}_r(G).$$

So $\Omega(G)_x = \Omega^{1/2}_{d(x)}(A(G)^*) \otimes \Omega^{1/2}_{r(x)}(A(G)^*)$. All of the line bundles of this paragraph are trivial. In particular, $\Omega(G) \cong G \times \mathbf{C}$. Hence we can identify $C^\infty_c(G)$ with the space of sections $C^\infty_c(G, \Omega(G))$. (Of course we have to interpret appropriately the latter space of sections as we did for $C^\infty_c(G)$ earlier to allow for the non-Hausdorffness of G.) Connes ([56, p.101]) then says that we can form the convolution $f * g$ in $C^\infty_c(G, \Omega(G))$ for $f, g \in C^\infty_c(G, \Omega(G))$ by defining

$$f * g(x) = \int_{G^{r(x)}} f(y) g(y^{-1}x) \tag{2.36}$$

"since it is the integral of a 1-density ... on the manifold $G^{r(x)}$". An interpretation of this integral will now be given. For another point of view, see Nistor, Weinstein and Xu ([191, p.21]).

Fix a nowhere vanishing section $u \to w_u$ of $\Omega^{1/2}(A(G)^*)$. For $y \in G^{r(x)}$ write $f(y) = t_{d(y)} \otimes w_{r(x)}$ and for $z \in G_{d(x)}$ write $g(z) = w_{d(x)} \otimes v_{r(z)}$. Note that both $t_{d(y)}$ and $v_{r(z)}$ are uniquely defined in $\Omega^{1/2}_{d(y)}(A(G)^*)$ and $\Omega^{1/2}_{r(z)}(A(G)^*)$ respectively. We also note that t, v are sections of the bundles $\Omega^{1/2}_d(G), \Omega^{1/2}_r(G)$ restricted to $G^{r(x)}, G_{d(x)}$ respectively.

---

geometry, the reader is referred to [274]. The Lie algebroid $A(G)$ is also important in the study of index theory for elliptic pseudodifferential operators on a Lie groupoid G, and is used in the construction of a natural asymptotic morphism associated with G ([191, 199, 198]).

Next, taking $z = y^{-1}x$, we have $g(y^{-1}x) = w_{d(x)} \otimes v_{d(y)}$. We interpret, using (F.10),

$$
\begin{aligned}
f(y)g(y^{-1}x) &= g(y^{-1}x) \otimes f(y) \\
&= w_{d(x)} \otimes v_{d(y)}t_{d(y)} \otimes w_{r(x)} \\
&= w_{d(x)} \otimes p_{d(y)} \otimes w_{r(x)}
\end{aligned}
$$

where p is a section of the *one*-density bundle $\Omega_d^1(G)$. We would like to interpret (2.36) to be "$w_{d(x)} \otimes \int p_{d(y)} \otimes w_{r(x)}$" but unfortunately, this does not make sense as $y \to p_{d(y)}$ is not a 1-section on the manifold $G^{r(x)}$.

To motivate what we should do, it is helpful to consider how a left Haar measure is defined on a k-dimensional Lie group H. One takes ([18]) a non-zero element $X \in \Lambda^k(T_e^*H)$ and translates it around to get the non-vanishing left invariant k-form ω: $y \to (L_{y^{-1}})^*X$ on G. This immediately gives that H is orientable and that the integral associated with ω is a left Haar measure on H.

In the situation above of the Lie groupoid G, n-forms are not adequate since G may not be orientable. (For example, consider the Moebius band which, as a line bundle, is (like all vector bundles) a Lie groupoid, but is not orientable.) So what we want to do is to use densities rather than forms (i.e. forms "mod" orientation). Indeed, for each $y \in G^{r(x)}$, the left translation map $L_{y^{-1}} : G^{r(x)} \to G^{d(y)}$ is smooth and is given by: $L_{y^{-1}}(a) = y^{-1}a$. We then obtain a section ω of $\Omega^1 T^* G^{r(x)}$ from the section $y \to p_{d(y)}$ by setting $\omega(y) = (L_{y^{-1}})^*(p_{d(y)})$. So ω is a 1-density on the manifold $G^{r(x)}$ and it has compact support (since the fact that f, g have compact support entails that the section p has as well). Hence we can integrate ω over $G^{r(x)}$ and the formula for convolution is:

$$
f * g(x) = \left(\int \omega \right)(w_{d(x)} \otimes w_{r(x)}) \in C_c^\infty(G, \Omega(G)). \qquad (2.37)
$$

There still remain some details to check, e.g. that $f * g \in C_c^\infty(G, \Omega)$ and that convolution is associative.

However, rather than pursuing this direction, we will instead develop a "left Haar system" approach to convolution where such details, as well as the basic representation theory for Lie groupoids, will follow from the general theory of left Haar systems on and the representations for locally compact groupoids (**3.1**). The advantage of using 1/2-densities to define convolution on a Lie groupoid is that it makes the convolution intrinsic to the groupoid, not dependent on a choice of left Haar system. However, the left Haar systems that we will discuss are equivalent to each other in all respects and so can be regarded effectively as unique on G. Indeed, anticipating the discussion of (**3.1**), the representation theory of a Lie groupoid G is independent of the choice of the smooth left Haar system. For example, given a quasi-invariant measure μ on G^0 and two such left Haar systems, the different versions of $\nu = \int_{G^0} \lambda^u \, d\mu(u)$ are equivalent since the two left Haar systems are locally equivalent. The same applies to the reduced and universal C^*-algebras of G. The effective uniqueness of smooth left Haar systems corresponds to the irrelevancy of left Haar systems in the approach of Connes discussed above.

The key idea in the smooth left Haar system approach is the same as that described above for the 1/2-density approach: that of translating a density on $A(G)^*$ around the groupoid to give a 1-density on the G^u's. Only this time, we use only 1-densities and these are *positive*. We first define the kind of left Haar systems appropriate to a Lie groupoid. As might be expected, there is a smoothness condition involved. We discuss some preliminaries.

Firstly, under natural measure theoretic conditions, if $T : X \to Y$ and μ is a measure on X, we can define a measure $\mu \circ T$ on Y by: $(\mu \circ T)(E) = \mu(T^{-1}E)$. Further for suitable functions g on Y, we have

$$\int g \, d(\mu \circ T) = \int (g \circ T) \, d\mu. \qquad (2.38)$$

(All this is elementary measure theory.) In the definition below, recall the map ψ_u of (2.35).

Definition 2.3.2 A *smooth left Haar system* for a Lie groupoid G is a family $\{\lambda^u\}$ $(u \in G^0)$ where each λ^u is a positive, regular Borel measure on the manifold G^u such that:

(i) if (V, ψ) is an r-fiberwise product open subset of G, $V \cong r(V) \times W$, and if λ_W is Lebesgue measure on $\mathbf{R}^k$ restricted to W, then for each $u \in r(V)$, the measure $\lambda^u \circ \psi_u$ is equivalent to λ_W, and the map $(u, w) \to d(\lambda^u \circ \psi_u)/d\lambda_W(w)$ belongs to $C^\infty(r(V) \times W)$ and is strictly positive;

(ii) for any $x \in G$ and $f \in C_c^\infty(G)$, we have

$$\int_{G^{d(x)}} f(xz) \, d\lambda^{d(x)}(z) = \int_{G^{r(x)}} f(y) \, d\lambda^{r(x)}(y). \qquad (2.39)$$

What (i) above is saying is that if we identify V with $r(V) \times W$, then for each $u \in r(V)$, λ^u is a strictly positive smooth measure on $G^u \cap V$ (see Appendix F) and the Radon-Nikodym derivatives $d(\lambda^u \circ \psi_u)/d\lambda_W$ vary smoothly on V. It is easy to prove that (2.39) is equivalent to the third left Haar system requirement (2.14) of Definition 2.2.2. Indeed, we only need to consider $f \in C_c(U)$ where U is an open Hausdorff subset of G, and in that case, we can obtain f as the uniform limit of a sequence in $C_c^\infty(G)$ with supports contained in a fixed compact subset of U. The first and second requirements of that Definition do not appear explicitly in Definition 2.3.2. However, the next proposition shows that every smooth left Haar system is indeed a left Haar system in the sense of Definition 2.2.2.

Proposition 2.3.1   *Every smooth left Haar system on G is a left Haar system in the sense of Definition 2.2.2.*

Proof.   Let $\{\lambda^u\}$ be a smooth left Haar system on G. By the above, property (iii) of Definition 2.2.2 is satisfied. It remains to show that the other two properties (i) and (ii) of that definition hold. Property (i) follows since each λ^u is locally equivalent to Lebesgue measure and so its support is the whole of G^u. It remains to prove (ii) of Definition 2.2.2, that if $g \in C_c(G)$ then $g^0 \in C_c(G^0)$. This is an easier version of the proof below giving the C^∞-version of property (ii), i.e. that $g^0 \in C_c^\infty(G^0)$ whenever $g \in C_c^\infty(G)$.

We can suppose for such a g that its support is inside V where (V, ψ) is an r-fiberwise product open subset of G, $V \cong r(V) \times W$. Then the function k, where $k(u, w) = d(\lambda^u \circ \psi_u)/d\lambda_W(w)$, belongs to $C^\infty(r(V) \times W)$. Next for $u \in r(V)$, we have

$$
\begin{aligned}
g^0(u) &= \int_{G^u} g \, d\lambda^u \\
&= \int_W (g \circ \psi_u^{-1}) \, d(\lambda^u \circ \psi_u) \\
&= \int_W g \circ \psi_u^{-1}(u, w) k(u, w) \, d\lambda_W(w). \quad (2.40)
\end{aligned}
$$

Now $(g \circ \psi_u^{-1})k \in C_c^\infty(r(V) \times W)$, and, using local charts for $r(V) \subset G^0$ to regard u as locally ranging over an open subset of $\mathbf{R}^p$, elementary analysis enables us to differentiate under the integral sign to obtain that $g^0 \in C_c^\infty(G^0)$ as required.   □

We now turn to the construction of smooth left Haar systems. Let $P_+(G)$ be the set of strictly positive 1-densities on $A(G)^*$. As discussed in Appendix F, $P_+(G)$ is non-empty. As in the earlier case, $P_+(G)$ is a cone. Locally, with respect to an r-fiberwise product open neighborhood V of a point in G^0, every $\alpha \in P_+(G)$ is of the form $g(x)dx_1 \cdots dx_k$ for a strictly positive C^∞-function g on $r(V)$. We will write $\alpha_u \in \Omega_u^1(A(G)^*)$ for the value of α at $u \in G^0$.

Also as in the earlier case, the family $\Lambda_s(G)$ of smooth left Haar systems on G is a cone. The following theorem says that this family can be identified with $P_+(G)$. (When G is an r-discrete Lie groupoid, the set $\Omega^1(A(G)^*)$ gets replaced, of course, by the set of strictly positive C^∞-functions on G^0, and the smooth left Haar systems are in one-to-one correpondence with this set as described in Proposition 2.2.5.)

Recall that $L_{x^{-1}}(z) = x^{-1}z$ $(z \in G^{r(x)})$.

Theorem 2.3.1   *Every $\alpha \in P_+(G)$ defines a smooth left Haar system $\{\Gamma^u(\alpha)\}$ where for each u, regarding $\Gamma^u(\alpha)$ as a 1-density on G^u,*

$$(\Gamma^u(\alpha))_x = (L_{x^{-1}})^*_{d(x)}(\alpha_{d(x)}). \tag{2.41}$$

Conversely, every smooth left Haar system $\{\lambda^u\}$ is of the form $\{\Gamma^u(\alpha)\}$ for some $\alpha \in P_+(G)$, and $\Gamma : P_+(G) \to \Lambda_s(G)$ is an isomorphism of cones.

Proof.   Let $z_0 \in G^u$ and let $V \cong r(V) \times W$ be an r-fiberwise product open subset of G with $d(z_0) \in r(V) \subset V$. By the continuity of d, there exists a chart Z for the manifold G^u containing z_0 such that $d(Z) \subset r(V)$. Let $\{z_i\}$ be coordinates for Z. Let $x \in G_u$. Then $d(x) = u$, and $L_{x^{-1}} : G^{r(x)} \to G^u$ is a diffeomorphism and $L_{x^{-1}}(x) = u$. Then $(L_{x^{-1}})^* : \Omega^1(G^u) \to \Omega^1(G^{r(x)})$. It follows by the change of variables formula for multiple integrals on an open subset of $\mathbf{R}^k$, that for $x \in Z$ and in terms of local coordinates,

$$(L_{x^{-1}})^*(g \, dw_1 \cdots dw_k) = (g \circ L_{x^{-1}}) \mid J(L_{x^{-1}}) \mid dz_1 \cdots dz_k$$

where, as usual, J stands for the Jacobian. In particular, for $z \in Z$, we have

$$(L_{z^{-1}})^*_{d(z)}(g(d(z)) \, dw_1 \cdots dw_k) = g(d(z)) \mid J(L_{z^{-1}})(z) \mid dz_1 \cdots dz_k. \tag{2.42}$$

Let $\alpha \in P_+(G)$. Then locally, we can write $\alpha = g \, dw_1 \cdots dw_k$ (with g strictly positive), and defining $\Gamma^u(\alpha)$ as in (2.41) and using (2.42), we obtain that $\Gamma^u(\alpha)$ is a positive 1-density on G^u. Let λ^u be the regular Borel measure on G^u associated with this density. Then in terms of local coordinates,

$$(d\lambda^u/dz)(z) = (g \circ d)(z) \mid J(L_{z^{-1}})(z) \mid . \tag{2.43}$$

Since g is strictly positive, we obtain (i) of Definition 2.3.2. It remains to show that (2.39) holds.

Let $x \in G$ and $f \in C_c^\infty(G^{r(x)})$. If $y \in G^{d(x)}$, then (identifying $\Gamma^u(\alpha)$ with the measure λ^u) we get

$$\begin{aligned}
(\lambda^{r(x)})_{xy} &= (L_{(xy)^{-1}})^*(\alpha_{d(xy)}) \\
&= (L_{x^{-1}})^*[(L_{y^{-1}})^*(\alpha_{d(y)})] \\
&= (L_{x^{-1}})^*(\lambda^{d(x)})_y. \tag{2.44}
\end{aligned}$$

From (2.44), we obtain

$$\lambda^{r(x)} = (L_{x^{-1}})^*(\lambda^{d(x)}) = \lambda^{d(x)} \circ L_x. \tag{2.45}$$

Applying (2.38) with $X = G^{d(x)}$, $Y = G^{r(x)}$, $f = g$ and $T = L_x$, we obtain from (2.45) that

$$\int f(y)\, d\lambda^{r(x)}(y) = \int f(xz)\, d\lambda^{d(x)}(z)$$

as required. It is clear from the definition of the Γ^u's that Γ is a one-to-one map that preserves the cone structures of $\Lambda_s(G), P_+(G)$. It remains to show that Λ is onto. So let $\{\lambda^u\}$ be a smooth left Haar system for G. From (i) of Definition 2.3.2, each λ^u defines a strictly positive density β^u on the manifold G^u. Take $\alpha_u = \beta^u_u$. Again using (i) of Definition 2.3.2 (to get, in particular, the smoothness of α along G^0) we obtain that $\alpha \in P_+(G)$. Reversing the argument of the preceding paragraph, we obtain ((2.45)) that for any $x \in G$ and any $y \in G^{d(x)}$,

$$\lambda^{r(x)}_{xy} = (L_{x^{-1}})^*(\lambda^{d(x)}_y).$$

In particular, taking $y = d(x)$, we get that

$$\lambda^{r(x)}_x = (L_{x^{-1}})^*(\lambda^{d(x)}_{d(x)}) = (L_{x^{-1}})^*(\beta_{d(x)}).$$

So for all $u \in G^0$, we have $\beta^u_x = (L_{x^{-1}})^*(\alpha_u)$ and ((2.41)) $\beta^u = \Gamma^u(\alpha)$. So Γ is onto as required.   □

The basic idea of the above proof is that one gets *every* smooth left Haar system on G by taking an element of $P_+(G)$ and left translating its values around the groupoid and then collecting these values on the G^u's to get the λ^u's. (It will be clear that what we are doing is entirely parallel to what was discussed earlier in the Lie group case where one translated a non-zero element of $\Lambda^k(T^*_e H)$ around the group to get a left Haar measure.) This procedure is illustrated later in the case of the tangent groupoid (Proposition 2.3.6).

Corollary 2.3.1   *Every Lie groupoid is a locally compact groupoid.*

This concludes our discussion of Lie groupoids in general, and the rest of the section concentrates on three important classes of these groupoids. These are the holonomy groupoid of a foliation, its r-discrete version and the tangent groupoid of a manifold.

Example 1 The holonomy groupoid of a foliation.

A brief introduction to foliations is given in Appendix F. To motivate the notion of *holonomy* we first look at the particular case of the *Kronecker foliation* of the torus **T** discussed in that appendix. The leaves of the foliation are diffeomorphic to **R**, and when represented as lines on the square (Fig. F.3), they have irrational slope θ. Let R be the leaf space equivalence relation on T so that M/R is the leaf space. As discusssed in Appendix F, the leaf space is of no value and instead, we stay with the groupoid R itself. The idea is to make R into a *Lie groupoid*, and use its C^*-algebra as a surrogate for the useless $C_0(M/R)$. (We saw a similar situation in **2.3** with the equivalence relation R_p, associated with Penrose tilings, on X_p.)

The problem is, then, to find charts which will make R into a Lie groupoid. To see how this can be done, consider Fig. 2.3.1 following, where the foliation is represented on the square D.

Here, $(a, b) \in R$, so that a and b are on the same leaf L. For simplicity, let a and b be in the interior of the square D. Let U be a parallelogram foliation chart with center a and α be the natural path from a to b in L. Now slide U along α parallel to the leaves to obtain another foliation chart V with center b. In the sliding, each plaque p in U stays on its own leaf and moves onto a plaque p' of V, with a sliding onto b.

Let $W_{a,b} = \cup_p(p \times p')$, the union being taken over all plaques p of U. Notice that each plaque p is paired up with *exactly one* plaque of the same leaf L in V. (There are *infinitely many* plaques of L in V, their union even being dense in V).

It is straightforward to show that the family of all such sets $\{W_{a,b}\}$ (allowing (a, b) to range over R and U to range over foliation charts centered at a) gives an atlas for a manifold topology on R for which R is a Lie groupoid. We will not stop to prove this as it is a special case of Theorem 2.3.2. This groupoid is the *holonomy groupoid* for the Kronecker foliation. To motivate the corresponding construction for a general foliation, observe that a *path* from a to b is involved in the Kronecker case. In that one-dimensional situation, there is effectively one path (i.e. up to homotopy) from a to b in the leaf L. For the general case, there are many paths available for consideration, and modifying the above argument for the Kronecker case leads to the fundamental notion of *holonomy* for a foliation.

As Lawson point out ([156, p.10]), holonomy is the *key concept in understanding the internal structure of foliations*. It can be regarded as giving information in groupoid[12] terms about how the leaves neighboring a given one are approaching or leaving that leaf.

Let $(M, \mathcal{F})$ be a foliated manifold. As usual, we set $n = \dim M$ and $k = \dim \mathcal{F}$. Let a, b be on the same leaf L of M, γ be a path in L from a to b and U, V be foliation charts containing a, b respectively with transverse sections A, B passing through a, b respectively. The idea of *holonomy*, due to Ehresmann (see, for example, Ehresmann and Shih [88], Ehresmann [84],

---

[12]In fact, Aof and Brown ([7]), developing ideas of Pradines ([208]), have shown that there is a version of the holonomy groupoid for *any* topological groupoid.

Haefliger [110], Reeb [228]), is to obtain paths from points of A close to a to points of B close to b by "following" γ, each path lying in the appropriate leaf. This pairs up points of A with points of B which turns out to be a diffeomorphism T_γ whose germ $[\gamma]$ at a is independent of the choices made. This germ is the *holonomy class* of γ. Instead of considering the set R of pairs (a, b) as we did in the special case of the Kronecker foliation, we make the set of *triples* $(a, [\gamma], b)$ into a Lie groupoid G called the *holonomy groupoid* or *graph* of $\mathcal{F}$. We now discuss this in more detail. Accounts of holonomy for foliations are given by Camacho and Neto ([44, p. 62ff.]), Molino ([177, p.22ff.]), Moore and Schochet ([178, p.54ff.]), Reinhart ([229, p.136ff.]) and Tamura ([259, p.113ff.]).

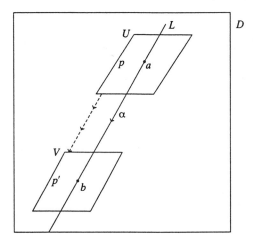

Figure 2.3.1

A straightforward compactness argument ([44, p.33]) shows that there is a sequence $U_0, \ldots, U_r$ of foliation charts and a partition $0 = t_0 < t_1 < \cdots < t_{r+1} = 1$ of $[0, 1]$ such that:
(a) if $U_i \cap U_j \neq \emptyset$ then $U_i \cup U_j$ is contained in a foliation chart;
(b) $\gamma([t_i, t_{i+1}]) \subset U_i$ $(0 \leq i \leq r)$.
Since $\gamma(t_{i+1}) \in U_i \cap U_{i+1}$, we have $U_i \cap U_{i+1} \neq \emptyset$. For $1 \leq i \leq r$, let $D_{i+1} = D'_i \cap U_{i+1}$ where D'_i is the transverse section of U_i passing through $\gamma(t_{i+1})$. We take $D(0) = A, D(r+1) = B$. Consider $x \in A$. For x close enough to a $(= \gamma(t_0))$, the plaque of U_0 containing x intersects $D(1)$ in exactly one point x_1. By taking x even closer to a (if necessary), the point x_1 becomes close enough to $\gamma(t_1)$ so that the plaque of x_1 in U_1 intersects $D(2)$ in exactly one point x_2. Continuing with this process, there is an open neighborhood Z_a of a in $A = D(0)$ such that for each $x \in Z_a$, the sequence $x_1, x_2, \ldots, x_{r+1} \in B$ exists. A simple inductive argument gives that the

In terms of local coordinates, the maps d, r are given by:

$$(t, s, u) \rightarrow (s, u), \qquad (t, s, u) \rightarrow (t, u)$$

and these are clearly submersions. Inversion in local coordinates is the map

$$(t, s, u) \rightarrow (s, t, u)$$

which is obviously smooth. Now let $(p, q) \in G^2$. We can write $p = (t, s, u), q = (s, s', u)$. It is readily checked that the map

$$(p, q) \rightarrow (t, s, s', u)$$

gives coordinates for G^2 as the submanifold $(d \times r)^{-1}(\Delta)$ of $G \times G$ ((2.34)). This concludes the proof that G is a Lie groupoid. □

From Theorem 2.3.1, the left Haar systems are identified with the cone $P_+(G^0)$. Identifying G^0 with M, we obtain that $P_+(G^0)$ is the cone of strictly positive sections of $\Omega^1(E)$ where E is the vector bundle T^*F and F is the subbundle of TM whose elements are tangent vectors along the leaves of the foliation.

When is $G = G(\mathcal{F})$ Hausdorff? The answer to this is strikingly similar to the corresponding result for the universal groupoid of an inverse semigroup given later in Proposition 4.3.6, and this emphasizes the close relationship between holonomy groupoids and the groupoids associated with inverse semigroups. The necessary and sufficient conditions for G to be Hausdorff, given below, are stated by Reinhart ([229, p.137]) and the proof is given by Moore and Schochet ([178, pp.59-60]). We reformulate it in a notation which brings out the parallel to Proposition 4.3.6.

Let $a, b \in L$ for some leaf L of M and U, V be foliation charts containing a, b respectively. Let A, B be the transverse sections for U, V containing a, b respectively, and γ, γ' be paths from a to b in L such that both $T_\gamma, T_{\gamma'}$ are defined from A to B. Let

$$D_{\gamma, \gamma'} = \{x \in A : T_\gamma x = T_{\gamma'} x\}.$$

Proposition 2.3.2   *The Lie groupoid G is Hausdorff if and only if for every such $D_{\gamma, \gamma'}$, its interior $D^o_{\gamma, \gamma'}$ in A is closed in A.*

Proof.   ⇐. Suppose that

$$(a_n, [\gamma_n], b_n) \rightarrow (a, [\gamma], b), (a', [\gamma'], b')$$

in G. By considering neighborhood bases at $(a, [\gamma], b), (a', [\gamma'], b')$ we obtain that $a_n \rightarrow a, a'$ and $b_n \rightarrow b, b'$. Since M is Hausdorff, we have $a = a', b = b'$. There exist U, A, V, B as in the paragraph preceding the proposition such that both $T_\gamma, T_{\gamma'}$ are defined from A to B. Then in the notation of (2.47), we can find $c_n \to a$ in A such that

$$[\gamma_n] = [s_{a_n c_n} \gamma_{c_n} s_{b_n c_n}^{-1}] = [s_{a_n c_n} \gamma'_{c_n} s_{b_n c_n}^{-1}].$$

It follows that $[\gamma_{c_n}] = [\gamma'_{c_n}]$ and hence that $T_\gamma = T_{\gamma'}$ on an open subset of A containing the sequence $\{c_n\}$. So $a = \lim_{n \to \infty} c_n \in \overline{D^o_{\gamma,\gamma'}}$, the closure of $D^o_{\gamma,\gamma'}$ in A. By assumption, $a \in D^o_{\gamma,\gamma'}$ and so $[\gamma] = [\gamma_a] = [\gamma'_a] = [\gamma']$. Hence G is Hausdorff.

$\Rightarrow$. Let $x \in \overline{D^o_{\gamma,\gamma'}}$ and $\{c_n\}$ be a sequence in $D^o_{\gamma,\gamma'}$ converging to x. Let $y = \lim_{n \to \infty} T_\gamma c_n$. Then $(c_n, [\gamma_{c_n}], T_\gamma c_n) = (c_n, [\gamma'_{c_n}], T_{\gamma'} c_n)$ converges in G to both $(x, [\gamma_x], y)$ and $(x, [\gamma'_x], y)$. Since G is Hausdorff, we obtain that $[\gamma_x] = [\gamma'_x]$, and so $T_\gamma x = T_{\gamma'} x$. So $x \in D^o_{\gamma,\gamma'}$, and $D^o_{\gamma,\gamma'}$ is closed in A. $\square$

We conclude our discussion of the holonomy groupoid by showing that the holonomy groupoid of the Reeb foliation of Fig. F.4 is not Hausdorff. Part of the foliation is shown below in Fig. 2.3.4.

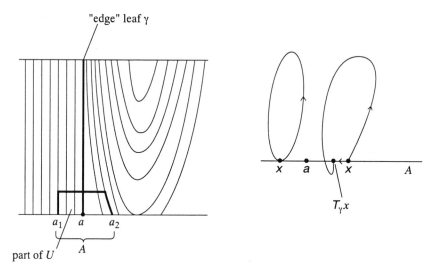

Figure 2.3.4

The point a is on an "edge" compact leaf L, i.e. a leaf that is a circle and from one direction the non-compact leaves flatten out against it while on the other side is an "interval's worth" of circular leaves like L. We take a foliation chart U containing a. Part of U is shown in Fig. 2.3.4. Next let $a = b$, $U = V$ and take $A = B$ to be the interval (a_1, a_2). Let γ be the path from a back to a going counter-clockwise once round the circle L. Let δ be the constant path at a. It is obvious that each δ_x is constant at x for

x close to a on A so that T_δ is the identity map on A. As for γ, for x to the left of a, γ_x just follows round the x-leaf counter-clockwise once, giving $T_\gamma x = x$. For x to the right of a (on a non-compact leaf) γ_x goes round the x leaf once counter-clockwise but ends up on A to the *left* of x. (See the right-hand diagram in Fig. 2.3.4.) So $D_{\gamma,\delta} = (a_1, a]$ giving $D^o_{\gamma,\delta} = (a_1, a)$ which is not closed in A. So the holonomy groupoid G is not Hausdorff by Proposition 2.3.2. Indeed, the elements $(a, [\gamma], a)$ and $(a, [\delta], a)$ cannot be separated.

Example 2 The r-discrete holonomy groupoid.

Until now in this section, we have discussed Lie groupoids in general and the holonomy groupoid of a foliation in particular. So far, there has been no mention of inverse semigroups. When the role of inverse semigroups is made explicit in the theory of holonomy groupoids, it gives an *r-discrete* version of the holonomy groupoid which we will call $G_r(\mathcal{F})$. The inverse semigroup involved is effectively generated by some of the T_γ's and is an example of what is called a *pseudogroup*. (A similar situation holds in the study of symplectic groupoids where finite dimensional pseudogroups play the role of a Lie group, and pseudogroup orbits give the symplectic leaves ([140, p.106ff.]).)

A great many C^*-algebras are obtained from pseudogroups. Indeed, in the paper by Arzumanian and Renault ([8]), the authors list seven classes of such C^*-algebras all obtained from the groupoids of germs of the pseudogroups: (i) foliation C^*-algebras, (ii) graph C^*-algebras, (iii) the C^*-algebra associated with an endomorphism, (iv) polymorphism C^*-algebras, (v) Ruelle and Smale algebras, (vi) C^*-algebras associated with quasicrystals, and finally (vii) Hecke algebras. A generalization of the construction of a C^*-algebra from a pseudogroup will be considered in **3.3** and used in Chapter 4. (i) is briefly considered below and in **3.3**, while (ii) and (vi) are discussed in **4.2**.

We now discuss the notion of *pseudogroup*.[14] This notion has a long history, going back as far as Lie in the 1880's. Cartan in the 1920's worked on the classification of Lie pseudogroups in $\mathbf{R}^n$. A sketch of the history of the notion of a *pseudogroup* is given by B. L. Reinhart ([229, p.9]). He points out that the term *pseudogroup* originated in the 1932 book [264] of Veblen and Whitehead. Haefliger ([110]) studied the relationship between pseudogroups and groupoids. Ehresmann ([84]) characterized pseudogroups in terms of groupoids, allowing products to be defined only when domain and range matched up. (This approach was used in Proposition 1.0.1.)[15]

---

[14]I am grateful to Alan Weinstein for helpful information about the history of pseudogroups and papers concerning them.

[15]See the first chapter of the book by Mark Lawson ([157]) for a recent discussion of

Definition 2.3.3 A *pseudogroup* on a topological space X is an inverse semigroup of homeomorphisms between open subsets of X.

There seems to be no standard definition of *pseudogroup* but the above definition is the simplest, and is satisfied by all other definitions of the terms known to the author. (This is also the definition taken by Mark Lawson ([157]) and Arzumanian and Renault ([8]).) J. Plante ([207]) does not require the domains of the elements of a pseudogroup to be open, but requires an additional *additive* condition on the pseudogroup. We will meet this additive condition in **3.3**. Also in the smooth context (such as that of the holonomy groupoid) we would naturally require the elements of a pseudogroup to be partial diffeomorphisms (rather than just partial homeomorphisms).

Indeed, to see how pseudogroups can be relevant to differential geometry, following Kobayashi and Nomizu ([147, p.2]), let us see how a manifold can be defined in terms of a pseudogroup. Let $\Gamma(\mathbf{R}^n)$ be the pseudogroup of partial diffeomorphisms between open subsets of $\mathbf{R}^n$. Then the "transition functions" condition (F.1) for the charts just comes down to saying that each $g_{\alpha\beta} \in \Gamma(\mathbf{R}^n)$. By restricting the pseudogroup, we obtain different kinds of manifolds (such as analytic or orientable).

Now a manifold is trivially a foliated manifold with only one leaf, and so one might expect there to be a notion of a Γ-*foliation* associated with a general foliation $\mathcal{F}$ on M in which the transition functions of (F.15) are required to be in a pseudogroup $\Gamma \subset \Gamma(\mathbf{R}^n)$. This is considered by Bott and Haefliger ([16]), see also Reinhart ([229]). As Bott and Haefliger point out, there is a classifying space $B\Gamma$ (corresponding to $B(G)$ in the group case) whose real cohomology gives the characteristic classes of Γ-foliations, and fits into the Godbillon-Vey, Gelfand-Fuks and Chern-Simons theories. We will not, of course, go into these theories here – even if the author were competent to do so – the point of mentioning them being to illustrate the importance of the pseudogroup concept.[16]

We now discuss an r-discrete version G_T of the holonomy groupoid which is often easier to handle than $G(\mathcal{F})$. As a very simple illustration, when $M = \mathbf{R}$ with only one leaf, we have that $G(\mathcal{F})$ is the trivial groupoid $\mathbf{R} \times \mathbf{R}$ while the r-discrete version can be taken to be the trivial groupoid $\mathbf{Z} \times \mathbf{Z}$. The general construction is described by Hilsum and Skandalis ([130]). See also Haefliger ([113]), Hurder ([134]) and Moore and Schochet ([178]). We will first discuss G_T as a reduction of the holonomy groupoid G to a subset T of G^0. Then we will examine G_T from the point of view of a pseudogroup

---

pseudogroups. I am grateful to Dr. Lawson for allowing me to consult this part of his book.

[16]The recognition of the importance of the pseudogroup concept is also implicit in the claim by J. H. C. Whitehead ([229, p.9]) that the work on pseudogroups may be Elie Cartan's best.

generated by the maps T_γ arising in the construction of the holonomy groupoid. (Recall that these are partial diffeomorphisms.) In terms of **3.3**, G_T is an example of a localization groupoid.

The definition of the reduction of a groupoid to a subset of its unit space was given in 1.12. If H is r-discrete (or ample) and T is a *closed* subset of H^0, then the reduction H_T by T is also r-discrete (or ample) in the relative topology. (There are a number of things to check with these claims but they are all easy.)

Let $(M, \mathcal{F})$ be a foliated manifold. The T by which we wish to reduce G needs to satisfy special conditions in order that G_T be an r-discrete Lie groupoid. This is ensured by the following lemma due to Hilsum and Skandalis ([130, (2. Lemme)]).

Lemma 2.3.1 *There exists a locally finite, countable cover $\{U_r\}$ ($r = 1, 2, 3, \ldots$) of foliation charts for M and transverse sections T_r of the U_r's such that for each r, $\overline{T_r} \cap \overline{\cup_{i \neq r} T_i} = \emptyset$.*

Proof. Let $\{O_i\}$ ($i = 1, 2, 3, \ldots$) be a locally finite cover of M by foliation charts. We can further suppose that for each i, there is a foliation chart (O_i', ϕ_i') such that $\overline{O_i} \subset O_i'$ and also that $\phi_i'(O_i') = U_2^k \times U_2^{n-k}$, $\phi_i'(O_i) = U_1^k \times U_1^{n-k}$ where U_r^p is the ball, center 0, radius r in $\mathbf{R}^p$. (That all of these requirements can be met follows by elementary manifold topology.) We will identify U_r^{n-k} with $\{0\} \times U_r^{n-k}$. For each j, let $S_j = \phi_j'^{-1}(\overline{U_1^{n-k}})$ (which contains a transverse section for O_j).

For each j, we will construct an open cover of O_j by foliation charts $\{\Omega_{j,i} : 1 \leq i \leq p_j\}$ of O_j and transverse sections $T_{j,i}$ of $\Omega_{j,i}$ such that each $\Omega_{j,i}$ is a union of plaques of O_j and $\overline{T_{j,i}} \cap \overline{T_{r,s}} = \emptyset$ if $(j,i) \neq (r,s)$. The lemma will then follow by taking the $\Omega_{j,i}, T_{j,i}$ to be respectively the sets U_r, T_r. Indeed, the family of sets $\{\Omega_{j,i}\}$ is a countable cover of the union M of the O_k's, and inherits from the O_k's the locally finite property. Also, $\overline{T_r} \cap \overline{\cup_{i \neq r} T_i} = \emptyset$. For if not, there exists a sequence $x_p \to x \in \overline{T_r}$ where $x_p \in T_{i_p}$ for some $i_p \neq r$. Since the family of U_r's is locally finite, there exists a neighborhood W of x for which there are only finitely many s's, $s_1, \ldots, s_l$, such that $W \cap U_s \neq \emptyset$. So eventually, $i_p \in \{s_1, \ldots, s_l\}$ and since $\overline{T_r} \cap \overline{\cup_{k=1}^l T_{s_k}} = \cup_{k=1}^l (\overline{T_r} \cap \overline{T_{s_k}}) = \emptyset$, we obtain a contradiction. It remains, then, to construct the $\Omega_{j,i}, T_{j,i}$.

Suppose that the $\Omega_{r,l}, T_{r,l}$ have been constructed for $r < j$. Then each such $\overline{T_{r,l}}$ is contained in a transverse section of O_r' and so meets a leaf in at most one point. It follows that $A = \cup \{\overline{T_{r,l}} : r < j, 1 \leq l \leq p_r\}$ is a closed subset of M which intersects any plaque of O_j in only finitely many points. Now $A \cap O_j'$ is a closed subset of O_j' and for each $z \in S_j$, the plaque Q_z of z in O_j' intersects $A \cap O_j'$ in finitely many points. Let $w = w_z \in (Q_z \sim A) \cap (U_2^k \times U_2^{n-k})$. Then there exists an open neighborhood V_w of w in O_j'

such that $\overline{V_w} \cap A = \emptyset$. We can suppose that $\phi'_j(V_w) = D_{1,w} \times D_{2,w}$ where $D_{1,w}$ is an open ball in U_2^k centered at the first component of $\phi'_j(w)$ and $D_{2,w}$ is an open ball in U_2^{n-k} centered at the second component of $\phi'_j(z)$. See Fig. 2.3.5. If $z' \in \overline{S_j}$, then there exists $z \in S_j$ close enough to z' so that we can enlarge D_{2,w_z} to ensure that $\phi'_j(z') \in D_{2,w_z}$. So the $D_{2,w}$'s cover $\overline{U_1^{n-k}}$.

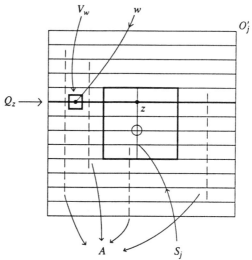

Figure 2.3.5

Since $\overline{U_1^{n-k}}$ is compact, there exist $z_1, \ldots, z_m \in S_j$ such that $\overline{U_1^{n-k}} \subset \cup_{i=1}^m D_{2,w_i}$ where $w_i = w_{z_i}$. Let $D_{2,i} = D_{2,w_i} \cap U_1^{n-k}$ and set

$$\Omega_{j,i} = (\phi'_j)^{-1}(U_1^k \times D_{2,i}).$$

Then $\cup_i \Omega_{j,i} = O_j$ and each $\Omega_{j,i}$ is, by definition, a union of plaques of O_j. Take $T_{j,i} = \phi_j^{-1}(\{u_i\} \times D_{2,i})$ where $u_i \in D_{1,w_i}$ and the u_i's are distinct. If $r < j$, then $\overline{T_{j,i}} \cap \overline{T_{r,s}} \subset \overline{T_{j,i}} \cap A = \emptyset$, and by the distinctness of the u_i's, the sets $\overline{T_{j,i}}$ are pairwise disjoint. This concludes the construction of the $\Omega_{j,i}, T_{j,i}$.   □

Let U_r, T_r be as in the above lemma. Let $T = \cup_r T_r$. By the lemma, it follows that the relative topology on T is the disjoint union topology, i.e. every open subset of T is of the form $\cup_r V_r$ where V_r is an open subset of T_r. Then T is a submanifold of dimension $(n - k)$ in $M = G^0$ since each T_r is such. Further, since the U_r's cover M and T_r is a transverse section of U_r, it follows that each leaf intersects T at least once. Also by local finiteness, every compact subset of M meets only a finite number of the U_r's.

Definition 2.3.4 The *r-discrete holonomy groupoid* G_T is the reduction of the holonomy groupoid G by T.

The elements of the groupoid G_T are those of the form $(x, [\gamma], y)$ where $x, y \in T$. Of course, the set T is not unique, but it is "up to Morita equivalence" ([113], [130, p.206], [178, p.294]). The next theorem proves that G_T is an r-discrete locally compact groupoid.

Theorem 2.3.3 *The groupoid G_T is a submanifold of the holonomy groupoid G, and is itself an r-discrete Lie groupoid.*

Proof. We first show that $H = G_T$ is a submanifold of G. Let $(a, [\gamma], b) \in H$. Let $a \in T_m, b \in T_n$. By the above lemma, we can find foliation charts U, V for M such that T_m, T_n are transverse sections of U, V and $U \cap T = T_m, V \cap T = T_n$. By contracting U, V, we can suppose that the domain and range of the local diffeomorphism T_γ are T_m, T_n respectively. If (as earlier) (t, s, u) are coordinates for the G-neighborhood $W_{[\gamma]}(U, V)$ of $(a, [\gamma], b)$, then the coordinates for $Z_{[\gamma]}(U, V) = W_{[\gamma]}(U, V) \cap H$ can be taken to be $(0, 0, u)$. It follows that H is a submanifold of G.

Obviously the unit space T of H is a Hausdorff manifold. With a as above, H^a is the set of triples $(a, [\delta], y)$ where y is in some T_j and δ is a path in the leaf of a from a to y. Since each T_j is a transverse section, the leaf of a intersects T_j in countably many points. So there are only countably many such y's in any T_j. By the Epstein-Millet-Tischler result earlier, there are only countably many holonomy classes $[\delta]$ to be considered. It follows that H^a is countable. Next, H^a is discrete. For let $(a, [\gamma], b) \in H^a$ be as above. Then with U, V as in the first paragraph of the proof, we obtain $Z_{[\gamma]}(U, V) \cap H^a = \{(a, [\gamma], b)\}$.

Next, using coordinates as in the case of G (see the proof of Theorem 2.3.2) the product and inversion maps for H are smooth, and the range and source maps are submersions. So H is a Lie groupoid, and hence (Corollary 2.3.1) is a locally compact groupoid. Now H is a union of the sets $Z_{[\gamma]}(U, V)$ as above. So to prove that H is r-discrete, it suffices to prove that $Z_{[\gamma]}(U, V) \in H^{op}$. Explicitly, $Z_{[\gamma]}(U, V) = \{(x, [\gamma_x], y) : x \in T_m \cap U, y \in T_n \cap V\}$. Now $r((x, [\gamma_x], y)) = x$ and r is bijection from $Z_{[\gamma]}(U, V)$ onto U. For every open subset U_1 of U, we have for some open subset V_1 of V that $r(Z_{[\gamma]}(U_1, V_1)) = U_1$. Since the family of such sets $Z_{[\gamma]}(U_1, V_1)$ is a basis for the topology of $Z_{[\gamma]}(U, V)$, it follows that r restricted to $Z_{[\gamma]}(U, V)$ is a homeomorphism (indeed, a diffeomorphism) onto U. Similarly considerations apply to d, and it follows that $Z_{[\gamma]}(U, V) \in H^{op}$ as required. $\square$

Let us now examine another approach to the construction of G_T. This approach is based on pseudogroups. As observed in the preceding proof, there is a countable basis for the topology of G_T consisting of sets of the form $Z_{[\gamma]}(U, V)$ where, for some m, n, some $a \in T_m$, some $b \in T_n$ in the leaf of a and some path γ from a to b in the leaf of a, we have

$$Z_{[\gamma]}(U, V) = \{(x, [\gamma_x], y) : x \in T_m \cap U, y \in T_n \cap V\}.$$

For each such γ, the map $T_\gamma : T_m \cap U \to T_n \cap V$ is a homeomorphism between open subsets of T, and if $(x, [\gamma], y) \in Z_{[\gamma]}(U, V)$, we have $T_\gamma(x) = y$. The set of T_γ's generates a countable pseudogroup S on T. The set of domains of the T_γ's is a basis for the topology of T. For each $s \in S$ and x in the domain of s, we will write $x.s$ for the action of s on x. (The reason for having S act on the right is that this fits in with the product $T_{\gamma_1 \circ \gamma_2} = T_{\gamma_2} \circ T_{\gamma_1}$.)

What is interesting is that the r-discrete holonomy groupoid can be recovered *from the pair (T, S) alone*. Indeed, if $s \in S$ and x is in the domain of s, define the pair $\overline{(x, s)}$ to be the germ of s at x. Then the set of such pairs G_S^{hol} is an r-discrete groupoid which can be identified with G_T. Indeed, Theorem 3.3.2 gives a general version of this. The product on G_S^{hol} is given, as in the transformation group, by: $\overline{(x, s)}\overline{(x.s, t)} = \overline{(x, st)}$ and inversion by: $\overline{(x, s)} \to \overline{(x.s, s^*)}$. Each non-zero $s \in S$ is a T_γ for some path γ, and the isomorphism between G_S^{hol} and G_T is given by $\overline{(x, s)} \to (x, [\gamma], T_\gamma x)$. We will not go into any more details about the identification of G_T with G_S^{hol} as they are straightforward.

However, the G_S^{hol} approach to the r-discrete holonomy groupoid puts it in the very general framework of an inverse semigroup acting on a locally compact Hausdorff space. This framework produces a large class of r-discrete groupoids, not just the r-discrete holonomy groupoid, and is the key to relating inverse semigroups to such groupoids through forming germs of the partial homeomorphisms which are the elements of S. This will be a central theme in **3.3** and Chapter 4.

Example 3   The tangent groupoid of a manifold.

The tangent groupoid G_M of a Hausdorff manifold M is an important Lie groupoid used in noncommutative geometry to study deformations of C^*-algebras. It is also useful in index theory. As Connes points out ([56, p.159]), the notion of deformation of algebras already plays a critical role in algebraic geometry ([99]), quantization ([159, 13]) and the construction of quantum groups ([76, 97]). Deformation of C^*-algebras gives rise to asymptotic morphisms with resultant implications for the E-theory of Connes and Higson ([60, 129]) and the concomitant KK-theory of Kasparov. In particular, the Connes-Higson asymptotic morphism $T^t : C_0(T^*M) \to \mathcal{K}(L^2(M))$ $(t \geq 1)$ gives at the K-theoretic level the Atiyah-Singer analytic index map from $K(T^*M)$ into $\mathbf{Z}(= K(\mathcal{K}(L^2M)))$ ([10], [56, p.104]).

We will not discuss these remarkable results further, but, at the intuitive level, try to motivate the definition of the tangent groupoid G_M from the deformation of C^*-algebras (cf. for example, [239]). As is well-known, classical mechanics can be regarded as a limiting case of quantum mechanics as the Planck constant $\hbar$ tends to zero. Phase space for the kinematics of classical mechanics is a Poisson manifold, which for the present discussion we will take to be the cotangent bundle T^*M of the manifold M. The observables are then scalar-valued functions on T^*M and a natural C^*-algebra to consider is the commutative C^*-algebra $C_0(T^*M)$. For quantum mechanics, the kinematics is given by a non-commutative C^*-algebra, a natural example of which is, of course, the algebra $\mathcal{K}(\mathcal{H})$ of compact operators on an infinite-dimensional, separable Hilbert space $\mathcal{H}$. It is natural in the present context to take this Hilbert space to be $L^2(M)$ (with respect to a strictly positive smooth measure on M (cf. Appendix F)). So we want to deform $\mathcal{K}(L^2(M))$ into $C_0(T^*M)$.

Rather than look at this from an algebra point of view, let us try to simplify matters by looking at the deformation from a groupoid perspective. From this perspective, both $C_0(T^*M)$ and $\mathcal{K}(L^2(M))$ have to be the C^*-algebras of locally compact, or even better, Lie groupoids, i.e. obtained from the representations of these groupoids. (We are here anticipating some of the discussion of **3.1**.) In the case of $C_0(T^*M)$, we take the required Lie groupoid G_1 to be the bundle of abelian groups TM, since using the Fourier transform on the individual T_xM's, the C^*-algebra of TM is $C_0(T^*M)$. Is there a natural groupoid G_2 whose C^*-algebra is $\mathcal{K}(L^2(M))$? Indeed there is! By Theorem 3.1.2 in Chapter 3, the C^*-algebra of the trivial Lie groupoid $M \times M$ is just $\mathcal{K}(L^2(M))$. (Intuitively, the two variable functions in $C_c(M \times M)$ just give the kernels for compact operators on $L^2(M)$.) The problem of deforming $\mathcal{K}(L^2(M))$ into $C_0(T^*M)$, then, is changed into that of deforming the trivial groupoid $M \times M$ into TM.

To do this, put a copy of $M \times M$ above each non-zero real number ϵ and a copy of TM above the real number 0, and take the union G of all these. As discussed below, G is a groupoid. All the deforming will take place within the compass of the one groupoid G. The ϵ is the deformation parameter, and as $\epsilon \to 0$, we want $M \times M \times \{\epsilon\}$ to "converge" in some sense to $TM \times \{0\}$. What we would like this to mean intuitively is that a sequence $(x_n, y_n, \epsilon_n) \to (x, X, 0)$ if $x_n \to x$, $y_n \to x$, $\epsilon_n \to 0$ and

$$\frac{x_n - y_n}{\epsilon_n} \to X.$$

(This formulation is given by Connes ([56, p.103]).) Of course we need to interpret $\frac{x_n - y_n}{\epsilon_n}$. Nonetheless, the preceding formulation is very helpful, since it shows that what is underlying the groupoid deformation is the simple idea of elementary calculus in which the derivative is the limit of a quotient.

The deformation of $M \times M$ into the bundle of Euclidean spaces TM fits in well with the notion of the "blowing up" of a subvariety in algebraic geometry (e.g. [102, 12]). (In the present case, the subvariety would be the diagonal $\Delta = \{(x, x) : x \in M\} \cong M$ in $M \times M$ and the tangent bundle is "fitted" on to Δ.) In the blowing up context, a "normal" projective space bundle is used in place of the tangent bundle. The normal bundle version is given, in the Riemannian context using the exponential map, by Hilsum and Skandalis ([131]). Another approach using local coordinates is given by Weinstein ([269]). The approach developed here is a variation of that of Weinstein, reducing the problem of the tangent groupoid G_M to the easier case where M is an open subset of $\mathbf{R}^n$.

So let M be an n-dimensional manifold and define

$$G_M = [(M \times M) \times \mathbf{R}^*] \cup (TM \times \{0\}).$$

Here $\mathbf{R}^* = \mathbf{R} \sim \{0\}$.[17] This is a disjoint union of the trivial groupoids $G_\epsilon = (M \times M) \times \{\epsilon\}$ and the tangent groupoid $G_0 = TM \times \{0\}$. As observed in Chapter 1, G_M is a groupoid with $G_M^2 = \cup_\epsilon G_\epsilon^2$, with product and inversion just the same as that given on the G_ϵ's. So there are two kinds of product, those of the form $(x, y, \epsilon)(y, z, \epsilon) = (x, z, \epsilon)$ for $\epsilon \neq 0$, and $(x, X, 0)(x, Y, 0) = (x, X + Y, 0)$ where X, Y are tangent vectors at $x \in M$. The elements of G^0 are of the form (x, x, ϵ) where $\epsilon \neq 0$ and $(x, 0, 0)$ where the second 0 is the 0 tangent vector at x. The next problem is to obtain charts for a smooth structure on G_M. Rather than deal with this directly, we consider first a special case.

Let U be an open subset of $\mathbf{R}^n$. We will embed G_U within $\mathbf{R}^{2n+1}$, with T_U being identified with $U \times \mathbf{R}^n$. Define a map $\psi : G_U \to \mathbf{R}^{2n+1}$ by requiring $\psi(x, y, \epsilon) = (x, \frac{x-y}{\epsilon}, \epsilon)$ if $\epsilon \neq 0$ while $\psi(x, X, 0) = (x, X, 0)$. Obviously, ψ is one-to-one. Let $\mathcal{V}_U = \psi(G_U)$. Then $\mathcal{V}_U$ is a groupoid under the structure inherited through ψ from G_U. An advantage of considering $\mathcal{V}_U$ rather than G_U is that in calculating with $\mathcal{V}_U$, we often do not need to distinguish between the cases $\epsilon \neq 0$, $\epsilon = 0$.

Proposition 2.3.3   *The groupoid $\mathcal{V}_U$ is the open subset*

$$\{(x, w, \epsilon) \in \mathbf{R}^{2n+1} : x \in U, x - \epsilon w \in U\} \tag{2.49}$$

of $\mathbf{R}^{2n+1}$, and its groupoid structure is given by:

$$(x, w, \epsilon)(x - \epsilon w, w', \epsilon) = (x, w + w', \epsilon) \tag{2.50}$$

$$(x, w, \epsilon)^{-1} = (x - \epsilon w, -w, \epsilon). \tag{2.51}$$

Proof.   If $\epsilon \neq 0$, then $(x, w, \epsilon) = \psi((x, x - \epsilon w, \epsilon))$, and conversely, if $(x, y, \epsilon) \in G_U$, then $\psi((x, y, \epsilon)) \in \mathcal{V}_U$ since $x - \epsilon \frac{x-y}{\epsilon} = y \in U$. If $\epsilon = 0$

---

[17]In [56], $\mathbf{R}^*$ is replaced by $(0, 1]$. This is convenient for deformations. On the other hand, it gives a Lie groupoid with boundary rather than simply a Lie groupoid.

then ψ carries G_ϵ onto the set $\{(x, w, 0) : (x, w, 0) \in \mathcal{V}_U\}$. This shows that $\mathcal{V}_U$ equals the set in (2.49). It is left to the reader to check that $\psi(G_U^2) = \mathcal{V}_U^2$ is the set of pairs $((x, w, \epsilon), (x - \epsilon w, w', \epsilon))$, and that the product $(x, w, \epsilon)(x - \epsilon w, w', \epsilon) = \psi(\psi^{-1}((x, w, \epsilon))\psi^{-1}((x - \epsilon w, w', \epsilon))) = (x, w + w', \epsilon)$ giving (2.50). Similar considerations apply to give (2.51). The openness of $\mathcal{V}_U$ in $\mathbf{R}^{2n+1}$ follows since it equals $F^{-1}(U \times U)$ where F is the continuous function from $\mathbf{R}^{2n+1}$ to $\mathbf{R}^{2n}$ given by:

$$F((x, w, \epsilon)) = (x, x - \epsilon w). \qquad \square$$

To ease notation, write $H = \mathcal{V}_U$. From (2.50) and (2.51), the range and source maps r, d of H are given by:

$$r((x, w, \epsilon)) = (x, 0, \epsilon), \quad d((x, w, \epsilon)) = (x - \epsilon w, 0, \epsilon). \qquad (2.52)$$

So the unit space H^0 of H is the set of triples $(x, 0, \epsilon)$ of H. Next, if $u = (x, 0, \epsilon) \in H^0$, then from (2.52), we have $H^u = \{(x, w, \epsilon) : (x, w, \epsilon) \in H\}$ and $H_u = \{(x + \epsilon w, w, \epsilon) : (x + \epsilon w, w, \epsilon) \in H\}$.

Now as an open subset of $\mathbf{R}^{n+1}$, the groupoid H is a manifold. The next result shows that it is a Lie groupoid with this smooth structure.

Proposition 2.3.4   *The groupoid H is a Lie groupoid.*

Proof.   It is obvious from the preceding that H^0 is a submanifold of $\mathbf{R}^{2n+1}$. (Indeed, it is identified with an open subset of $\mathbf{R}^{n+1}$.) It is obvious from (2.51) and (2.52) that the inversion map is smooth and that r and d are submersions. Now H^2 can be identified with

$$Z = \{(x, w, w', \epsilon) : x, x - \epsilon w, x - \epsilon(w + w') \in U\}$$

where a pair $((x, w, \epsilon), (x - \epsilon w, w', \epsilon)) \in H^2$ is associated with (x, w, w', ϵ). The multiplication map $m : Z \to H$ and the map $d \times r : Z \to H^0$ ((2.34)) are given by:

$$m((x, w, w', \epsilon)) = (x, w + w', \epsilon), \quad (d \times r)((x, w, w', \epsilon)) = (x - \epsilon w, 0, \epsilon).$$

Also Z is an open subset of $\mathbf{R}^{3n+1}$ since it equals $f^{-1}(U \times U \times U)$ where $f : Z \to \mathbf{R}^{3n}$ is the continuous function given by:

$$f(x, w, w', \epsilon) = (x, x - \epsilon w, x - \epsilon(w + w')).$$

Clearly with the manifold topology on Z that it inherits as an open subset of $\mathbf{R}^{3n+1}$, multiplication on Z $(= H^2)$ is smooth. It remains to show that this "open subset" smooth structure on Z coincides with the submanifold smooth structure that was defined earlier on Z using the submersion $d \times r$. This is intuitively obvious. It can be shown by changing coordinates on Z using the diffeomorphism $(x, w, w', \epsilon) \to (x', w, w', \epsilon)$ where $x' = x - \epsilon w$, and observing that in those new coordinates, $d \times r$ is a projection map so that the discussion of submersions in Appendix F applies. $\square$

For a general manifold, we will use G_U's as charts for G_M. To this end, we need to lift coordinate changes at the level of U to coordinate changes at the G_U level. To do this, we require the following lemma. I believe that this lemma is well-known but have been unable to find a reference for it. In the lemma, the elements of $\mathbf{R}^n$ are regarded as column vectors.

Lemma 2.3.2   *Let U, V be open subsets of $\mathbf{R}^n$ and $\Phi : U \to V$ be smooth. Let $a_0 \in U$. Then there exists an open neighborhood U_1 of a_0 in U and C^∞-functions $\phi_i : U_1 \times U_1 \to \mathbf{R}^n$ such that for all $a, b \in U_1$,*

$$\Phi(a) - \Phi(b) = \sum_{i=1}^{n} (a_i - b_i) \phi_i(a, b). \tag{2.53}$$

Proof.   We will prove the result in the case $n = 2$ leaving the obvious adaptation for general n to the reader. Let U_1 be an open ball contained in U and centered at a_0. For $a = (a_1, a_2)$ and $b = (b_1, b_2)$ in U_1, let

$$\phi_1(a, b) = \int_0^1 \frac{\partial \Phi}{\partial x_1}(ta_1 + (1 - t)b_1, a_2) \, dt$$

and

$$\phi_2(a, b) = \int_0^1 \frac{\partial \Phi}{\partial x_2}(b_1, ta_2 + (1 - t)b_2) \, dt.$$

If $a_1 \neq b_1$ and $x_1 = ta_1 + (1 - t)b_1$, then

$$\frac{\partial \Phi}{\partial x_1} = \frac{\partial \Phi}{\partial t} \frac{\partial t}{\partial x_1} = (a_1 - b_1)^{-1} \frac{\partial \Phi}{\partial t}$$

and a trivial integration gives

$$(a_1 - b_1)\phi_1(a, b) = \Phi(a) - \Phi(b_1, a_2).$$

Similarly

$$(a_2 - b_2)\phi_2(a, b) = \Phi(b_1, a_2) - \Phi(b)$$

and (2.53) immediately follows. $\square$

Corollary 2.3.2   *In the notation of the proposition,*

$$d\Phi_a = [\phi_1(a, a), \dots, \phi_n(a, a)].$$

For the remainder of this section, we will, when convenient, identify G_U with $\mathcal{V}_U$, the context clarifying any ambiguity. Let U, V be open subsets of $\mathbf{R}^n$ and $g : U \to V$ be a diffeomorphism. Define $T_g : G_U \to G_V$ by setting

$$T_g((x, w, \epsilon)) = \left(g(x), \frac{g(x) - g(x - \epsilon w)}{\epsilon}, \epsilon \right)$$

if $\epsilon \neq 0$ and

$$T_g((x, w, 0)) = (g(x), dg_x(w), 0).$$

Proposition 2.3.5   *The map T_g is a diffeomorphism.*

Proof.   From the definition of T_g, we have $(T_g)^{-1} = T_{g^{-1}}$. So by symmetry, we just have to show that T_g is smooth. It is obvious from the above definition that T_g is smooth on $G_U \sim \{(x, w, 0) : (x, w, 0) \in G_U\}$. Let $(x_0, w_0, 0) \in G_U$. By Lemma 2.3.2, there exists an open neighborhood U_1 of x_0 in U and C^∞-functions $\phi_i : U_1 \times U_1 \to \mathbf{R}^n$ such that for $x, y \in U_1$, $g(x) - g(y) = \sum_{i=1}^n (x_i - y_i)\phi_i(x, y)$. Let $W = F^{-1}(U_1 \times U_1) \subset G_U$ where $F(x, w, \epsilon) = (x, x - \epsilon w)$. Then W is an open neighborhood of $(x_0, w_0, 0)$, and if $(x, w, \epsilon) \in W$ with $\epsilon \neq 0$, then

$$
\begin{aligned}
T_g((x, w, \epsilon)) &= (g(x), (g(x) - g(x - \epsilon w))/\epsilon, \epsilon) \\
&= \left(g(x), \sum_{i=1}^n w_i \phi_i(x, x - \epsilon w), \epsilon \right). \quad (2.54)
\end{aligned}
$$

By Corollary 2.3.2, the equation (2.54) also holds when $\epsilon = 0$. Since the right-hand side of (2.54) obviously defines a C^∞ function on W, the proof is complete.   □

Let $\mathcal{C}$ be a maximal atlas for M. For $(U, \phi) \in \mathcal{C}$, define $\tilde{\phi} : G_U \to G_{\phi(U)}$ in the obvious way: if $\epsilon \neq 0$, then $\tilde{\phi}((x, y, \epsilon)) = (\phi(x), \phi(y), \epsilon)$, while if $\epsilon = 0$, then $\tilde{\phi}((x, X, 0)) = (\phi(x), d\phi_x(X), 0)$. By Proposition 2.3.3, $G_{\phi(U)} = \mathcal{V}_{\phi(U)}$ is a Lie groupoid which is an open subset of $\mathbf{R}^{2n+1}$.

Theorem 2.3.4   *The set $\mathcal{A} = \{(G_U, \tilde{\phi}) : (U, \phi) \in \mathcal{C}\}$ defines a Hausdorff $(2n + 1)$-dimensional manifold structure on G_M for which G_M is a Lie groupoid.*

Proof.   The family of G_U's cover G_M – note that the U's do not need to be connected – and each $\tilde{\phi} : G_U \to G_{\phi(U)}$ is bijective. If V is an open subset of U, then $\tilde{\phi}(G_V)$ itself is, of course, an open subset of $\mathbf{R}^{2n+1}$ and equals $\tilde{\phi}_{|V}(G_V)$. Next, if $(V, \psi) \in \mathcal{C}$ and $g = \psi\phi^{-1} : \phi(U \cap V) \to \psi(U \cap V)$, then by Proposition 2.3.5, the map T_g is a diffeomorphism from $G_{\phi(U \cap V)}$ onto $G_{\psi(U \cap V)}$. In addition, from the definition involved, we have

$$T_g = \tilde{\psi}\tilde{\phi}^{-1}.$$

Using the "alternative" definition of a manifold given earlier in the section, the family $\mathcal{A}$ is an atlas for a Hausdorff $(2n + 1)$-dimensional manifold structure on G_M. It remains to show that G_M is a Lie groupoid.

Firstly, G_M^0 is a submanifold of G_M since $G_U^0 = G_U \cap G_M^0$ is a submanifold of the Lie groupoid G_U. A similar argument applies to give that every G_M^u is a submanifold of G_M. Next, the inversion map is smooth on each G_U and so is smooth on G_M. Then the maps d, r are submersions since they are submersions on every G_U. An atlas for the submanifold G_M^2 of $G_M \times G_M$ is given by the family of sets of the form $G_M^2 \cap G_U^2$. Next, if I is an interval in $\mathbf{R}$ not containing 0, then the open subgroupoid $M \times M \times I$ of G_M has its product manifold structure, and trivially the product map on $M \times M \times I$ is smooth. Let $b = ((x, y, \epsilon), (y, z, \epsilon)) \in G_M^2$. If $\epsilon \neq 0$, then some $(M \times M \times I) \cap G_M^2$ is an open neighborhood of b in G_M^2, and on that neighborhood, the product map is smooth. If $b = ((x, X, 0), (x, Y, 0)) \in G_M^2$, then some G_U^2 is an open neighborhood of b in G_M^2, and on G_U^2, the product map is smooth. These are the only two possibilities for an element $b \in G_M^2$ so that the product map for G_M is smooth. So G_M is a Lie groupoid. $\square$

We conclude this section by explicitly constructing, using the theory developed earlier in the section, a smooth left Haar system $\{\lambda^u\}$ on G_U where U is an open subset of $\mathbf{R}^n$. Note that even though $\lambda^{(x,x,\epsilon)}$ "$\to \infty$" as $\epsilon \to 0$, nevertheless it still tends to Lebesgue measure λ on $\mathbf{R}^n$! In the following proposition, we distinguish G_U from $\mathcal{V}_U$.

Proposition 2.3.6   *Let λ be Lebesgue measure on $\mathbf{R}^n$. Then a smooth left Haar system λ^u for $G_U = (U \times U \times \mathbf{R}^*) \cup (U \times \mathbf{R}^n \times \{0\})$ is given by:* $\lambda^{(x,x,\epsilon)} = \epsilon^{-n} \lambda_{|U}$ *when $\epsilon \neq 0$ and $\lambda^{(x,X,0)} = \lambda$.*

Proof.   As above, let $\psi : G_U \to H = \mathcal{V}_U$. Then H^0 is the set of elements of the form $(x, 0, \epsilon)$ $(x \in U, \epsilon \in \mathbf{R})$. Also, $H^{(x,0,\epsilon)} = \{(x, w, \epsilon) : (x, w, \epsilon) \in H\}$. Clearly $(x, 0, \epsilon) \to |\, dw\,|_{(x,0,\epsilon)}$ belongs to $P_+(H)$. Note that from (2.50),

$$L_{(x-\epsilon t, -t, \epsilon)}(x, w, \epsilon) = (x - \epsilon t, w - t, \epsilon).$$

Using this, Theorem 2.3.1 and (F.14), we obtain a left Haar system $\{\mu^{(x,0,\epsilon)}\}$ for H where $\mu_{(x,t,\epsilon)}^{(x,0,\epsilon)}$ equals

$$(L_{(x-\epsilon t, -t, \epsilon)})^* (|\, dw\,|_{(x,0,\epsilon)}) = |\, J(L_{(x-\epsilon t, -t, \epsilon)})\,| |\, dw\,|_{(x,t,\epsilon)} = |\, dw\,|_{(x,t,\epsilon)} \; .$$

So $\mu^{(x,0,\epsilon)} = \lambda_{|U_{x,\epsilon}}$ where $U_{x,\epsilon} = \{w : (x, w, \epsilon) \in H\}$ and $\mu^{(x,0,0)} = \lambda$. So we get a smooth left Haar system $\{\lambda^u\}$ on G_U by setting $\lambda^u = \psi^*(\mu^u)$. Recall that $\psi((x, y, \epsilon)) = (x, (x - y)/\epsilon, \epsilon)$ if $\epsilon \neq 0$ and $\psi((x, X, 0)) = (x, X, 0)$. A simple computation gives the smooth left Haar system of the statement of the proposition. $\square$

CHAPTER 3

Groupoid C^*-Algebras and Their Relation to Inverse Semigroup Covariance C^*-Algebras

3.1 Representation theory for locally compact groupoids

Let G be a locally compact groupoid. We recall (Definition 2.2.2) that G is equipped with a left Haar system $\{\lambda^u\}$.

In the representation theory of locally compact groups, unitary representations of such a group G are integrated up with respect to left Haar measure against $C_c(G)$ (or even $L^1(G)$) functions to give a representation of the convolution algebra $C_c(G)$. In the locally compact groupoid case, however, there are many measures involved in the Haar system. Each unit u has "its own" measure, λ^u, which lives on the "little piece" $G^u = r^{-1}(\{u\})$ of the groupoid. Leaving aside exactly what a representation L of the groupoid G should be for the moment and thinking very roughly, for any $f \in C_c(G)$ and $u \in G^0$, we will have to integrate up $x \rightarrow f(x)L(x)$ in some sense over G^u with respect to λ^u and then combine together what we get over $u \in G^0$.

The natural mathematical framework for this is that in which we have a (probability) measure μ and a Hilbert bundle $\{H_u\}$ $(u \in G^0)$. Each $L(x)$ will then be an operator from $H_{d(x)}$ to $H_{r(x)}$ and the above "combining" will be done by integrating with respect to μ.[1]

The requirement that groupoid representation theory should involve a probability measure μ on the unit space G^0 seems initially unfamiliar from a group point of view, since representation theory for a locally compact group does not explicitly use such a measure. However, in that case, there is such a measure implicitly involved – the point mass at the single unit of the group, the identity! Of course in that context, it is not worth the trouble to mention it! By contrast, a groupoid with a large unit space will have *many* measures μ relevant for representation theory.

However, as one would expect, not *every* probability measure on G^0 will be relevant for representation theory. By considering the representation theory of transformation groupoids ([179, Ch.1])), such a measure should satisfy some kind of invariance condition. The condition that we need is that of *quasi-invariance*. We now describe in more detail representation theory for locally compact groupoids. This description is based on Renault's account in [230].[2]

If X is locally compact Hausdorff, then the set of probability measures on X is denoted by $P(X)$. Let G be a locally compact groupoid. Since G (and G^2) need not be Hausdorff, a little care is needed in places with the measure theory.

A *positive Borel measure* on G is a $[0, \infty]$-valued measure on the Borel σ-algebra $\mathcal{B}(G)$. Regularity for a positive Borel measure on G is defined exactly as for the locally compact Hausdorff case ([120, p.127]). (There are many compact (Hausdorff) subsets of G available for the purposes of inner regularity by (ii) of Definition 2.2.1.)

Let $\mu \in P(G^0)$. Then μ and the left Haar system determine a regular Borel positive measure ν on $\mathcal{B}(G)$ conveniently written:

$$\nu = \int_{G^0} \lambda^u \, d\mu(u). \tag{3.1}$$

More precisely, let U be an open subset of G contained in some open Hausdorff subset U' of G such that the closure of U in U' is compact. By Urysohn's lemma, there exists $F \in C_c(G)$ such that $F \geq \chi_U$. Define a linear functional ϕ_U on $C_c(U)$ by defining

$$\phi_U(f) = \int d\mu(u) \int f(x) \, d\lambda^u(x).$$

---

[1]Strictly, a "modular function" needs also to be taken into account.

[2]In the approaches of Renault ([230]) and Muhly ([179]), results of Bourbaki ([19]) and Effros ([82]) respectively are used to facilitate the measure theory. However, in the present (non-Hausdorff) context, it seemed preferable to work out the details from scratch. This also gives a self-contained account, and indeed, as we will see, the details, while requiring care, use only basic measure theory.

(The preceding repeated integral makes sense and is finite because of (ii) of Definition 2.2.2.) Now for such an f,

$$| \phi_U(f) | \leq \int d\mu(u) \int | f(x) | F(x) \, d\lambda^u(x)$$

$$\leq M\|f\|_\infty$$

where $M = \int d\mu(u) \int F(x) \, d\lambda^u(x) < \infty$. So the functional ϕ_U is continuous on $C_c(U)$, and by the Riesz representation theorem, there exists a regular Borel measure ν_U on U such that

$$\int_G f \, d\nu_U = \int_{G^0} d\mu(u) \int_{G^u} f(x) \, d\lambda^u(x)$$

for $f \in C_c(U)$.

Now if U, V are open Hausdorff subsets of G as in the preceding paragraph, then the measures ν_U, ν_V, restricted to $U \cap V$ coincide. Simple measure theory then gives a measure ν defined on the σ-ring generated by such open Hausdorff subsets of G, where ν restricted to any such U coincides with ν_U. Since G is covered by countably many such subsets U (by using (ii) of Definition 2.2.1), this σ-ring is just $\mathcal{B}(G)$, and so we obtain ν on $\mathcal{B}(G)$. The measure ν is regular since all of the ν_U's are. By reducing to the case of the ν_U's where the classical theory applies, it follows that $C_c(G)$ is dense in $L^p(G, \nu)$ ($1 \leq p < \infty$) (and in the corresponding L^p-spaces for the associated measures ν^{-1}, ν_0 defined below).

Associated with ν are the regular Borel measures ν^{-1} on G and ν^2 on G^2.

For the first of these, we define $\nu^{-1}(W) = \nu(W^{-1})$ ($W \in \mathcal{B}(G)$). Alternatively, we can write

$$\int f \, d\nu^{-1} = \int \check{f} \, d\nu. \tag{3.2}$$

where $\check{f}$ is as in (2.12). Since $(\lambda^u)^{-1} = \lambda_u$, we have

$$\int f \, d\nu^{-1} = \int d\mu \int f(x^{-1}) \, d\lambda^u(x)$$

$$= \int d\mu(u) \int f(x) \, d\lambda_u(x). \tag{3.3}$$

So parallel to (3.1) we have

$$\nu^{-1} = \int_{G^0} \lambda_u \, d\mu(u). \tag{3.4}$$

The measure ν^2 on the closed subset G^2 of $G \times G$ is defined by setting for $f \in C_c(G^2)$,

$$\nu^2(f) = \int_{G^0} d\mu(u) \int\!\!\int f(x, y) \, d\lambda_u(x) \, d\lambda^u(y). \tag{3.5}$$

To clarify (3.5), we first make sense of it for functions f of the form $g \otimes h \in C_c(U) \otimes C_c(V)$ restricted to $U \dot\times V$ ((2.13)), where U, V are open Hausdorff subsets of G as in the ν-case earlier. Then

$$\nu^2(g \otimes h) = \int (\check{g})^0(u) h^0(u) \, d\mu(u)$$

is well-defined. Using (ii) of Definition 2.2.2, the right-hand side of (3.5) then defines a continuous linear functional on the span of such restrictions in $C_c(U \dot\times V)$. Since G^2 is closed in $G \times G$, it follows that $U \dot\times V$ is an open, locally compact Hausdorff subset of G^2. Using the Stone-Weierstrass theorem, we obtain a measure $\nu^2_{U \dot\times V}$ on $U \dot\times V$. Arguing as in the case of ν, we obtain the required measure ν^2 determined by its restrictions $\nu^2_{U \dot\times V}$ on every $U \dot\times V$.

The (probability) measure μ is called ([230, p.23]) *quasi-invariant* if ν is equivalent to ν^{-1}. (A simple but instructive example which computes the quasi-invariant measures on a trivial groupoid is given in the proof of Theorem 3.1.2.)

The above definition of quasi-invariance for μ may appear a little strange at first sight. To motivate it, let us consider the transformation groupoid $X \times H$ associated with the tranformation group (X, H) discussed in Chapter 1. (So H is a countable discrete group with a right action on a locally compact Hausdorff space X.)

In that situation (and indeed for the continuous case) the representations of $C_c(X \times H)$ were shown by Effros and Hahn ([83]) to be given by covariant pairs (ρ, π) where ρ is a representation of $C_0(X)$ and π is a unitary representation of H. (See, for example, [179, Chapter 1]. This will be extended to inverse semigroup actions in **3.3**.) The "basic measure" μ on X for ρ (see below) is then characterized as *quasi-invariant*. This means that if $h \in H$ and $h\mu$ is the measure on X given by

$$(h\mu)(E) = \mu(Eh),$$

then μ and $h\mu$ are equivalent. Let us show that this notion of quasi-invariance, when translated into ν-terms, coincides with the groupoid definition in terms of ν above.

Let $\mu \in P(X)$. Then $\nu = \int_X \lambda^x \, d\mu(x) = \mu \times \lambda$ where λ is counting measure on H. Thinking of the locally compact groupoid $X \times H$ as the union of the countable family of "horizontal slices" $A_h = X \times \{h\}$ as in Chapter 1, we see that $\nu \sim \nu^{-1}$ if and only if $\nu \mid_{A_h} \sim \nu^{-1} \mid_{A_h}$ for all h. If $E \in \mathcal{B}(X)$, then $\nu(E \times \{h\}) = \mu(E)$ while $\nu^{-1}(E \times \{h\}) = \nu(\{(x, h)^{-1} : x \in E\}) = \nu(\{(xh, h^{-1}) : x \in E\}) = h\mu(E)$. It follows immediately that the two versions of quasi-invariance for μ coincide. The advantage of its formulation in terms of ν is, of course, that it applies to locally compact groupoids in complete generality.

If μ is quasi-invariant then the Radon-Nikodym derivative

$$D = \frac{d\nu}{d\nu^{-1}} \tag{3.6}$$

is called the *modular function* of μ.

In the case of a locally compact group H with $\mu = \delta_e$, then $\nu = \lambda$, left Haar measure on H, and $\nu^{-1} = \lambda^{-1}$. In that case, by [120, Theorem 15.15], the right-hand side of (3.6) does give the modular function Δ for H. A useful property of Δ in the group context is that it is a homomorphism, i.e. for all $x, y \in H$, we have $\Delta(x^{-1}) = \Delta(x)^{-1}$ and $\Delta(xy) = \Delta(x)\Delta(y)$.

In the case of a locally compact groupoid G, a similar result holds. As noted by Renault ([230, p.23]), this follows from the work of P. Hahn on measure groupoids ([115]). Firstly, with μ quasi-invariant as above, it is easy to prove that

$$D(x^{-1}) = D(x)^{-1} \ \nu\text{-a.e..} \tag{3.7}$$

Indeed, if $g \in C_c(G)$, then

$$
\begin{aligned}
\int g \, d\nu &= \int gD \, d\nu^{-1} \\
&= \int g(x^{-1})D(x^{-1}) \, d\nu(x) \\
&= \int g(x^{-1})D(x^{-1})[D(x) \, d\nu^{-1}(x)] \\
&= \int g(x)D(x)D(x^{-1}) \, d\nu(x),
\end{aligned}
$$

so that

$$D(x)D(x^{-1}) = 1$$

for ν-a.e. x. Proving the equality

$$D(xy) = D(x)D(y) \ \nu^2\text{-a.e.} \tag{3.8}$$

requires a lengthy argument due to P. Hahn ([115]). As noted by P. Muhly ([179]), a result of A. Ramsay ([220, Theorem 3.20]) combined with the results of Hahn gives that D can actually be taken to be a Borel homomorphism of G (so that D is a Borel map for which $D(x^{-1}) = D(x)^{-1}$ for all $x \in G$ and $D(xy) = D(x)D(y)$ for all $(x, y) \in G^2$).

In Hahn's approach, as noted in Chapter 1, G^2 is itself a groupoid, with inversion given by $(x, y) \to (xy, y^{-1})$. So we can define the measure $(\nu^2)^{-1}$ on G^2 by:

$$\int F(x, y) \, d(\nu^2)^{-1}(x, y) = \int F(xy, y^{-1}) \, d\nu^2(x, y).$$

Using the quasi-invariance of μ, it is not difficult to show that $\nu^2 \sim (\nu^2)^{-1}$ so that we can write

$$\int F(x,y)\rho(x,y)\,d\nu^2(x,y)$$

$$= \int F(x,y)\,d(\nu^2)^{-1}(x,y) \qquad (3.9)$$

$$= \int F(xy,y^{-1})\,d\nu^2(x,y) \qquad (3.10)$$

for some function ρ on G^2. Alternatively expressed, we have

$$\int F(x,y)\,d\nu^2(x,y) = \int F(xy,y^{-1})\rho^{-1}(xy,y^{-1})\,d\nu^2(x,y). \qquad (3.11)$$

One might expect there to be a relation between $\rho^{-1} = \frac{d\nu^2}{d(\nu^2)^{-1}}$ and $D = \frac{d\nu}{d\nu^{-1}}$, and indeed Hahn shows that in fact there is. As noted by Renault, Hahn's results give that for ν^2-a.e. (x,y), we have

$$D(y) = \rho^{-1}(x,y) = D(xy)D(x)^{-1}. \qquad (3.12)$$

Of course, (3.8) follows from (3.12). The details of the remarkable arguments of ([115]) which give (3.8) will be omitted. (Is there a straightforward direct proof of (3.8) as in the locally compact group case ([120, p.195])?)

Following Renault ([230, p.24]), it is useful for representation theory (see, for example, (3.17)) to have available a measure equivalent to ν that is symmetrical with respect to inversion on G. To this end, let ν_0 be the measure on $\mathcal{B}(G)$ given by

$$d\nu_0 = D^{-1/2}\,d\nu.$$

Then $\nu_0(E) = \nu_0(E^{-1})$ for all $E \in \mathcal{B}(G)$. Indeed, using (3.7), we have

$$\nu_0(E) = \int \chi_E(x)D^{-1/2}(x)\,d\nu(x)$$

$$= \int \chi_E(x^{-1})D^{-1/2}(x^{-1})\,d\nu^{-1}(x)$$

$$= \int \chi_{E^{-1}}(x)D^{1/2}(x)D^{-1}(x)\,d\nu(x))$$

$$= \nu_0(E^{-1}).$$

It is well-known that if H is a locally compact Hausdorff group, then the unitary representations of H correspond to the non-degenerate $\|.\|_1$-continuous *-representations of $C_c(H)$ on a Hilbert space. (Usually, $L^1(H)$ is used in place of $C_c(H)$, but the latter is a dense subalgebra of $L^1(H)$ so that the C_c-version, which fits in much better with the groupoid context, also holds.) We now discuss the counterpart of this result (due to Renault) for the locally compact groupoid G, groupoid representations corresponding to unitary representations in the group case. We will use the theory of Hilbert bundles. Accounts of this theory are given by Dixmier ([72, Part 2]), Ramsay ([219]) and Muhly ([179]).

A *Hilbert bundle* will be denoted by a triple such as $(X, \mathcal{K}, \mu)$. Here, X is a (second countable) locally compact Hausdorff space and μ is a probability measure on X. The notation $\mathcal{K}$ stands for a collection of Hilbert spaces $\{H_u\}$ where u ranges over X. A *section* of the bundle is a function $f : X \to \cup_{u \in X} H_u$ where $f(u) \in H_u$. Left implicit in this notation, there is also given a sequence of such sections f_n which is *fundamental* in the sense that for each pair m, n, the function $u \to \langle f_m(u), f_n(u) \rangle$ is μ-measurable on X, and for each $u \in X$, the $f_n(u)$'s span a dense subspace of H_u. Use of Gram-Schmidt shows that we can take $\{f_n\}$ to be *orthonormal* in the sense that for each $u \in X$, the sequence $\{f_n(u)\}$ is an orthonormal basis for H_u. The fundamental sequence $\{f_n\}$ defines the notion of *measurability* for sections in the sense that a section f is called measurable if each function $u \to \langle f(u), f_n(u) \rangle$ is μ-measurable. The Hilbert space $L^2(X, \{H_u\}, \mu)$ is defined in the obvious way as the space of (equivalence classes) of measurable sections f for which the function $u \to \|f(u)\|_2^2$ is μ-integrable. The norm of $f \in L^2(X, \{H_u\}, \mu)$ is given by: $\|f\|_2 = (\int_X \|f(u)\|^2 \, d\mu(u))^{1/2}$ and the inner product by: $\langle f, g \rangle = \int_X \langle f(u), g(u) \rangle \, d\mu(u)$.

Given X, μ and a (separable) Hilbert space H, the *constant* Hilbert bundle is the triple $(X, \{K_u\}, \mu)$ where $K_u = H$ for all u and we take for a fundamental sequence the constant sections $u \to e_n$ where $\{e_n\}$ is an orthonormal basis for H. A simple computation shows that in the constant bundle case, $L^2(X, \{K_u\}, \mu)$ is canonically identified with $L^2(X, \mu) \otimes H$ ([72, p.175]). Any Hilbert bundle $(X, \{H_u\}, \mu)$ where $\dim H_u = p$ $(1 \le p \le \aleph_0)$ is isomorphic to the constant bundle associated with the Hilbert space H of dimension p. (For each u, simply send $f_n(u)$ to e_n for any fundamental sequence $\{f_n\}$ for $(X, \{H_u\}, \mu)$ ([72, p.167]).) Such a bundle is called *trivial* and in that case, we have

$$L^2(X, \{H_u\}, \mu) \cong L^2(X, \mu) \otimes H. \tag{3.13}$$

Each $F \in L^\infty(X, \mu)$ identifies isometrically with a multiplication or *diagonalizable* operator $T_F \in B(L^2(X, \{H_u\}, \mu))$: define for f in the Hilbert space $L^2(X, \{H_u\}, \mu)$ and $u \in X$,

$$T_F(f)(u) = F(u)f(u). \tag{3.14}$$

A fact that we will need later is that if $\{f_n\}$ is a fundamental orthonormal sequence for the Hilbert bundle, then ([72, p.172]) the span of the functions $\{T_F(f_n) : F \in L^\infty(X, \mu), n \ge 1\}$ is dense in $L^2(X, \{H_u\}, \mu)$.

Hilbert bundles arise in a standard way when a commutative C^*-algebra $C_0(X)$ is represented non-degenerately on a Hilbert space. In that case ([72, Part 1, Ch. 7], [146, 4.4]), there is a probability measure μ (called a *basic* measure) on X and a representation of the Hilbert space as a Hilbert bundle $L^2(X, \{H_u\}, \mu)$ with respect to which the elements of $C_0(X)$ act as diagonalizable operators. So the operator associated with $g \in C_0(X)$ on the original Hilbert space is now identified as an operator on the L^2-space of sections of the bundle, and given explicitly by $u \to g(u)I_u$ where I_u is the identity operator on H_u. (Think of the spectral theorem!)

The idea behind a groupoid representation L on a Hilbert space is that we think of the Hilbert space as the L^2-space $L^2(G^0, \{H_u\}, \mu)$ of a Hilbert bundle over the unit space G^0 of the groupoid. Each x in the groupoid then acts as a unitary element $L(x)$ from the Hilbert space $H_{d(x)}$ onto the Hilbert space $H_{r(x)}$. This fits in with the multiplication on G. For if $r(y) = d(x)$, so that the product xy is defined in G, then the composition $L(x) \circ L(y)$ is also defined and is required to be $L(xy)$ (at least "almost everywhere"). While the basic idea is natural, the precise definition, given below, becomes a little involved owing to the necessary measure theoretic requirements. The measure μ on G^0 associated with the bundle is required to be quasi-invariant. We also have the measures ν, ν^2 on G, G^2 available to deal with the measurability requirements for inversion and product.

Definition 3.1.1 *A representation ([230, p.52]) of the locally compact groupoid G is defined by a Hilbert bundle $(G^0, \{H_u\}, \mu)$ where μ is a quasi-invariant measure on G^0 (with associated measures $\nu, \nu^{-1}, \nu^2, \nu_0$) and, for each $x \in G$, a unitary element $L(x) \in B(H_{d(x)}, H_{r(x)})$ such that:*

(i) *$L(u)$ is the identity map on H_u for all $u \in G^0$;*

(ii) *$L(x)L(y) = L(xy)$ for ν^2-a.e. $(x, y) \in G^2$;*

(iii) *$L(x)^{-1} = L(x^{-1})$ for ν-a.e. $x \in G$;*

(iv) *for any $\xi, \eta \in L^2(G^0, \{H_u\}, \mu)$, the function*

$$x \to \langle L(x)\xi(d(x)), \eta(r(x)) \rangle \qquad (3.15)$$

is ν-measurable on G.

Our notation for the representation of G will be the triple $(\mu, \{H_u\}, L)$ or (with $\mu, \{H_u\}$ left implicit) simply L. Note that the inner product in (3.15) makes sense since $\xi(d(x)) \in H_{d(x)}$ and so $L(x)\xi(d(x))$, like $\eta(r(x))$, belongs to $H_{r(x)}$, the inner product in (3.15) being then that of $H_{r(x)}$.

There are two natural examples of groupoid representations suggested by the locally compact group case, both of which involve no measure theoretic complications. Indeed, two easily defined examples of representations of a locally compact *group* M are the *trivial representation* and the *left regular representation*. The trivial representation L_{triv} on M is the one dimensional representation given by: $L_{triv}(x) = 1$ for all $x \in M$. The left regular representation L_{red} of M on $L^2(G)$ is given by:

$$L_{red}(x)(F)(t) = F(x^{-1}t) \quad (t \in G).$$

Putting the group into the groupoid context, the quasi-invariant measure μ for these representations is, of course, the point mass at the identity e. For a general locally compact groupoid G and *any* given quasi-invariant measure μ on G^0 there are natural versions of L_{triv} and L_{red}.

For the trivial representation case, take the Hilbert bundle over G^0 to be given by the trivial bundle $G^0 \times \mathbf{C}$. (So each H_u is just the one-dimensional Hilbert space $\mathbf{C}_u = \mathbf{C}$.) Then the trivial representation (for μ) is given by: $L_{triv}(x) = 1$, the identity map on $\mathbf{C} = \mathbf{C}_{d(x)} = \mathbf{C}_{r(x)}$. It is (indeed!) trivial to check that L_{triv} is a representation of G.

For the left regular representation ([230, p.55]), take

$$H_u = L^2(G^u).$$

Consider $C_c(G)$ as a space of sections of $\{H_u\}$ by identifying $F \in C_c(G)$ with the section: $u \to F_{|G^u} \in C_c(G^u) \subset L^2(G^u)$. For $F_1, F_2 \in C_c(G)$ the requirement that $u \to \langle F_1(u), F_2(u) \rangle$ be μ-measurable follows since $F_1 \overline{F_2}$ is in $B_c(G)$ and hence $(F_1 \overline{F_2})^0 \in B_c(G^0)$ as discussed in **2.2**. For $C \in \mathcal{C}$ in the notation of Definition 2.2.1, there exists a sup-norm dense countable subset $\mathcal{A}_C$ of $C_c(C^o)$, and a fundamental sequence of sections is obtained by using sums of the functions in $\mathcal{A}_C$ as C ranges over $\mathcal{C}$. So $\{H_u\}$ is a Hilbert bundle. Then we define $L_{red}(x) : H_{d(x)} \to H_{r(x)}$ by:

$$(L_{red}(x))(h)(z) = h(x^{-1}z) \tag{3.16}$$

where $h \in L^2(G^{d(x)})$ and $z \in G^{r(x)}$. So $L_{red}(x)$ is just the extension of the map $f \to x * f$ of (2.18) to $L^2(G^{d(x)})$, and this extension is a bijective isometry and hence is unitary. It is easy to check that L_{red} satisfies the other representation conditions of Definition 3.1.1.

As commented earlier, we are interested in linking up representations of G with representations of $C_c(G)$. Recall (**2.2**) that $C_c(G)$ is a normed *-algebra under the I-norm, and that all representations of $C_c(G)$ (on a Hilbert space) considered in the book are assumed to be I-norm continuous. Since $C_c(G)$ is separable (Theorem 2.2.1), every representation of $C_c(G)$ generates a separable C^*-algebra, and such C^*-algebras can always be realized on a separable Hilbert space. We can assume then that every representation of $C_c(G)$ under consideration is on a separable Hilbert space.

We will now state and prove the result of Renault ([230, 233]) that a representation L of the locally compact groupoid G "integrates up" to give a representation $\pi_L : C_c(G) \rightarrow B(L^2(\mathcal{H}))$, where $\mathcal{H} = L^2(G^0, \{H_u\}, \mu)$, and where π_L is given by:

$$\langle \pi_L(f)\xi, \eta \rangle = \int_G f(x) \langle L(x)(\xi(d(x))), \eta(r(x)) \rangle \, d\nu_0(x). \qquad (3.17)$$

We will see in the proof of Proposition 3.1.1 that the integrand in (3.17) is in $L^1(G, \nu_0)$.

It will be useful to have available an explicit formula which says what $\pi_L(f)\xi \in \mathcal{H}$ is for μ-a.e. $u \in G^0$. To this end, assume for the moment that $\pi_L(f) \in B(\mathcal{H})$. (We will prove that this is the case in the theorem below.) To determine this formula, recall that $d\nu_0 = D^{-1/2} d\nu$ and that $\nu = \int \lambda^u \, d\mu(u)$. Using these formulae in (3.17) gives that $\langle \pi_L(f)\xi, \eta \rangle$ is equal to

$$\int_{G^0} d\mu(u) \int_{G^u} f(x) \langle L(x)(\xi(d(x))), \eta(r(x)) \rangle D(x)^{-1/2} \, d\lambda^u(x). \qquad (3.18)$$

Then passing the λ^u-integral through the $\langle , \rangle$ of the inner product in (3.18) gives that $\langle \pi_L(f)\xi, \eta \rangle$ is equal to

$$\int_{G^0} \langle \int_{G^u} f(x) L(x)(\xi(d(x))) D(x)^{-1/2}(x) \, d\lambda^u(x), \eta(u) \rangle \, d\mu(u). \qquad (3.19)$$

Note in the above that we were able to substitute u for $r(x)$ in the expression $\eta(r(x))$ since for $x \in G^u$, we have $r(x) = u$.

So we can define, for μ-a.e. $u \in G^0$,

$$\pi_L(f)\xi(u) = \int_{G^u} f(x) L(x)(\xi(d(x))) D^{-1/2}(x) \, d\lambda^u(x). \qquad (3.20)$$

Here, the right-hand side of (3.20) is interpreted as a vector-valued integral in the weak topology of $\mathcal{H}$ using the Riesz representation theorem, so that by definition $\pi_L(f)\xi = \xi' \in \mathcal{H}$, where $\langle \xi', \eta \rangle$ is the right-hand side of (3.18). With this interpretation, (3.20) gives the required formula for $\pi_L(f)\xi$.

Proposition 3.1.1   ([230, Proposition 1.7, p.52]) *The equation* (3.17) *defines a representation* π_L *of* $C_c(G)$ *of norm* ≤ 1 *on* $\mathcal{H} = L^2(G^0, \{H_u\}, \mu)$.

Proof.   We show first that the integrand in (3.17) belongs to $L^1(G, \nu_0)$ and that

$$| \langle \pi_L(f)\xi, \eta \rangle | \leq \|f\|_I \|\xi\|_2 \|\eta\|_2. \qquad (3.21)$$

The proof depends on an ingenious use of the Cauchy-Schwartz inequality (which goes back to an argument of Hahn [115]). Recalling that $d\nu^{-1} = D^{-1} d\nu$, we have, using (3.1), (3.3), (3.4), (2.24) and (2.23),

$$\int | f(x) \langle L(x)(\xi(d(x))), \eta(r(x)) \rangle | \, d\nu_0(x)$$

$$\leq \int [|\, f(x)\,|^{1/2}\, \|\xi(d(x))\| D(x)^{-1/4}] \times$$

$$[|\, f(x)\,|^{1/2}\, \|\eta(r(x))\| D(x)^{1/4}]\, d\nu_0(x)$$

$$\leq \ [\int |\, f(x)\,|\, \|\xi(d(x))\|^2 D(x)^{-1/2}\, d\nu_0(x)]^{1/2} \times$$

$$[\int |\, f(x)\,|\, \|\eta(r(x))\|^2 D(x)^{1/2}\, d\nu_0(x)]^{1/2}$$

$$= \ [\int |\, f(x)\,|\, \|\xi(d(x))\|^2\, d\nu^{-1}(x)]^{1/2} \times$$

$$[\int |\, f(x)\,|\, \|\eta(r(x))\|^2\, d\nu(x)]^{1/2}$$

$$\leq \ [\int_{G^0} \|\xi(u)\|^2\, d\mu(u) \int_{G_u} |\, f(x)\,|\, d\lambda_u(x))]^{1/2} \times$$

$$[\int_{G^0} \|\eta(u)\|^2\, d\mu(u) \int_{G^u} |\, f(x)\,|\, d\lambda^u(x)]^{1/2}$$

$$\leq \ [\|f\|_{I,d}^{1/2}\|\xi\|][\|f\|_{I,r}^{1/2}\|\eta\|]$$

$$\leq \ \|f\|_I \|\xi\| \|\eta\|.$$

This gives (3.21) and that there is a bounded linear map $\pi_L : C_c(G) \to B(\mathcal{H})$ with $\|\pi_L\| \leq 1$ given by (3.17).

Now let $f, g \in C_c(G)$ and $\xi, \eta \in \mathcal{H}$. We want to show that

$$\pi_L(f * g) = \pi_L(f)\pi_L(g).$$

Before considering the rather technical argument below establishing this, the reader may find it helpful to motivate it by giving an easy direct argument for this result *ignoring null sets* by showing, using (3.20), that

$$\pi_L(f * g)\xi(u) = \pi_L(f)(\pi_L(g)\xi(u)).$$

The problem then lies with the need to consider null sets, and since both the product formula in (ii) of Definition 3.1.1 and the equality of (3.8) hold only ν^2-a.e., the measure ν^2 has to come into the calculation. We start with (3.17), which does not hold "a.e.", instead of (3.20).

Using (2.21), (3.6), (3.20), (3.11), (3.12), (ii) of Definition 3.1.1, (3.8), the fact that $d\nu = D\, d\nu^{-1}$ and Fubini's theorem, we have, with

$$F(x,t) = f(xt)g(t^{-1})\langle L(x)\xi(d(x)), \eta(r(x))\rangle D^{1/2}(x),$$

that

$$\langle \pi_L(f * g)\xi, \eta\rangle$$

$$= \int f * g(x)\langle L(x)\xi(d(x)), \eta(r(x))\rangle D^{-1/2}(x)\, d\nu(x)$$

$$= \int (\int f(xt)g(t^{-1})\, d\lambda^{d(x)}(t))\langle L(x)\xi(d(x)), \eta(r(x))\rangle \times$$
$$D^{-1/2}(x)[D(x)\, d\nu^{-1}(x)]$$
$$= \int d\mu(u) \int d\lambda_u(x) \int F(x,t)\, d\lambda^{d(x)}(t)$$
$$= \int F(x,t)\, d\nu^2(x,t)$$
$$= \int F(xt,t^{-1})\rho^{-1}(xt,t^{-1})\, d\nu^2(x,t)$$
$$= \int d\mu(u) \iint f(x)g(t)\langle L(xt)\xi(d(xt)), \eta(r(xt))\rangle \times$$
$$D^{1/2}(xt)[D(x)D(xt)^{-1}]\, d\lambda_u(x)\, d\lambda^u(t)$$
$$= \int d\mu(u) \int f(x)\langle L(x)(\int g(t)L(t)\xi(d(t)) \times$$
$$D^{-1/2}(t)\, d\lambda^{d(x)}(t)), \eta(u)\rangle D^{-1/2}(x)[D(x)\, d\lambda_u(x)]$$
$$= \int d\mu(u) \int f(x)\langle L(x)(\int g(t)L(t)\xi(d(t))D^{-1/2}(t)\, d\lambda^{d(x)}(t)),$$
$$\eta(u)\rangle[D^{-1/2}(x)\, d\lambda^u(x)]$$
$$= \int f(x)\langle L(x)\pi_L(g)\xi(d(x)), \eta(r(x))\rangle\, d\nu_0(x)$$
$$= \langle \pi_L(f)\pi_L(g)\xi, \eta\rangle.$$

So π_L is a homomorphism.

Next we show that $\pi_L(f^*) = \pi_L(f)^*$. Indeed, using the symmetry of ν_0 and (iii) of Definition 3.1.1, we have

$$\langle \pi_L(f^*)\xi, \eta\rangle$$
$$= \int \overline{f(x^{-1})}\langle L(x)\xi(d(x)), \eta(r(x))\rangle\, d\nu_0(x)$$
$$= \int \overline{f(x)}\langle L(x^{-1})\xi(d(x^{-1})), \eta(r(x^{-1}))\rangle\, d\nu_0(x)$$
$$= \overline{\int f(x)\langle L(x)\eta(d(x)), \xi(r(x))\rangle\, d\nu_0(x)}$$
$$= \overline{\langle \pi_L(f)\eta, \xi\rangle}$$

so that $\pi_L(f^*) = \pi_L(f)^*$.

Lastly, we need to show that π_L is non-degenerate. Suppose that this is not so. Then the span of the vectors $\pi_L(f)\xi$, where $f \in C_c(G)$ and $\xi \in \mathcal{H}$, is not dense in $\mathcal{H}$, and so there exists non-zero $\eta \in \mathcal{H}$ such that $\langle \pi_L(f)\xi, \eta\rangle = 0$ for all $f \in C_c(G)$ and all $\xi \in \mathcal{H}$, i.e.

$$\int f(x)\langle L(x)\xi(d(x)), \eta(r(x))\rangle\, d\nu_0(x) = 0.$$

By (ii) of Definition 2.2.1, there is a countable basis for G consisting of open, Hausdorff subsets U. For such a set U, let $f \in C_c(U)$. By approximating the $L^1(G, \nu_0)$-function

$$x \longrightarrow |\, f(x)\langle L(x)\xi(d(x)), \eta(r(x))\rangle\,|$$

by functions of the form

$$x \longrightarrow g(x)f(x)\langle L(x)\xi(d(x)), \eta(r(x))\rangle$$

with $g \in C_c(U)$, we have

$$\int |\, f(x)\langle L(x)\xi(d(x)), \eta(r(x))\rangle\,|\; d\nu_0(x) = 0.$$

By varying f and U, this gives that $\langle L(x)\xi(d(x)), \eta(r(x))\rangle = 0$ ν_0-a.e.. Let $\{\xi_n\}$ be a fundamental sequence for $(G^0, \{H_u\}, \mu)$. Then there exists a ν_0-null set E in G such that for all $x \in G \sim E$ and all n, we have

$$\langle L(x)\xi_n(d(x)), \eta(r(x))\rangle = 0. \tag{3.22}$$

Now for $x \in G \sim E$, the vectors $\xi_n(d(x))$ span a dense subspace of $H_{d(x)}$ and since $L(x)$ is unitary, it follows that the vectors $L(x)\xi_n(d(x))$ span a dense subspace of $H_{r(x)}$. From (3.22), we obtain $\eta(r(x)) = 0$ ν_0-a.e.. Since ν_0 is equivalent to ν, we have $0 = \int \|\eta(r(x))\|\, d\nu(x) = \int d\mu(u) \int \|\eta(u)\|\, d\lambda^u(x) = \int \|\eta(u)\|\lambda^u(G^u)\, d\mu(u)$, and since $\lambda^u(G^u) > 0$ for all $u \in G^0$, we have $\eta(u) = 0$ μ-a.e.. Hence $\eta = 0$ in $\mathcal{H}$ and we have a contradiction. $\qquad\square$

As a simple but instructive example of a π_L, let us consider the case discussed earlier where $L = L_{triv}$ with respect to some quasi-invariant measure μ on G^0. Recalling that in this case, $H_u = \mathbf{C}$, we have $L^2(G^0, \{\mathbf{C}_u\}, \mu) = L^2(G^0, \mu)$. With this identification, use of (3.20) gives that for $f \in C_c(G)$ and $\xi \in L^2(G^0, \mu)$, we have

$$\pi_{L_{triv}}(f)\xi(u) = \int_{G^u} f(x)\xi(d(x))D^{-1/2}(x)\, d\lambda^u(x). \tag{3.23}$$

In the locally compact group case where there is only one unit, we have $L^2(G^0, \mu) = \mathbf{C}$, and we obtain

$$\pi_{L_{triv}}(f) = \int_G f(x)D^{-1/2}(x)\, d\lambda(x). \tag{3.24}$$

The usual formula when G is a locally compact group is:

$$\pi_{L_{triv}}(f) = \int_G f(x)\, d\lambda(x) \tag{3.25}$$

where λ is left Haar measure on G. (The $D^{-1/2}(x)$ term present in (3.24) but not in (3.25) reflects, as discussed in **2.2**, the fact that we are using a different involution on $C_c(G)$: from a groupoid point of view, $f^*(x) = \overline{f(x^{-1})}$, but from the usual locally compact group point of view, $f^*(x) = D(x^{-1})\overline{f(x^{-1})}$ where D is the modular function on G.) We will consider later in the section the representation $\pi_{L_{red}}$ for the left regular representation L_{red}.

The fundamental theorem of analysis on locally compact groupoids, part of which has been proved in Proposition 3.1.1, is due to Renault ([233]), and is stated in the following theorem. The really difficult part of the proof is to show that every representation of $C_c(G)$ is some π_L. A complete, detailed account of the proof of the fundamental theorem in the Hausdorff case is given by Paul Muhly in his book [179]. This proof can be adapted to cover the general non-Hausdorff case.

We note that for the r-discrete groupoids an earlier proof[3] by Renault ([230, p.65ff.]), which will be refined to give Theorem 3.2.1, also gives Theorem 3.1.1, in that case. The proof of that will be sketched after that of Corollary 3.2.1.

Theorem 3.1.1   *Let G be a locally compact groupoid. Then every representation of $C_c(G)$ is of the form π_L for some representation L of G, and the correspondence $L \rightarrow \pi_L$ preserves the natural equivalence relations on the representations of G and the representations of $C_c(G)$.*

An important step in the proof of Theorem 3.1.1 is that of extending a representation Π of $C_c(G)$ on a Hilbert space $\mathcal{H}$ to a representation of $B_c(G)$ on $\mathcal{H}$. For later use, we discuss this extension here. The discussion is based on [230, p.62]. Let $U, F \geq \chi_U$ be as in the paragraph following (3.1), and let $f \in C_c(U)$. Then for any $u \in G^0$, we have

$$\int_{G^u} | f(x) | F(x) \, d\lambda^u(x) \leq \|f\|_\infty \|F\|_{I,r}.$$

A similar inequality holds for the (I, d)-norm, and we have

$$\|f\|_I \leq \|f\|_\infty \|F\|_I.$$

So

$$\|\Pi(f)\| \leq \|\Pi\| \|f\|_I \leq \|\Pi\| \|f\|_\infty \|F\|_I.$$

So Π, restricted to $C_c(U)$, is $\|.\|_\infty$-continuous, and by the Riesz representation theorem, there exists, for each $\xi, \eta \in \mathcal{H}$, a complex regular Borel measure $\mu_{U,\xi,\eta}$ on U such that

$$\langle \Pi(f)\xi, \eta \rangle = \int f \, d\mu_{U,\xi,\eta}. \tag{3.26}$$

---

[3]The author is grateful to Paul Muhly for helpful discussion about these two proofs of Renault.

Arguing as we did in defining ν from the ν_U's, there exists a unique regular Borel measure $\mu_{\xi,\eta}$ on the ring of Borel subsets of G contained in some compact subset of G such that

$$\langle \Pi(f)\xi, \eta \rangle = \int f \, d\mu_{\xi,\eta}. \tag{3.27}$$

for all $f \in C_c(G)$. The right-hand side of (3.27) is defined for any $f \in B_c(G)$, and this gives a linear map Π' from $B_c(G)$ into $B(\mathcal{H})$.

Next, we claim that Π' is a *-homomorphism. To prove this, for fixed $g \in C_c(G)$, we apply Lemma 2.2.1 first to show that the space of functions $f \in B_c(G)$ with $\Pi'(f * g) = \Pi'(f)\Pi'(g)$ is the same as $B_c(G)$, and second to show that the space of functions $h \in B_c(G)$ such that $\Pi'(f * h) = \Pi'(f)\Pi'(h)$ for all $f \in B_c(G)$ is also $B_c(G)$. (Compare the proof of Corollary 2.2.1.) Similarly, one can prove that Π' preserves the involution on $B_c(G)$.

For later use, if G is r-discrete, then the representation Π of $C_c(G)$ determines a natural representation π_1 of the inverse semigroup G^{op}. Indeed, if $B \in G^{op}$ is contained in a compact subset of G, then $\chi_B \in B_c(G)$, and we define $\pi_1(B) = \Pi'(\chi_B)$. Any $A \in G^{op}$ is the union of an increasing sequence $\{A_n\}$ of such B's, and we can take $\pi_1(A)$ to be the strong operator limit of the sequence $\{\pi_1(A_n)\}$. (See Proposition 3.2.6.) It is left to the reader to show that π_1 is indeed a representation of G^{op}. We will continue our discussion of π_1 after Proposition 3.2.2 and also after Corollary 3.2.1.

Given Theorem 3.1.1 and a general locally compact groupoid, it is easy to specify Π' on $B_c(G)$. Indeed, just write $\Pi = \pi_L$ for some representation L of G and define Π' as we did for π_L allowing $B_c(G)$-functions rather than just $C_c(G)$ functions. So

$$\langle \Pi'(f)\xi, \eta \rangle = \int_G f(x)\langle L(x)(\xi(d(x))), \eta(r(x)) \rangle \, d\nu_0(x). \tag{3.28}$$

Using Lemma 2.2.1 and (3.27), it is easy to show that Π' is indeed given by (3.28). In future we will usually write Π in place of Π'.

An immediate advantage of using (3.28) is that exactly the same argument used in case of $C_c(G)$ in the proof of Proposition 3.1.1 gives that Π is continuous on the seminormed algebra $(B_c(G), \|.\|_I)$.

Recall (2.2) that when G is r-discrete, the space $B_c(G)$ is a normed *-algebra under convolution with norm $\|.\|_I$, and contains $C_c(G)$ as a normed *-subalgebra. By the above, Π is a continuous representation of $B_c(G)$.

It is also easy to check that $\phi = \Pi_{|C_c(G^0)}$ is a representation of $C_c(G^0) \subset B_c(G)$ on $\mathcal{H}$. Indeed, the only thing needing to be checked is that ϕ is non-degenerate. This is immediate if G^0 is compact, since then χ_{G^0} is the identity of $B_c(G)$. For the general case, argue as in the proof of the existence of a bounded approximate identity in $C_c(G)$ for the r-discrete case (given after (2.30)).

We now give a simple illustration of the transition from L to π_L based on ([230, pp.55-58]) which will be used later. Here, L is the *left regular representation of G on μ* where μ is quasi-invariant.

In this case, the Hilbert bundle fibers H_u ($u \in G^0$) are given by: $H_u = L^2(G^u, \lambda^u)$. As usual, let $\nu = \int \lambda^u \, d\mu(u)$. The Hilbert space $L^2(G^0, \{H_u\}, \mu)$ of square integrable sections is then canonically identified with $L^2(G, \nu)$, each $\xi \in L^2(G^0, \{H_u\}, \mu)$ being identified with the section $u \to \xi(u) = \xi_{|G^u} \in L^2(G^u, \lambda^u)$.

A fundamental sequence for the bundle $\{H_u\}$ is determined μ-a.e. by any orthonormal sequence of $L^2(G, \nu)$ in $C_c(G)$. To prove this, one can replace G by an open Hausdorff subset U of G, restricting the Hilbert bundle appropriately, and ν by ν_U. One uses the fact that any $f \in C_c(G^u)$ extends to a function in $C_c(G)$ (see **2.2**). The details are left to the reader. We now specify the representation L.

For $x \in G$, the unitary map

$$L_{red}(x) \in B(L^2(G^{d(x)}, \lambda^{d(x)}), L^2(G^{r(x)}, \lambda^{r(x)}))$$

is given by (3.16).

Let us calculate π_L in this case from (3.17) leaving explanations until after the details. Let $f, \xi, \eta \in C_c(G)$. As discussed above, ξ, η are regarded as sections, so that, for example, $\xi(u) = \xi_{|G^u}$. Now

$$\langle \pi_L(f)\xi, \eta \rangle =$$

$$\int_G f(x)\langle L(x)\xi(d(x)), \eta(r(x)) \rangle \, d\nu_0(x) \tag{3.29}$$

$$= \int_G f(x) \, d\nu_0(x) \int_{G^{r(x)}} \xi(x^{-1}y)\overline{\eta(y)} \, d\lambda^{r(x)}(y) \tag{3.30}$$

$$= \int_{G^0} d\mu(v) \int_{G^v} f(x)D^{-1/2}(x) \, d\lambda^v(x)$$

$$\times \int_{G^v} \xi(x^{-1}y)\overline{\eta(y)} \, d\lambda^v(y) \tag{3.31}$$

$$= \int_{G^0} d\mu(v) \int_{G^v} \overline{\eta(y)} \, d\lambda^v(y) \times$$

$$\int_{G^v} f(x)D^{1/2}(x^{-1}y)D^{-1/2}(y)\xi(x^{-1}y) \, d\lambda^v(x) \tag{3.32}$$

$$= \int_{G^0} d\mu(v) \int_{G^v} \overline{\eta(y)}D^{-1/2}(y) \, d\lambda^v(y) \times$$

$$\int_{G^v} f(x)(D^{1/2}\xi)(x^{-1}y) \, d\lambda^v(x) \tag{3.33}$$

$$= \int_{G^0} d\mu(v)(f * (D^{1/2}\xi))(y)\overline{\eta(y)}D^{-1/2}(y) \, d\lambda^v(y) \tag{3.34}$$

$$= \langle f * (D^{1/2}\xi), D^{1/2}\eta \rangle_{L^2(\nu^{-1})}. \tag{3.35}$$

The complicated look of an argument like that above can be disconcert-ing. However, most of it is basic measure theory. In particular, Fubini's theorem is used whenever appropriate. Let us briefly discuss each step in turn. The equation (3.29) is just (2.14), while in (3.30), we use (3.16) and the fact that the fiber of the Hilbert bundle $L^2(G, \nu)$ at $u = r(x)$ is $L^2(G^{r(x)}, \lambda^{r(x)})$. In (3.31), we substitute $d\nu_0(x) = D^{-1/2}(x) \, d\nu(x)$ and use $\nu = \int_{G^0} \lambda^u \, d\mu(u)$. In (3.32), we take D to be a Borel homomorph-ism, so that $D(x) = (D(x^{-1}y))^{-1} D(y)$. Then (3.33) rearranges (3.32) to involve a convolution formula that is calculated (using (2.20)) in (3.34). Finally, the $L^2(G, \nu^{-1})$ inner product in (3.35) is obtained by writing $D^{-1/2}(y) = D^{1/2}(y) D^{-1}(y)$ and using $d\nu^{-1}(y) = D^{-1}(y) \, d\nu(y)$.

The above representation π_L is defined on the Hilbert space $L^2(G, \nu)$. Now the map $\xi \to D^{1/2} \xi$ is an isometry from $L^2(G, \nu)$ into $L^2(G, \nu^{-1})$. In-deed,

$$\|D^{1/2} \xi\|^2_{L^2(\nu^{-1})} = \int_G D(x) \mid \xi(x) \mid^2 D(x)^{-1} \, d\nu(x) = \|\xi\|^2_{L^2(\nu)}.$$

What (3.35) shows is that when we use this isometry to identify $L^2(G, \nu)$ with $L^2(G, \nu^{-1})$ then π_L, the left regular representation on μ, becomes both elegant and natural; it is just *function convolution* (extended to the convolving of a $C_c(G)$-function with an $L^2(G, \nu^{-1})$-function):

$$\pi_L(f)(F) = f * F. \tag{3.36}$$

Not only that, as one can check directly, (3.36) gives a representation of $C_c(G)$ on $L^2(G, \nu^{-1})$ for *any* probability measure μ – the measure μ no longer needs to be quasi-invariant! We will return to the left regular repre-sentations later in this section.

For the present, we discuss the *universal* C^*-algebra associated with a general locally compact groupoid G. Since representations of $C_c(G)$ are assumed to be I-norm continuous (and indeed have norm ≤ 1), we can define a C^*-seminorm $\|.\|$ on $C_c(G)$ by setting

$$\|f\| = \sup_\pi \|\pi(f)\|.$$

This seminorm is actually a norm (since it dominates the reduced C^*-norm on $C_c(G)$ discussed below), and the completion of $C_c(G)$ under this norm is defined to be ([230, p.58]) the C^*-*algebra of G*, denoted $C^*(G)$. The representation theory of $C^*(G)$ is, by Theorem 3.1.1, identifiable with that of G. This is the groupoid parallel to the well-known result that for a locally compact group H, the representations of $C^*(H)$ are given by the unitary representations of H.

We now pause to discuss two important examples of $C^*(G)$. In the first example, G is a trivial groupoid, while in the second, $G = R_p$, the r-discrete equivalence relation groupoid associated with the Penrose tilings. Both of these groupoids were discussed in **2.2**.

Recall that a trivial groupoid G is defined to be $X \times X$, where X is a locally compact Hausdorff space on which is given a positive regular Borel measure μ on X with support equal to X, and that the left Haar system of G is given by μ. By scaling μ, we can take μ to be a probability measure. (It is obvious that this will make no difference to the representation theories of G or of $C^*(G)$.) An interesting and important result is that $C^*(G)$ *is isomorphic to the C^*-algebra of compact linear operators on a separable Hilbert space.*[4] This follows from the paper of Williams and Muhly ([182]) and is also proved by Muhly in [179]. The proof given below is a version of his proof.

Theorem 3.1.2 *Let X be a locally compact Hausdorff space, $G = X \times X$ be the trivial groupoid with left Haar system defined by some probability measure μ on X. Then $C^*(G) \cong \mathcal{K}(L^2(X, \mu))$.*

Proof. Let $\pi : C_c(G) \to C^*(G)$ be the canonical homomorphism and realize $C^*(G)$ on a (separable) Hilbert space $\mathcal{H}$. Then by Theorem 3.1.1, $\pi = \pi_L$ for some representation $(\kappa, \{H_u\}, L)$ of G. Recall that for $(x, y) \in G$, we have $r((x, y)) = x, d((x, y)) = y$. Fix $x_0 \in X$, Then by Definition 3.1.1, the map $L_{(x,x_0)} : H_{x_0} = H_{d((x,x_0))} \to H_{r((x,x_0))} = H_x$ is unitary. Hence $\dim H_x = \dim H_{x_0}$ is the same for all x. By (3.13), $L^2(G^0, \{H_u\}, \kappa) \cong L^2(G^0, \kappa) \otimes H$ where $H = H_{x_0}$.

We now claim that κ is equivalent to μ. Indeed, let ν be the measure on G associated with κ. Then $\nu = \int_{G^0} \lambda^u \, d\kappa(u)$, and since $\lambda^u = \mu$ for all u, we have for $f \in C_c(G)$,

$$\int_G f(x, y) \, d\nu(x, y) = \int_X d\kappa(x) \int_X f(x, y) \, d\mu(y),$$

so that $\nu = \kappa \times \mu$. Similarly, $\nu^{-1} = \mu \times \kappa$. By quasi-invariance, we have that the measures $\kappa \times \mu, \mu \times \kappa$ are equivalent, and elementary measure theory gives that κ, μ are absolutely continuous with respect to each other, so that κ, μ are equivalent. Obviously, μ is itself quasi-invariant, and it is simple to check that two equivalent quasi-invariant measures for a locally compact groupoid give the same groupoid representations and isomorphic C^*-algebras. So we can take $\kappa = \mu$. (That is, effectively, there is only *one* quasi-invariant measure on X and it is μ itself.) The modular function for μ is 1.

---

[4]Using an easier version of the proof of Theorem 3.1.2, one can also show that the reduced $C^*_{red}(G)$, defined below, is isomorphic to $\mathcal{K}(L^2(X, \mu))$. (Of course, this also follows from the theorem, since $C^*_{red}(G)$ is a non-zero homomorphic image of the simple C^*-algebra $C^*(G)$.)

We now calculate π_L on $C_c(G)$. From the first paragraph of the proof, the Hilbert space of this representation is

$$L^2(G^0, \mu) \otimes H = L^2(X, \mu) \otimes H \cong L^2(X, H),$$

the latter being the L^2-space of H-valued functions with respect to μ.

For $f \in C_c(G)$, $x \in G$ and $F \in L^2(X, H)$, we have by (3.20):

$$
\begin{aligned}
\pi_L(f)(F)(x) &= \int_X f(x, y)(L_{(x,y)} F(y)) \, d\mu(y) \\
&= \int_X f(x, y) L_{(x,x_0)} (L_{(x_0,y)} F(y)) \, d\mu(y) \\
&= \int_X f(x, y) L_{(x,x_0)} (T(F)(y)) \, d\mu(y), \qquad (3.37)
\end{aligned}
$$

where for $F \in L^2(X, H) \cong L^2(X, \mu) \otimes H$, we define the section TF of the trivial bundle $X \times H$ by:

$$TF(y) = L_{(x_0,y)} F(y).$$

We now want to show that $TF \in L^2(X, H)$ and that T is unitary on $L^2(X, H)$. Clearly, T is linear. To show that TF is a measurable section, we need to prove that for $\xi \in H$, the map

$$y \to \langle TF(y), \xi \rangle = \langle L_{(x_0,y)} F(y), \xi \rangle \qquad (3.38)$$

is μ-measurable. To this end, by (3.15), the two-variable map

$$(x, y) \to \langle L_{(x,y)} F(y), g(x)\xi \rangle$$

is $\mu \times \mu$-measurable for all $g \in C_c(X, \mu)$. Fixing $x = x_0$ and taking $g(x_0) = 1$ gives (by standard results on product measure spaces) the measurability of the function in (3.38). To obtain that $TF \in L^2(X, H)$ and that T is isometric, one argues:

$$
\begin{aligned}
\int_X \|TF(y)\|^2 \, d\mu(y) &= \int_X \|L_{(x_0,y)} F(y)\|^2 \, d\mu(y) \\
&= \int_X \|F(y)\|^2 \, d\mu(y) \\
&= \|F\|^2 \qquad (3.39)
\end{aligned}
$$

since (by (i) of Definition 3.1.1) the map $L_{(x_0,y)}$ is unitary. A similar argument shows that the map T', where $T'(F)(y) = L_{(y,x_0)} F(y)$, is a linear isometry on $L^2(X, H)$. Since $T' = T^{-1}$, it follows that T is unitary.

Let $\pi = T\pi_L T^{-1}$, a representation equivalent to π_L. Then $\pi(C_c(G))$ also generates a copy of $C^*(G)$. Let $F' \in L^2(X, H)$. Calculating $\pi(f)$

for $f \in C_c(G)$, we get, using (3.37) and the facts that L is a groupoid homomorphism and that $L(x_0, x_0)$ is the identity:

$$
\begin{aligned}
\pi(f)(F')(x) &= (T\pi_L(f)(T^{-1}(F')))(x) \\
&= L_{(x_0,x)}\pi_L(f)(T^{-1}F')(x) \\
&= L_{(x_0,x)} \int_X f(x,y) L_{(x,x_0)}(T(T^{-1}F'))(y)\, d\mu(y) \\
&= \int_X f(x,y) F'(y)\, d\mu(y).
\end{aligned}
$$

With $F' = g \otimes \xi$ and comparing with (3.23) – recall that the modular function D is 1 – we have

$$
\pi(f)(g \otimes \xi) = \pi_{L_{triv}}(f)(g) \otimes \xi.
$$

Running ξ through an orthonormal basis for H gives that π is a multiple of $\pi_{L_{triv}}$ – in fact it is $\cong (\dim H)\pi_{L_{triv}}$. So

$$
\|\pi(f)\| = \|\pi_{L_{triv}}(f)\|.
$$

For $g \in L^2(X, \mu)$, we have

$$
\pi_{L_{triv}}(f)(g)(x) = \int_X f(x,y) g(y)\, d\mu(y).
$$

In other words, the range of $\pi_{L_{triv}}$ consists of the kernel operators coming from the continuous functions with compact support on $X \times X$. So this range is an algebra of compact operators which is dense in $\mathcal{K}(L^2(X,\mu))$.[5] The theorem now follows.   □

A very easy case of the theorem is the case where $G = X \times X$ with X a finite set with n elements. In that case, we can take μ to be $(n^{-1})\sum_{x \in X} \delta_x$ and obtain that $C^*(G)$ is just the algebra of compact operators on n-dimensional space, i.e. $C^*(G) \cong M_n$, the algebra of $n \times n$ complex matrices. (Of course, this is easy to check in a direct, elementary way since $C^*(G) = C_c(G)$.)

As an aside, the above theorem has an interesting application in the study of deformation quantization using the tangent groupoid of **2.3**. (See, for example, [56, p.103-104].) (Details of proofs are omitted.) Indeed, let M

---

[5]This is well known but easy to prove directly. Indeed, if $f(x,y) = h(x)k(y) = (h \otimes k)(x,y)$ for $h, k \in C_c(X)$, then $\pi_{L_{triv}}(f)$ is obviously of rank 1, and the span of such operators is norm dense in the algebra of finite rank operators and hence also in the algebra of compact operators. That every $\pi_{L_{triv}}(f)$ is compact can be shown by approximating f by linear combinations of functions of the form $h \otimes k$ using the Stone-Weierstrass theorem.

be a smooth manifold, and recall that the tangent groupoid $G_M = G_1 \cup TM$, where G_1 is the locally compact groupoid $(M \times M) \times \mathbf{R}^*$. Now G_1 is an open subgroupoid of G, and it is easily seen that $C_c(G_1)$ is an ideal in $C_c(G)$. We also have a short exact sequence

$$0 \to C_c(G_1) \to C_c(G) \to C_c(TM) \to 0.$$

This can be shown to lift to the C^*-level. It is not surprising that

$$C^*(G_1) \cong C^*(M \times M) \otimes C_0(\mathbf{R}^*) = \mathcal{K} \otimes C_0(\mathbf{R}^*),$$

where $\mathcal{K}$ is the algebra of compact operators on an infinite dimensional separable Hilbert space and we have used the preceding theorem. On the other hand, using the Fourier transform (each $T_x M \cong \mathbf{R}^n$) one can show that $C^*(TM) \cong C_0(T^*M)$, so that one obtains a short exact sequence of C^*-algebras:

$$0 \to \mathcal{K} \otimes C_0(\mathbf{R}^*) \to C^*(G) \to C_0(T^*M) \to 0.$$

Using this, an asymptotic morphism, called the *Connes-Higson* asymptotic morphism, is constructed from $C_0(T^*M)$ into $\mathcal{K}$. (A direct construction of the Connes-Higson asymptotic morphism at the level of the tangent groupoid is given in [198].) Connes also gives, using this short exact sequence, a proof of the Atiyah-Singer index theorem.

The above description has been a diversion from the main purpose of this section, which is to give an account of the representation theory of locally compact groupoids. We have included it to indicate the remarkable mathematics with which groupoids are involved in noncommutative geometry, as well as to give an application of Theorem 3.1.2 in the context of the tangent groupoid. We now return to the representation theory theme and examine another example of a groupoid C^*-algebra. In this case, the C^*-algebra can be calculated directly without having to consider *any* quasi-invariant measures!

Here the groupoid is the equivalence relation R_p associated with Penrose tilings. Recall (**2.2**) that X_p is the compact space of sequences $\{x_n\}$ where x_n is either 0 or 1 and $x_{n+1} = 0$ if $x_n = 1$. The equivalence relation R_p on X_p is defined by: xR_py whenever $x_i = y_i$ eventually. The topology on R_p was the "inductive limit" topology associated with the sets $R_p^{(N)}$ defined in (2.19). We noted that R_p is an r-discrete groupoid. Connes shows ([56, p.91]) that the reduced C^*-algebra $C^*_{red}(R_p)$ is an AF-algebra. (Reduced groupoid C^*-algebras will be considered later.) We will give effectively Connes's argument below with an adaptation to the case of the universal C^*-algebra $C^*(R_p)$.[6]

---

[6]In fact, it can be shown that the reduced and universal C^*-algebras for R_p are actually isomorphic.

Recall (e.g. [186, p.183]) that an AF-algebra is a C^*-algebra A that contains an increasing sequence $\{A_n\}$ of finite-dimensional C^*-algebras such that $\cup_{i=1}^{\infty} A_n$ is dense in A. A fundamental theorem of Elliott (see, for example, [15, Ch.7]) gives that K-theory distinguishes between different AF-algebras.

For each n, let K_n be the set of $0, 1$-sequences $(x_1, \ldots x_n)$ of length n such that $x_{i+1} = 0$ if $x_i = 1$ $(1 \leq i < n)$. Let $P_n : X_p \to K_n$ be the projection onto the first n-terms of a sequence. Let W_n be the equivalence relation on K_n given by: $(x_1, \ldots, x_n) W_n (y_1, \ldots, y_n)$ if and only if $x_n = y_n$. Clearly, if $(x, y) \in R_p^{(n)}$ then $(P_n x, P_n y) \in W_n$. Also, if $w W_n w'$ in K_n, then there exist $x, y \in R_p^{(n)}$ such that $w = P_n x, w' = P_n y$. For any function $f : W_n \to \mathbf{C}$, we can define a function $f_n \in C(R_p^{(n)}) \subset C_c(R_p)$ by setting: $f_n((x, y)) = f((P_n x, P_n y))$. Define $Q_n(f) = f_n$. Let C_n be the vector space of such functions f_n. We claim that C_n is a *-subalgebra of $C_c(R_p)$ isomorphic under Q_n to the finite-dimensional algebra $C_c(W_n)$.

Clearly, Q_n is linear and one-to-one and preserves the involution. It remains to check that C_n is closed under convolution and that Q_n is a homomorphism. Recalling that the left Haar system on each of the r-discrete groupoids W_n, R_p is the counting measure system, we have

$$
\begin{aligned}
f_n * g_n((x, y)) &= \int f_n((x, s)) g_n((s, x)(x, y)) \, d\lambda^x(s) \\
&= f * g(P_n x, P_n y)
\end{aligned}
$$

since f_n, g_n vanish outside $R_p^{(n)}$ and the number of s's for which $(x, s) \in R_p^{(n)}$ is the same as the number of corresponding r's in K_n for which $(P_n x, r) \in W_n$. The desired result now follows.

We now calculate $C_c(W_n)$. Let $K_n(0), K_n(1)$ be the subsets of K_n whose elements end in 0 and 1 respectively. Then W_n is the disjoint union of the trivial groupoids $(K_n(0) \times K_n(0))$ and $(K_n(1) \times K_n(1))$. It follows easily that $C_c(W_n)$, and hence C_n, is isomorphic to $M_{k_n} \times M_{k'_n}$ where k_n, k'_n are respectively the number of elements in $K_n(0), K_n(1)$.

There is a natural embedding i_n of C_n into C_{n+1}. Indeed, in an obvious notation, $K_{n+1}(0) = (K_n(0)0) \cup (K_n(1)0)$ and $K_{n+1}(1) = K_n(0)1$ since a 1 must be followed by a 0. So $k_{n+1} = k_n + k'_n$ and $k'_{n+1} = k_n$, and C_n can be regarded as a subalgebra of C_{n+1}. Let $C = \cup_{n=1}^{\infty} C_n$. Then C is a *-subalgebra of $C_c(R_p)$ and is a finite dimensional C^*-subalgebra of $C^*(R_p)$. To show that $C^*(R_p)$ is an AF-algebra, it is sufficient to show that C is I-norm dense in $C_c(R_p)$. For by Proposition 3.1.1 and Theorem 3.1.1, it follows that C is dense in $C^*(R_p)$ so that the latter is an AF-algebra.

To this end, let $F \in C_c(R_p)$. Since the open sets $R_p^{(n)}$ cover R_p, we have $F \in C_c(R_p^{(N)})$ for some N. We construct a sequence $\{F_r\}$ in C such that $\|F_r - F\|_I \to 0$. Let $\epsilon > 0$. By the uniform continuity of F on the compact space $R_p^{(N)}$, there exists $q \geq N$ and a function $f : W_q \to \mathbf{C}$ such that, in the above notation, $| f_q(xz, yz) - F(xz, yz) | < \epsilon$ for all $(x, y) \in W_q, (xz, yz) \in R_p$. Since F vanishes outside $R_p^{(N)}$, we can take $f((x, y)) = 0$ if for any $i \geq N$, $x_i \neq y_i$. So f_q also belongs to $C(R_p^{(N)})$, and for any $t \in X_p$, the sets $(R_p^{(N)})^t, (R_p^{(N)})_t$ have $\leq m$ elements, where m is the number of elements of K_N. It follows that $\|f_q - F\|_I < m\epsilon$. The existence of the desired sequence now follows, and $C^*(R_p)$ is an AF-algebra.

The explicit information about the embedding of C_n in C_{n+1} makes it easy to calculate the dimension group (i.e. the pair $(K_0(A), K_0(A)^+)$) for the C^*-algebra $A = C^*(R_p)$. Connes does this in [56, p.92], and uses the information to interpret K-theoretically tile density results for Penrose tilings. We will discuss Penrose tilings in **4.2**, Example 5. We now return to the representation theory of a general locally compact groupoid.

As in the group case, there is a *reduced* C^*-algebra $C^*_{red}(G)$. This C^*-algebra $C^*_{red}(G)$ is defined below.

Firstly, let $\mu \in P(G^0)$, the set of probability measures on G^0, and ν be the measure on G associated with μ as discussed earlier. (Note that μ is not assumed to be quasi-invariant.) Replacing π_L of (3.36) by $Ind\,\mu$, the latter is a representation of $C_c(G)$ on $L^2(\nu^{-1})$ given by

$$Ind\,\mu(f)(F) = f * F. \tag{3.40}$$

The justification for the terminology Ind in this context is, that as Renault shows ([230, p.81f.]), the representation $Ind\,\mu$ is induced (in the sense of Rieffel) from the multiplication representation of $C_c(G^0)$ on $L^2(G^0, \mu)$. We would expect left regular representations to be induced representations, since, in the locally compact group case, the left regular representation is well-known to be induced from the trivial one dimensional representation of the subgroup $\{e\}$ and hence, in the present notation, is just $Ind\,\delta_e$. The interpretation of $Ind\,\mu$ as an induced representation is illuminating, and for that reason, a brief discussion of it is given in Appendix D.

The norm of the C^*-algebra generated by $Ind\,\mu(C_c(G))$ will be denoted by $\|.\|_\mu$. For $v \in G^0$ let $Ind\,v = Ind\,\delta_v$. For our purposes, it will be useful to have available the explicit formulae for the representations $Ind\,\mu$ and $Ind\,v$.

Note first that with $\mu = \delta_v$, we have (using (3.1)) that $\nu = \lambda^v, \nu^{-1} = \lambda_v$, these measures being regarded as defined in the obvious way on $\mathcal{B}(G)$. So for $v \in G^0$ and $F, H \in C_c(G)$, we have, using (3.40) and (2.21) (c.f. [230, p.82]), that

$$\langle Ind\,v(f)(F), H \rangle$$
$$= \int f * F(x)\overline{H(x)}d\lambda_v(x) \tag{3.41}$$

$$= \int d\lambda_v(x) \int f(xt)F(t^{-1})\overline{H(x)}d\lambda^v(t). \qquad (3.42)$$

Note that using (iii) of Definition 2.2.2, the function $v \to \langle Ind\, v(f)(F), H \rangle$ is in $C_c(G^0)$. For a general $\mu \in P(G^0)$, we therefore have:

$$\langle Ind\, \mu(f)(F), H \rangle = \int_{G^0} \langle Ind\, v(f)(F), H \rangle \, d\mu(v). \qquad (3.43)$$

For $f \in C_c(G)$, we define

$$\|f\|_{red} = \sup_{\mu \in P(G^0)} \|Ind\, \mu(f)\|. \qquad (3.44)$$

Then ([230, p.82]) $\|.\|_{red}$ is a C^*-norm on $C_c(G)$ whose completion is defined to be the *reduced C^*-algebra* $C^*_{red}(G)$ of G. (Indeed, if $\|f\|_{red} = 0$, then $Ind\, v(f) = 0$ for all $v \in G^0$ and, using the density of the support of λ_v on G_v and the equality (3.41), for all $F \in C_c(G)$, we have $f * F = 0$. Using Lemma 2.2.1, we have $f * F = 0$ for all $F \in B_c(G)$. Taking $F = \chi_K$ where K is a large enough compact subset of G^0 then gives $f = f * \chi_K = 0$. So $\|.\|_{red}$ is a norm on $C_c(G)$.)

An alternative way of defining $\|.\|_{red}$ (e.g. [179, 56]) specifies

$$\|f\|_{red} = \sup_{v \in G^0} \|Ind\, v(f)\|.$$

The equivalence of the two definitions of $\|f\|_{red}$ seems well-known, but it follows immediately from the following proposition which we will need later.

Proposition 3.1.2   *Let $\mu \in P(G^0)$ and have support C. Then for $f \in C_c(G)$,*

$$\|f\|_\mu = \sup_{v \in C} \|Ind\, v(f)\|. \qquad (3.45)$$

Proof.   Let $f, F \in C_c(G)$ and M be the right-hand side of (3.45). Then using (3.43) and the fact that $\mu \in P(G^0)$,

$$\begin{aligned}
\|Ind\, \mu(f)(F)\|^2 &= \int_{G^0} \langle Ind\, v(f^* f)(F), F \rangle \, d\mu(v) \\
&\leq \int_{G^0} \|Ind\, v(f)(F)\|^2 \, d\mu(v) \\
&\leq M^2 \int_G \|F\|^2_{L^2(\lambda_v)} \, d\mu(v) \\
&\leq M^2 \int_G |F|^2 \, d\nu^{-1} \qquad (3.46)
\end{aligned}$$

where (as usual) $\nu = \int \lambda^v \, d\mu(v)$. It follows that $\|f\|_\mu \leq M$. (As noted in the discussion of ν earlier, the space $C_c(G)$ is dense in the Hilbert spaces involved.)

For the converse, let $v_0 \in C$. We will show that $Ind\, v_0$ is weakly contained in $Ind\, \mu$. It is sufficient to show (by [73, Theorem 3.4.4]) that for $F \in C_c(G)$, there exists a sequence $F_n \in C_c(G)$ such that for all $f \in C_c(G)$,

$$\langle Ind\, \mu(f)(F_n), F_n \rangle \to \langle Ind\, v_0(f)(F), F \rangle. \tag{3.47}$$

To this end, let $f, F \in C_c(G)$. Let $\{V_n\}$ be a neighborhood basis of open sets for v_0 in G^0 such that for all n, $V_{n+1} \subset V_n$. Since $v_0 \in C$, we have $\mu(V_n) > 0$ for all n.

Let $f_n \geq 0$ be in $C_c(V_n) \subset C_c(G^0)$ and be such that $\int f_n^2 \, d\mu = 1$. Let $F_n = (f_n \circ d)F \in C_c(G)$. Then for each $v \in G^0$,

$$\langle Ind\, v(f)(F_n), F_n \rangle = f_n^2(v)\langle Ind\, v(f)(F), F \rangle \tag{3.48}$$

using (3.42). Let $\epsilon > 0$. Since the map $v \to \langle Ind\, v(f)(F), F \rangle$ is continuous on G^0, there exists an N such that for all $n \geq N$ and all $v \in V_n$,

$$| \langle Ind\, v(f)(F), F \rangle - \langle Ind\, v_0(f)(F), F \rangle | < \epsilon.$$

Using (3.43) and (3.48), we have, for $n \geq N$,

$$| \langle Ind\, \mu(f)(F_n), F_n \rangle - \langle Ind\, v_0(f)(F), F \rangle | =$$
$$| \int_{V_n} f_n(v)^2 [\langle Ind\, v(f)(F), F \rangle - \langle Ind\, v_0(f)(F), F \rangle] d\mu(v) |$$
$$< \epsilon.$$

Then (3.47) now follows. □

3.2 Representation theory for groupoids that are r-discrete, and their inverse semigroups of open G-sets

In this section, G will be assumed to be an r-discrete groupoid.

Recall that $B(X)$ is the space of complex-valued, bounded Borel functions on a topological space X. In the notation of **2.2**, let $A \in G^{op}$ and $f \in B(A)$. Recall also that r_A, d_A are the restrictions of the range and source maps r, d to A. Define $f \circ r_A^{-1} \in B(G^0)$ by: $f \circ r_A^{-1}(x) = f(a)$ if $x = r(a)$ for some $a \in A$ and is zero otherwise. The function $f \circ d_A^{-1} \in B(G^0)$ is defined similarly. Note that if $f \in C_0(A)$, then $f \circ r_A^{-1} \in C_0(G^0)$. The continuity follows since r_A is a homeomorphism from A onto $r(A)$.

The following proposition will ease certain later calculations.

Proposition 3.2.1 Let $A, B \in G^{op}$, $\mu \in P(G^0)$ and $\nu = \int_{G^0} \lambda^u \, d\mu(u)$.

(i) A function $f \in B(A)$ is ν-integrable if and only if $f \circ r_A^{-1}$ is μ-integrable, and in that case,

$$\int f \, d\nu = \int f \circ r_A^{-1} \, d\mu, \quad \int f \, d\nu^{-1} = \int f \circ d_A^{-1} \, d\mu;$$

(ii) If $C \in B(A)$ then $\nu(C) = \mu(r(C))$ and $\nu^{-1}(C) = \mu(d(C))$;

(iii) if $E \in B(A)$ and $F \in B(B)$, then

$$\nu^2(E \dot\times F) = \mu(d(E) \cap r(F)). \tag{3.49}$$

Proof. (i) This is an immediate consequence of the fact that $\int_{G^u} f(x) \, d\lambda^u = f \circ r_A^{-1}(u)$ if $u \in r(A)$ and is zero otherwise. (Recall that λ^u is counting measure and that $A \in G^{op}$.) A similar argument applies to the last equality of (i).

(ii) Apply (i) with $f = \chi_C$ and $f = \chi_{C^{-1}}$.

(iii) Recall ((2.13)) that $E \dot\times F = (E \times F) \cap G^2$. Since $u \in d(E) \cap r(F)$ if and only if there exists $x \in E, y \in F$ such that $d(x) = u = r(y)$, and as G is r-discrete, we have (cf. (3.5))

$$\begin{aligned}
\nu^2(E \dot\times F) &= \int_{G^0} d\mu(u) \int_{E \dot\times F} d\lambda_u(x) d\lambda^u(y) \\
&= \mu(d(E) \cap r(F)).
\end{aligned}$$

This concludes the proof. □

Let X be a locally compact Hausdorff space and S an inverse semigroup. We wish to define the inverse semigroup version of a right action of a group on X. As in Chapter 1, a *right action* of S on X is an antihomomorphism $s \rightarrow \alpha_s$ of S into the inverse semigroup $\mathcal{I}(X)$ such that each α_s is a homeomorphism from one *open* subset of X onto another. When S is countable and the family of domains of the α_s's forms a basis for the topology of X, we will call the pair (X, S) a *localization*. The study of localizations and the C^*-algebras which they generate will be taken up in **3.3**.

There is (Chapter 1) a natural right action $s \rightarrow \beta_s$ of any inverse semigroup S on its idempotent semilattice E. (Of course we treat E as discrete.) (This action will be extended to the "filter completion" of E in the course of constructing the universal groupoid of S in **4.3**.) Here, the domain D_s of β_s is the set $\{e \in E : e \leq ss^*\}$, and the action is given by:

$$\beta_s(e) = s^*es. \tag{3.50}$$

It is easy to check that $s \rightarrow \beta_s$ is a right action of S. For example, to check that β_s is one-to-one on D_s, one argues that if $e, f \in D_s$ and $s^*es = s^*fs$, then $e = s(s^*es)s^* = s(s^*fs)s^* = f$.

In the context of the r-discrete groupoid G, there is a related canonical right action of the inverse semigroup G^{op} on the locally compact space G^0. Precisely, for $A \in G^{op}$ and for $u \in r(A)$ define

$$\alpha_A(u) = A^{-1}uA. \tag{3.51}$$

To make sense of this, $\alpha_A(u)$ is only defined if $u \in r(A)$, for otherwise, there would be no element of A that u could multiply on the left. So if we take $u = r(a)$ with $a \in A$, then, using the fact that A is a G-set, a is unique, and since we have $uA = \{a\}$, we obtain

$$A^{-1}uA = \{b^{-1} : b \in A\}a = \{a^{-1}a\} = \{d(a)\}.$$

Dropping the set braces off the $d(a)$, we have that α_A is a homeomorphism from the open set $r(A)$ onto the open set $d(A)$, and for each $a \in A$,

$$\alpha_A(r(a)) = d(a). \tag{3.52}$$

(The homeomorphism property follows since r_A, d_A are homeomorphisms.) Alternatively expressed, α_A is just the map $d \circ r_A^{-1}$.

The map $A \to \alpha_A$ defines a *right action* of G^{op} on G^0. Indeed, for $A, B \in G^{op}$, we have $u \in r(AB)$ if and only if for some $a \in A$, both $u = r(a), d(a) \in r(B)$ if and only if u belongs to the domain of $\alpha_B \circ \alpha_A$. Then for such a u, $\alpha_{AB}(u) = (AB)^{-1}u(AB) = B^{-1}(A^{-1}uA)B = \alpha_B(\alpha_A(u))$, so that the map $A \to \alpha_A$ is indeed an antihomomorphism as claimed. The reader needs to be aware of the potentially confusing fact that the *domain* of each α_A is the "range" $r(A)$ of A in G!

The next proposition gives a convenient formulation of quasi-invariance for a probability measure on the unit space of an r-discrete groupoid G. It generalizes the characterization of quasi-invariance for the transformation groupoid $X \times H$ discussed in **3.1**. This result is effectively contained in [230, pp.31-32].

For $A \in G^{op}$ and $E \in \mathcal{B}(d(A))$, we define $(\mu \circ \alpha_A)(E) = \mu(\alpha_A^{-1}(E))$. Note that $\alpha_A^{-1} = \alpha_{A^{-1}}$, and that for appropriate functions g on $d(A)$, we have, using (2.38), that

$$\int g\, d(\mu \circ \alpha_A) = \int g \circ \alpha_A\, d\mu.$$

The modular function D for a quasi-invariant measure was defined in (3.6).

Proposition 3.2.2 *Let G be an r-discrete groupoid and μ be a probability measure on G^0. Then μ is quasi-invariant if and only if, for every $A \in G^{op}$, $\mu \circ \alpha_A \sim \mu$ on $d(A)$. Further, in those circumstances,*

$$(D(r_A^{-1}(u)))^{-1} = \frac{d(\mu \circ \alpha_A^{-1})}{d\mu}(u) \tag{3.53}$$

for μ-a.e. $u \in r(A)$, where D is the modular function of μ.

Proof.   Since G is a countable union of G^{op}-sets, it follows that $\nu \sim \nu^{-1}$ if and only if $\nu_{|A} \sim (\nu^{-1})_{|A}$ for all $A \in G^{op}$. Let $A \in G^{op}$ and let $f : A \to [0, \infty)$ be measurable. Then using the equality $\alpha_A^{-1} = r_A \circ d_A^{-1}$ and (i) of Proposition 3.2.1,

$$
\begin{aligned}
\int_G f \, d\nu^{-1} &= \int_{G^0} f(d_A^{-1}(u)) \, d\mu(u) \\
&= \int_{G^0} f \circ r_A^{-1}(\alpha_A^{-1}(u)) \, d\mu(u) \\
&= \int_{G^0} f(r_A^{-1}(u)) \, d(\mu \circ \alpha_A^{-1})(u).
\end{aligned}
\tag{3.54}
$$

Since, again by (i) of Proposition 3.2.1,

$$
\int_G f \, d\nu = \int_{G^0} f(r_A^{-1}(u)) \, d\mu(u)
\tag{3.55}
$$

comparison with (3.54) gives $\nu_{|A} \sim (\nu^{-1})_{|A}$ if and only if $\mu \circ \alpha_A^{-1} \sim \mu$ on $r(A)$ for all $A \in G^{op}$ (if and only if $\mu \circ \alpha_A \sim \mu$ on $d(A)$ for all $A \in G^{op}$). The equality

$$
\int_G f \, d\nu = \int_G f D \, d\nu^{-1}
$$

together with (3.54) and (3.55) then gives (3.53).   □

Let L be a representation of G associated with a quasi-invariant probability measure μ on G^0 and a Hilbert bundle $(G^0, \{H_u\}, \mu)$ as in **3.1**. Recall that there is a representation π_L of the normed *-algebras $C_c(G)$ and $B_c(G)$ on $L^2(G^0, \{H_u\}, \mu)$ given by (3.17).

Recall also Theorem 3.1.1 which asserts that conversely, every representation of $C_c(G)$ is the integrated form of a groupoid representation, i.e. is a π_L for some L. The main result Theorem 3.2.1 of this section is that not only does a representation of $C_c(G)$ determine a G-representation, but so also do certain intrinsically specifiable representations of a wide class of inverse subsemigroups of G^{op}. This is a central result of the present work.

In preparation for this, we first discuss how a G-representation L determines a representation of G^{op}. We saw earlier (after Theorem 3.1.1) that every representation Π of $C_c(G)$ determines a representation π_1 of G^{op}. Starting with L, however, a simpler "multiplier algebra" argument can be given to produce π_1, and as this technique should be useful in other contexts, we will describe it here. (Of course, we could directly use the earlier argument to obtain $\pi_1(A) = \pi_L(A)$ where $\pi_L(A)$ is given as in (3.57) below.)

We regard G^{op} as a subsemigroup of the multiplier algebra $M(B_c(G))$ of the normed *-algebra $B_c(G)$, and then we can extend π_L in a standard way to $M(B_c(G))$ and finally restrict it to G^{op}. To this end, any $A \in G^{op}$ defines a multiplier of the normed algebra $B_c(G)$ using the convolution formula (2.21), with χ_A inserted in as the left or right function of the convolution. Indeed, it is easily checked that for $f \in B_c(G)$,

$$\chi_A * f(x) = f(A^{-1}x), f * \chi_A(x) = f(xA^{-1}), \qquad (3.56)$$

and these define functions in $B_c(G)$ whose I-norm is $\leq \|f\|_I$. (Of course, $f(A^{-1}x)$ is defined to be 0 if $r(x)$ does not belong to $r(A) = d(A^{-1})$, and to be $f(y)$ when $r(x) \in r(A)$, where $A^{-1}x = \{y\}$. Similar considerations apply to $f(xA^{-1})$.) The requisite multiplier properties and isomorphic embedding of G^{op} into the multiplier algebra are also readily verified. (For this result in a more general context, see [230, pp.62-65].)

Since $B_c(G)$ has a bounded approximate identity (**2.2**), general Banach algebra theory gives that $B_c(G)$ can be regarded as a subalgebra of $M(B_c(G))$ and that π_L extends to a continuous representation, also denoted by π_L, of $M(B_c(G))$, which restricts to give a representation of the subsemigroup G^{op} of $M(B_c(G))$. (The non-degeneracy of π_L regarded as a homomorphism on G^{op} follows from Proposition 3.2.3 below.) Let $\{\chi_{U_n}\}$ be the bounded approximate identity of $B_c(G)$ discussed in **2.2**. Then $\chi_A * \chi_{U_n} = \chi_{AU_n} \to \chi_A$ strictly, and applying (3.17) with $f = \chi_{AU_n} \in B_c(G)$ and a measure theoretic convergence theorem gives

$$\pi_L(A) = \pi_L(\chi_A)$$

where the right-hand side is defined as in (3.17). Substituting in the latter gives:

$$\langle \pi_L(A)\xi, \eta \rangle = \int_A \langle L(x)(\xi(d(x))), \eta(r(x)) \rangle \, d\nu_0(x). \qquad (3.57)$$

In practice, we are not usually confronted with the large inverse semigroup G^{op} but rather with a (countable) inverse subsemigroup S of G^{op} which determines the topology of G in the sense that *it is a basis for the topology of G*. Such inverse semigroups S always exist, since G^{op} contains a countable basis $\mathcal{A} \subset G^{op}$ for G (using (ii) of Definition 2.2.1) and we can take S to be the inverse semigroup generated by $\mathcal{A}$. When S is a countable inverse subsemigroup of G^{op} which is a basis for the topology of G, then the G^{op}-action of (3.51), restricted to S, gives a localization (G^0, S). For G^0 is open in G and so S contains a basis for G^0, and each U in this basis is the domain of the identity map α_U.

There is a often a natural S available. For example, if G is ample (**2.2**), it would be natural to take $S = G^a$. Also, as we will see in **3.3**, when a localization is realized in terms of r-discrete groupoids, it is natural to take for S the original inverse semigroup of the localization realized as a subsemigroup of G^{op}.

However, until further notice, S will be an inverse subsemigroup of G^{op} which is a basis for the topology of G *but is not necessarily countable*. Ignoring what, for our purposes, is the trivial case when G is a group, we can (and will) suppose that the empty set $0 \in S$. (Indeed, if S is not a group, then G^0 is not a singleton, and by the basis condition on S and the fact that G^0 is open and Hausdorff, there exist $A, B \in S$ with $A \cup B \subset G^0$ and $A \cap B = 0 = AB \in S$.)

Since $S \subset G^{op}$, the above representation π_L of G^{op} restricts to give a homomorphism of S on $L^2(G^0, \{H_u\}, \mu)$ which will also be denoted by π_L. (We are getting dangerously close to overusing the notation π_L, but the context should make clear what the domain of a particular occurrence of π_L is, and the ambiguity in the notation π_L seems preferable to a proliferation of terminology.)

Recall that for us, an inverse semigroup representation has to be *non-degenerate*; in order to show that the homomorphism π_L on S is a representation, we therefore need to show that it is non-degenerate.

Proposition 3.2.3   *Let L be a representation of G, S be an inverse subsemigroup of G^{op} which is a basis for the topology of G. Then π_L, regarded as a homomorphism on S, is non-degenerate (so that π_L is a representation of S).*

Proof.   Let $f \in L^2(G^0, \{H_u\}, \mu)$ have compact support C in G^0. Since G^0 is open in G and S is a basis for the topology of G, we can cover C by a finite number $U_1, \ldots, U_n$ of open sets belonging to S, each of which has compact closure in G^0. Let $g = \sum_{i=1}^n \chi_{U_i} \in B_c(G)$ and $F = g^{-1}f \in L^2(G^0, \{H_u\}, \mu)$. Note that $\pi_L(g)$ is in the span of $\pi_L(S)$ and is the multiplication operator by g, while $f = \pi_L(g)F$. Since the space of such sections f is dense in $L^2(G^0, \{H_u\}, \mu)$, it follows that π_L is non-degenerate on S. □

We now discuss a useful extra structure possessed by G^{op}, that of *addition of orthogonal elements*. For motivation, recall (Proposition 2.1.4) that every inverse semigroup can be realised as *-semigroup of partial isometries on a Hilbert space $\mathcal{H}$. Now if T_1, T_2 are partial isometries with orthogonal initial subspaces and orthogonal final subspaces, then $T_1 + T_2$ is also a partial isometry. Algebraically, the preceding conditions on T_1, T_2 are equivalent to:

$$T_1 T_2^* = 0 = T_1^* T_2. \qquad (3.58)$$

Note that the maps T_1 and T_2 in (3.58) can be interchanged by applying the adjoint operation to both sides of (3.58). Under the circumstances of (3.58), it is reasonable to say that T_1 is *orthogonal* to T_2. This notion of orthogonality readily reformulates in the context of G^{op} with $+$ replaced by $\cup$.

So for $A, B \in G^{op}$, we say that A is *orthogonal* to B $(A \perp B)$ if

$$AB^{-1} = 0 = A^{-1}B. \tag{3.59}$$

By taking inverses in (3.59) we obtain that $A \perp B$ if and only if $B \perp A$. The equalities of (3.59) are equivalent to:

$$d(A) \cap d(B) = 0 = r(A) \cap r(B). \tag{3.60}$$

In particular, if $A \perp B$, then $A \cap B = 0$. We then have the following easy proposition.

Proposition 3.2.4 *If $A \perp B$ in G^{op}, then $C = A \cup B \in G^{op}$. The same conclusion holds with G^a in place of G^{op} when G is ample.*

Proof. Of course, C is an open subset of G. We show that C is Hausdorff. Let $x, y \in C$ with $x \neq y$. If both x, y are in A or in B then since A, B are Hausdorff, we can separate x and y by disjoint open neighborhoods in C. So we can suppose that $x \in A$ and $y \in B$. But then A and B are disjoint neighborhoods of x and y. So C is Hausdorff. Finally, C is a G-set by (3.60).

The corresponding result for G^a follows from the G^{op} case since the union of two compact sets is compact. $\square$

A useful condition on S is that of additivity. Here, the inverse subsemigroup S of G^{op} is called *additive* if it is a basis for the topology of G and, whenever $A, B \in S$ with $A \cup B \in G^{op}$, then $A \cup B \in S$. Trivially, G^{op} is additive. Also for the ample case, the inverse semigroup G^a is additive if G is ample.

Associated with S is the inverse semigroup S^0 of elements $A \in S$ with $A \subset G^0$. From the comments following Proposition 2.2.3, we have $S^0 = E(S)$. So S^0 is a semilattice of open G-sets for the groupoid of units G^0 and is a basis for the topology of G^0. Of course, the additivity of S immediately implies the additivity of S^0 as a subsemigroup of $(G^0)^{op}$. (The latter is just the family of open subsets of G^0.) Note also that, if π is a representation of S, then the restriction π^0 of π to S^0 is also a representation. (Nondegeneracy follows since if $s \in S$ and ξ belongs to the Hilbert space of π, then $\pi(s)(\xi) = \pi(ss^*)(\pi(s)\xi)$ and $ss^* \in S^0$.)

For the rest of this section, the inverse semigroup S will be assumed to be additive (as well as being a basis for the topology of G).

The inverse semigroup S can be "large" (e.g. uncountable as in the G^{op}-case) and we impose an additional "regularity" condition on its representations. This will help to make them manageable. For $U \in S$, let $\mathcal{A}_U$ be the family of elements $V \in S$ which are contained in a compact subset of U. Since S is additive, $\mathcal{A}_U$ is directed upwards under inclusion and so is a net. Using the basis property of S and (ii) of Definition 2.2.1, we have $\cup \mathcal{A}_U = U$.

For partial isometries T_1, T_2 on a Hilbert space $\mathcal{H}$, we write

$$T_1 \leq T_2 \quad (T_2 \geq T_1)$$

if $T_1 = T_2 T_1^* T_1$. When T_1, T_2 belong to an inverse semigroup of partial isometries on $\mathcal{H}$, then $\leq$ is just the natural partial ordering on the semigroup (discussed after Proposition 2.2.3). Geometrically, $T_1 \leq T_2$ means that the initial subspace of T_1 is contained in the initial subspace of T_2, and T_1 and T_2 coincide on the initial subspace of T_1. So in the appropriate sense, T_1 is a "restriction" of T_2. From this perspective, it is clear that $\leq$ gives a partial ordering on the set of partial isometries on $\mathcal{H}$.

Now suppose that π is a representation of S. Suppose that $V, V' \in \mathcal{A}_U$ and $V \subset V'$. Then $V = V'(V)^{-1}V$ and applying π to this equation gives $\pi(V) = \pi(V')\pi(V)^*\pi(V)$ so that $\pi(V) \leq \pi(V')$. It easily follows from the "restriction" viewpoint of the preceding paragraph that the net of partial isometries $\pi(V)$ ($V \in \mathcal{A}_U$) converges in the strong operator topology to a partial isometry T. (The case when the partial isometries $\pi(V)$ are projections is an immediate consequence of Vigier's theorem.) In terms of the partial ordering, we could write $T = \sup_{\mathcal{A}_U} \pi(V)$. The initial subspace of T is the closure of the union of the initial subspaces of the $\pi(V)$'s. Also, since $\pi(V) \leq \pi(U)$ for all $V \in \mathcal{A}_U$, we have $T \leq \pi(U)$.

We will call a representation π of S on a Hilbert space $\mathcal{H}$ *regular* if whenever $U \in S$ then

$$\pi(U) = \lim\{\pi(V) : V \in \mathcal{A}_U\}, \tag{3.61}$$

the limit being taken in the strong operator topology. The use of the term *regular* is of course motivated by the notion of *inner regularity* for Borel measures. We will see below that regularity (and also additivity below) for a representation can in fact be formulated in terms of the idempotent semilattice $E(S)$ of S.

A representation π of S is said to be *additive* if $\pi(0) = 0$ and whenever $A \perp B$ in S, then $\pi(A \cup B) = \pi(A) + \pi(B)$. (Recall that $\pi(A), \pi(B)$ are partial isometries on $\mathcal{H}$ and since $\pi(0) = 0$, it follows that $\pi(A) \perp \pi(B)$ in the sense of (3.58) so that the partial isometry $\pi(A) + \pi(B)$ makes sense.) A finite sequence $A_1, A_2, \ldots, A_n$ in S is called *orthogonal* if $A_i \perp A_j$ for $i \neq j$. Note that given such a sequence $\{A_i\}$, $A_1 \perp \cup_{i=2}^n A_i$ and it follows by induction that a representation π of S is additive if and only if for every such sequence,

$$\pi(A) = \sum_{i=1}^n \pi(A_i). \tag{3.62}$$

A representation π of S on $\mathcal{H}$ is called (finitely) *subadditive* if whenever $U_1, \ldots, U_n \in S^0$, then

$$\pi(\cup_{i=1}^n U_i) \leq \sum_{i=1}^n \pi(U_i). \tag{3.63}$$

The operators $\pi(U_i), \pi(\cup_{i=1}^n U_i)$ are projections since $U_i, \cup_{i=1}^n U_i \in S^0$. (Here, recall that if $\mathcal{H}$ is a Hilbert space and $T_1, T_2 \in B(\mathcal{H})$ are Hermitian, then $T_1 \leq T_2$ if and only if for all $\xi \in \mathcal{H}$, we have $\langle T_1 \xi, \xi \rangle \leq \langle T_2 \xi, \xi \rangle$.) In the ample case with $S = G^a$, additivity of π actually implies (Corollary 3.2.1) its regularity and subadditivity.

There is also the cognate notion of a *countably subadditive* representation in which the finite sequence $\{U_i\}$ is replaced by an infinite sequence (with $\cup_{i=1}^\infty U_i \in S^0$). (The corresponding sum of non-negative terms $\sum_{i=1}^\infty \langle \pi(U_i)\xi, \xi \rangle$ for the right-hand side of (3.63) is allowed to equal ∞.)

The next proposition gives the useful result that the regular and additive conditions for a representation π of S is equivalent to the corresponding conditions for the restriction π^0 of π to S^0.

Proposition 3.2.5   *A representation π of S has the property of being regular or additive if and only if the representation π^0 of S^0 has the same property.*

Proof.   It is trivial that if π has any one of the two properties then π^0 also has that property. It remains to establish the converse.

Suppose that π^0 is regular and let $U \in S$. Since π^0 is regular, there exists a sequence $\{K_n\}$ in S^0 with each $K_n \subset r(U)$ and having compact closure in $r(U)$ such that $\pi^0(K_n) \to \pi^0(r(U))$ strongly. Then $K_n U \in \mathcal{A}_U$, and in the strong operator topology, $\pi(K_n U) = \pi^0(K_n)\pi(U) \to \pi^0(r(U))\pi(U) = \pi(U)$. Since $\lim_{V \in \mathcal{A}_U} \pi(V) \leq \pi(U)$, the regularity of π follows.

Suppose now that π^0 is additive. Then $\pi(0) = 0$. Suppose that $U_1 \perp U_2$ in S. Then by (3.60) and the additivity of π^0, we have $\pi(U_1 \cup U_2) = \pi(r(U_1 \cup U_2)(U_1 \cup U_2)) = \pi^0(r(U_1) \cup r(U_2))\pi(U_1 \cup U_2) = (\pi^0(r(U_1)) + \pi^0(r(U_2)))\pi(U_1 \cup U_2) = \pi(r(U_1)(U_1 \cup U_2)) + \pi(r(U_2)(U_1 \cup U_2)) = \pi(U_1) + \pi(U_2)$. So π is additive. $\square$

The following proposition reduces the theory of regular, additive, subadditive representations of S to that of the regular, additive, countably subadditive representations of G^{op}. (The result [148, Corollary, p.162] can be regarded as a "localization" version of this proposition.) We first extend the notion of the net $\mathcal{A}_U$ above to the case where $U \in G^{op}$ (rather than just $U \in S$). The definition is just the same as before: $\mathcal{A}_U$ is the set of elements $V \in S$ where V is contained in a compact subset of U. Since S is a basis for the topology of G, we have $\cup_{V \in \mathcal{A}_U} V = U$. Also, as for the limit in (3.61), the corresponding strong operator limit in (3.64) below exists and is a partial isometry.

Proposition 3.2.6   *Let S be an additive inverse subsemigroup of G^{op} and π be a regular, additive, subadditive representation of S on the Hilbert space $\mathcal{H}$. For each $A \in G^{op}$, define*

$$\pi(A) = \lim_V \pi(V) \qquad (3.64)$$

where V runs over the net $\mathcal{A}_U$ and the limit is taken in the strong operator topology. Then π is a regular, additive, countably subadditive representation of G^{op} and restricts to the original π on S.

Proof.   Using the facts that the weak operator topology on $B(\mathcal{H})$ is *-continuous and that the inversion on G is a homeomorphism, it follows that $\pi(A)^* = \pi(A^{-1})$. Let $A, B, C \in G^{op}$ with $AB = C$. We show that $\pi(C) = \pi(A)\pi(B)$. Firstly, $\pi(A)\pi(B)$ is a partial isometry. For the product net of elements $\pi(U)\pi(V)$, where $U \in \mathcal{A}_A, V \in \mathcal{A}_B$, converges to $\pi(A)\pi(B)$ by the joint continuity of the strong operator topology on bounded sets, and since this net is increasing, its limit is also a partial isometry. Let $W \in \mathcal{A}_C$. Let $K \in \mathcal{C}(C)$ be such that $W \subset K$. If U, V are contained in compact subsets of A, B respectively, then $UV \in S$ is contained in a compact subset of $C = AB$. The set of UV's, where $U \in \mathcal{A}_A, V \in \mathcal{A}_B$, gives an open cover for C. So we can cover K by a finite number of such sets $U_i V_i$ ($1 \le i \le n$) with each $U_i \subset K_i \subset A$ and $V_i \subset L_i \subset B$ for some compact sets K_i, L_i. Let $U = \cup U_i, V = \cup V_i$. Then by the additivity of S, we have $U \in \mathcal{A}_A$, $V \in \mathcal{A}_B$, $UV \in \mathcal{A}_C$ and $W \subset K \subset UV$. Since π is a homomorphism on S, we get $\pi(A)\pi(B) \ge \pi(U)\pi(V) = \pi(UV) \ge \pi(W)$. It follows by taking sup's over W that $\pi(A)\pi(B) \ge \pi(C)$. The reverse inclusion follows since if $U' \in \mathcal{A}_A$ and $V' \in \mathcal{A}_B$, then $U'V' \in \mathcal{A}_C$. So $\pi(A)\pi(B) = \pi(C)$.

The additivity and regularity of π on G^{op} is straightforward. The restriction of this π to S coincides with the original π by the latter's regularity. To prove the countable subadditivity of π on G^{op}, let $\{A_i\}$ be a sequence in $E(G^{op})$ and $A = \cup_{i=1}^{\infty} A_i$. Let $U \in S$ and $K \in \mathcal{C}(A)$ be such that $U \subset K \subset A$. Since K is compact in A, there exists N such that $U \subset K \subset B = \cup_{i=1}^{N} A_i$. By regularity, we only have to show that π is finitely subadditive on G^{op}. To this end, cover K by a finite number of open sets V_{α,i_α} such that for each α, $V_{\alpha,i_\alpha} \in S$ and is contained in a compact subset of A_{i_α}. Now put $U_j = \cup_{i_\alpha = j} V_{\alpha,i_\alpha} \subset A_j$. Then $U_j \in S$ (by the additivity of S), is contained in a compact subset of A_j and $U \subset \cup_{j=1}^{N} U_j$. Since $U(\cup_{j=1}^{N} U_j) = U$, we have $\pi(U) \le \pi(\cup_{j=1}^{N} U_j)$. Using the finite additivity of π on S, we obtain $\pi(U) \le \pi(\cup_{j=1}^{N} U_j) \le \sum_{j=1}^{N} \pi(U_j) \le \sum_{j=1}^{N} \pi(A_j)$ and finite subadditivity follows by taking the sup over $\pi(U)$.   □

We recall what is meant by a *resolution of the identity* on a locally compact space X with respect to a Hilbert space $\mathcal{H}$ ([243, 12.17]). This is mapping P from $\mathcal{B}(X)$ into the set of self-adjoint projections in $B(\mathcal{H})$ such that:

(i) $P(\emptyset) = 0$ and $P(X) = I$;

(ii) for all $E_1, E_2 \in \mathcal{B}(X)$, we have $P(E_1 \cap E_2) = P(E_1)P(E_2)$, and if $E_1 \cap E_2 = \emptyset$, then $P(E_1 \cup E_2) = P(E_1) + P(E_2)$;

(iii) for each $\xi, \eta \in \mathcal{H}$, the map $E \to \langle P(E)\xi, \eta \rangle$ is a regular Borel (finite) measure $\mu_{\xi,\eta}$ on $\mathcal{B}(X)$.

If P is a resolution of the identity on X with respect to $\mathcal{H}$, then there is a representation Φ_P on $\mathcal{H}$ of $B(X)$ given by

$$\Phi_P(f) = \int_X f \, dP$$

where $\langle \Phi_P(f)\xi, \eta \rangle = \int_X f \, d\mu_{\xi,\eta}$.

The restriction of Φ_P to $C_0(X)$ is a representation Q_P of $C_0(X)$ on $\mathcal{H}$. Conversely (cf. [243, p.306]), if Q is a representation of $C_0(X)$ on $\mathcal{H}$, then there exists a unique resolution of the identity P on X, such that $Q = Q_P$.

We will also need the following result ([242, pp.346-347]) which enables us to obtain a regular Borel measure on a locally compact Hausdorff space from an "inner content" on its family of open sets.

Proposition 3.2.7   *Let X be a locally compact Hausdorff space. Suppose that μ is a positive, real-valued function on the family $\mathcal{U}$ of open subsets of X such that μ is increasing, additive, countably subadditive and for which*

$$\mu(U) = \sup\{\mu(V) : V \text{ is open and has compact closure in } U\}.$$

Then μ extends to a regular Borel measure, also denoted by μ, on X where for $E \in \mathcal{B}(X)$, we define

$$\mu(E) = \inf\{\mu(U) : E \subset U\}. \tag{3.65}$$

The next proposition is the key to relating the regular, additive, subadditive representations of S to the covariant representations to be discussed in **3.3**. Recall that any $e \in E(S)$ is an open subset of G^0 so that $\chi_e \in B(G^0)$.

Proposition 3.2.8   *Let S be an additive inverse subsemigroup of G^{op}. Then a representation π of S on a Hilbert space $\mathcal{H}$ is regular, additive and subadditive if and only if there exists a resolution of the identity P on $\mathcal{B}(G^0)$ such that for all $e \in E(S)$, $P(\chi_e)(= P(e)) = \pi(e)$.*

Proof. Let $\pi : S \rightarrow B(\mathcal{H})$ be a regular, additive, subadditive representation of S. By Proposition 3.2.6, π extends to a representation, also denoted by π, of G^{op} given by (3.64). This representation is regular, additive and countably subadditive. Then the restriction π^0 of π to the family $\mathcal{U}'$ of open subsets of G^0 is a regular, additive, countably subadditive projection-valued representation. Since the extension π is non-degenerate and G^0 is the identity of G^{op}, we have $\pi^0(G^0) = I$.

Define a projection-valued map $P : \mathcal{B}(G^0) \rightarrow \mathbf{R}$ by:

$$P(E) = \inf_U \pi^0(U) \tag{3.66}$$

the inf (which is also a strong limit) being taken over the (decreasing) net of those open subsets U of G^0 containing E. We will see below that P is a resolution of the identity on G^0.

To prove this, note first that π^0 and P coincide on $\mathcal{U}'$ so that for all $e \in E(S)$, $P(\chi_e) = \pi(e)$.

Next, define, for $\xi, \eta \in \mathcal{H}$, the set function $P_{\xi,\eta}$ on $\mathcal{B}(G^0)$ by: $P_{\xi,\eta}(E) = \langle P(E)\xi, \eta \rangle$. Then $P_{\xi,\eta}$ is a complex-valued, regular Borel measure on $\mathcal{B}(G^0)$. To prove this, we can suppose that $\xi = \eta$. Put $P_{\xi,\xi} = \mu$. From the corresponding properties for π^0, the set function μ', where μ' is the restriction of μ to $\mathcal{U}'$, is finite, increasing, additive and countably subadditive. Further, for every $A \in \mathcal{U}'$, $\mu'(A)$ is the sup of $\mu'(U)$ over all $U \in \mathcal{U}'$ whose closure in G^0 is compact and contained in A. By Proposition 3.2.7, μ' extends to a regular Borel measure on G^0, and comparing (3.65) with (3.66), we see that this extension is just μ. So every $P_{\xi,\eta}$ is a regular Borel finite measure on G^0.

We now claim that P is a resolution of the identity on $\mathcal{B}(G^0)$. To this end, we check (i),(ii) and (iii) of the definition given before Proposition 3.2.7. It is obvious that (i) holds and (iii) was proved above. So it remains to prove (ii). Let $E_1, E_2 \in \mathcal{B}(G^0)$. We show first that $P(E_1 \cap E_2) = P(E_1)P(E_2)$. Note that since π^0 is a representation, we have

$$P(U_1 \cap U_2) = P(U_1)P(U_2)$$

for any open sets U_1, U_2 in G^0. Let $\xi \in \mathcal{H}$. It follows that $P(E_1 \cap E_2) \leq P(E_1)P(E_2)$, using the joint continuity of the strong operator topology on bounded sets. To get equality, we just have to show that with $\mu = P_{\xi,\xi}$ as above, for any open $U \subset G^0$ with $E_1 \cap E_2 \subset U$ and $\epsilon > 0$, there exist open sets U_1, U_2 in G^0 such that $E_1 \subset U_1, E_2 \subset U_2$ and $\mu(U_1 \cap U_2) < \mu(U) + \epsilon$. (For then we obtain the inequality $\langle P(E_1)P(E_2)\xi, \xi \rangle \leq \langle P(E_1 \cap E_2)\xi, \xi \rangle$ giving $P(E_1)P(E_2) = P(E_1 \cap E_2)$.) This is easy to do : let V_i be open subsets of G^0 such that $E_i \sim (E_1 \cap E_2) \subset V_i$ and $\mu(E_i \sim (E_1 \cap E_2))$ is sutiably close to $\mu(V_i)$. Then take $U_i = V_i \cup U$. Lastly, we have to show that if $E_1 \cap E_2 = \emptyset$, then $P(E_1 \cup E_2) = P(E_1) + P(E_2)$. This follows from the fact that each $P_{\xi,\eta}$ is a measure. So P is a resolution of the identity on $\mathcal{B}(G^0)$ extending π^0.

Conversely, let π be a representation of S for which there exists a resolution of the identity P on $\mathcal{B}(G^0)$ such that for all $e \in E(S)$, $P(e) = \pi(e)$. Then π is additive using Proposition 3.2.5 and the fact that $\pi(\emptyset) = P(\emptyset) = 0$. Let $U \in S$ with $U \subset G^0$. Since S contains a countable basis of relatively compact sets for the topology of G^0 and S is additive, there exists an increasing sequence $\{U_n\}$ in S such that each U_n has compact closure in U and $\cup_{n=1}^{\infty} U_n = U$. It follows by the countable additivity of the measure $A \rightarrow \langle P(A)\xi, \xi \rangle$ on $\mathcal{B}(G^0)$ ($\xi \in \mathcal{H}$) (cf. [243, p.302]) that $\pi(U)$ is the strong operator limit of $\{\pi(U_n)\}$. The regularity of π now follows using Proposition 3.2.5. The subadditivity of π follows from the finite additivity property of P. So π is regular, additive and subadditive. $\qquad \square$

Theorem 3.2.1 below shows that in the appropriate sense, the representation theory of S is exactly equivalent to the representation theory of $C_c(G)$. The method of proof of this theorem is modelled on Renault's original proof of Theorem 3.1.1 for a large class of groupoids, including those r-discrete groupoids that we are discussing which are Hausdorff. In this proof, the central technique is a disintegration theorem of Guichardet as applied in the context of Borel G-sets which mediates between the G- and $C_c(G)$-representations. Later, of course, Renault in [233] was able to dispense completely with G-sets. However Renault's first proof [230, pp.65-69] can be modified to give Theorem 3.2.1, a result which does not seem to follow from Theorem 3.1.1.

Guichardet's theorem is stated and proved in Appendix E. The theorem will ennable us to "disintegrate" the partial isometries arising from certain S-representations. The following theorem is the key to relating inverse semigroup and groupoid representations.

Theorem 3.2.1   *Let S be an additive inverse subsemigroup of G^{op}. If L is a representation of G, then the representation π_L, where π_L is defined by (3.57), is a regular, additive, subadditive representation of S. Conversely, every regular, additive, subadditive representation of S is of the form π_L for some representation L of G.*

Proof.   Suppose that L is a representation of G associated with the Hilbert bundle $(G^0, \{H_u\}, \mu)$. Then π_L is a representation of S on $\mathcal{H} = L^2(G^0, \{H_u\}, \mu)$. Suppose that $A \perp B$ in S. Then $A \cap B = 0$, and using (3.57), we have

$$\pi_L(A \cup B) = \pi_L(A) + \pi_L(B).$$

So π_L is additive. The subadditivity of π_L follows from:

$$\langle \pi_L(A)\xi, \xi \rangle \leq \int_A \|\xi(u)\|_2^2 \, d\mu(u)$$

for $A \in S^0$.

We now show that π_L is regular on S. Let $A \in S^0$ and $\mathcal{A}_A$ be as in the proof of Proposition 3.2.6. Then for any $U \in \mathcal{A}_A$ and any $\xi \in \mathcal{H}$, we have $UA = U = AU$, and so

$$\|(\pi_L(A) - \pi_L(U))\xi\|^2 =$$
$$\langle \pi_L(A)\xi, \xi \rangle - \langle \pi_L(U)\xi, \xi \rangle. \tag{3.67}$$

Now every compact subset K of A is contained in a member of $\mathcal{A}_A$. Using (3.57), (3.67) and the monotone convergence theorem, we obtain that $\pi_L(U) \to \pi_L(A)$ strongly. So π_L is regular on S by Proposition 3.2.5. So π_L is a regular, additive, subadditive representation of S.

We now prove the converse result. Let $\pi : S \to B(\mathcal{H})$ be a regular, additive, subadditive representation of S. Then by Proposition 3.2.6, π extends to a regular, additive, countably subadditive representation, also denoted by π, on G^{op}, and by Proposition 3.2.8, there exists a resolution of the identity P on $\mathcal{B}(G^0)$ such that $P(U) = \pi(U)$ for all $U \in S^0$. Let ϕ be the representation $f \to \int f \, dP$ of $B(G^0)$ on $\mathcal{H}$.

Let μ be a basic measure on the spectrum Z of $\phi(C_0(G^0))$ with associated Hilbert bundle $(Z, \{H_z\}, \mu)$ with fibers $H_z \neq \{0\}$ for all $z \in Z$. Then $\mathcal{H}$ is naturally identified with $L^2(Z, \{H_z\}, \mu)$ and each $\phi(f)$ ($f \in B(G^0)$) with the diagonalizable operator T_f. Identifying Z with a locally compact subset of G^0, we can extend the Hilbert bundle $(Z, \{H_z\}, \mu)$ trivially to G^0 and (abusing notation a little) call the extended bundle $(G^0, \{H_u\}, \mu)$. We then have $\mathcal{H}$ identified with $L^2(G^0, \{H_u\}, \mu)$ and μ identified with a probability measure on G^0. We shall also identify $\phi(f)$ with the multiplication operator T_f for $f \in B(G^0)$. Let $U_A = \pi(A)$ for $A \in G^{op}$.

We now show that μ is quasi-invariant. To prove this, it is sufficient by Proposition 3.2.2 to show that for $A \in G^{op}$, $\mu \circ \alpha_A \sim \mu$ on $d(A)$. Let $A \in G^{op}$. Let B be an open subset of G^0 and $f = \chi_B$. Since $\pi(A)\pi(B)\pi(A)^{-1} = \pi(ABA^{-1}) = T_{\chi_{ABA^{-1}}}$ and $u \in ABA^{-1}$ if and only if $f(A^{-1}uA) = 1$, we have

$$U_A T_f U_A^* = T_{f \circ \alpha_A}. \tag{3.68}$$

(Here, $f \circ \alpha_A(u) \in B(G^0)$ is defined to be zero if u does not belong to $r(A)$.)

So for $\xi \in \mathcal{H}$, we have

$$\langle U_A T_f U_A^* \xi, \xi \rangle = \langle T_{f \circ \alpha_A} \xi, \xi \rangle.$$

Hence

$$\int f(u) \|U_A^* \xi(u)\|^2 \, d\mu(u) = \int f \circ \alpha_A(u) \|\xi(u)\|^2 \, d\mu(u), \tag{3.69}$$

and substituting $f = \chi_B$ in (3.69) gives

$$\int_B \|U_A^* \xi(u)\|^2 \, d\mu(u) = \int_{ABA^{-1}} \|\xi(u)\|^2 \, d\mu(u). \tag{3.70}$$

Applying the monotone class result of measure theory, we see that (3.68) holds with $f = \chi_E$ for all $E \in \mathcal{B}(G^0)$. Since the span of such characteristic functions χ_E is norm dense in $B(G^0)$ we have that for all $f \in B(G^0)$,

$$U_A T_f U_A^* = T_{f \circ \alpha_A}. \tag{3.71}$$

Now if $C \in \mathcal{C}(A)$ and ξ is chosen so that $\|\xi(u)\| = 1$ for all $u \in d(A)$ for which $H_u \neq \{0\}$, then from (3.70) with C in place of B,

$$\mu \circ \alpha_A(C) = \int_C \|U_A^* \xi(u)\|^2 \, d\mu(u).$$

It follows using the regularity of $\mu \circ \alpha_A$ that if $E \in \mathcal{B}(d(A))$ and $\mu(E) = 0$, then $\mu \circ \alpha_A(E) = 0$. Replacing E by $\alpha_{A^{-1}}(E)$ gives the reverse implication. So μ is quasi-invariant. It also follows that (3.71) holds for all $f \in L^\infty(G^0, \mu)$.

For any $D \in G^{op}$, the unitary operator U_D has the restriction $\mathcal{H}_{d(D)}$ of $\mathcal{H} = L^2(G^0, \{H_u\}, \mu)$ to $d(D)$ as its initial subspace and the restriction $\mathcal{H}_{r(D)}$ of $\mathcal{H} = L^2(G^0, \{H_u\}, \mu)$ to $r(D)$ as its final subspace. (Consider the projections $U_{D^{-1}D}, U_{DD^{-1}}$.) We now apply Theorem E.0.4 with $X = d(A), X' = r(A), \phi = \alpha_A^{-1}, U = U_A$ and the measures on X, X' just the restrictions of μ. (Since μ is quasi-invariant, Proposition 3.2.2 gives $\mu \circ \alpha_A^{-1} \sim \mu$ on $X = d(A)$.) Then there exist for μ a.e. $v \in r(A)$, unitary maps $u^A(v) : H_{\alpha_A(v)} \to H_v$, unique μ a.e., such that for all $\xi \in \mathcal{H}_{d(A)}$,

$$U_A \xi(v) = \delta_A(v)^{1/2} u^A(v)(\xi(\alpha_A(v))) \tag{3.72}$$

for μ a.e. $v \in r(A)$, where

$$\delta_A(v) = \frac{d(\mu \circ \alpha_A^{-1})}{d\mu}(v) \tag{3.73}$$

on $r(A)$. Simple "Radon-Nikodym" calculations show that for $A, B \in S$,

$$\delta_{AB}(u) = \delta_B(\alpha_A(u))\delta_A(u) \tag{3.74}$$

for μ a.e. $u \in r(AB)$. In particular, since $\alpha_{AA^{-1}}$ is the identity map, we have

$$1 = \delta_{A^{-1}}(\alpha_A(u))\delta_A(u) \tag{3.75}$$

for μ a.e. $u \in AA^{-1}$.

It follows from (3.72) that the map

$$v \to \langle u^A(v)(\xi(\alpha_A(v))), \eta(v)\rangle \tag{3.76}$$

is μ-measurable on $r(A)$ for all $\xi, \eta \in L^2(G^0, \{H_u\}, \mu)$.

Let $\nu = \int_{G^0} \lambda^u \, d\mu(u)$, and for $D \in G^{op}$ and $x \in D$, define $L^D(x) = u^D(r(x))$. Let $A, B \in G^{op}$ and $C = A \cap B \in G^{op}$. We claim that $L^A(x) = L^B(x)$ ν a.e. on C. Let $\xi' \in \mathcal{H}_{d(C)}$. Note that $A(d(C)) = C$. Also, α_C is the restriction of α_A to $r(C)$, $\delta_A \mid_{r(C)} = \delta_C$ μ a.e. and $\mathcal{H}_{d(C)} \subset \mathcal{H}_{d(A)}$. Since $\chi_{d(C)} \xi' = \xi'$, we have $U_C \xi' = U_A U_{d(C)} \xi' = U_A \xi'$. The same result holds with B in place of A, and by the uniqueness of the $u^A(v)$, $u^B(v)$ and $u^C(v)$, we have for μ a.e. $v \in r(C)$ that $u^A(v) = u^C(v) = u^B(v)$. Since $\nu(E) = \mu(r(E))$ for any measurable $E \subset A \cap B$ (Proposition 3.2.1, (ii)), we have $L^A(x) = L^B(x)$ ν a.e. on $A \cap B$.

Since G is a countable union of G^{op}-elements A_i, we obtain a map L on G by expressing G as a disjoint union of sets $E_i \in \mathcal{B}(A_i)$ and setting $L(x) = L^{A_i}(x)$ if $x \in E_i$. We take G^0 to be one of the E_i's (which will be an A_i also). By the result of the preceding paragraph, if different A_i's and E_i's are used, then the new L coincides with the first off a ν-null set. Thus L is essentially unique. We now show that we can arrange for L to be a representation of G, namely, that the required properties (i)-(iv) of Definition 3.1.1 can be arranged to hold.

To prove (i), it follows from (3.72) and the facts that both U_{G^0} and α_{G^0} are identity maps that $L(u)$ is the identity map μ-a.e. on G^0. By changing L on a null set, we can suppose that (i) holds. Using (3.76), Proposition 3.2.1 and the fact that $\alpha_A(r(x)) = d(x)$ for $x \in A$, we see that the function in (3.15) is measurable on every $A \in G^{op}$. The condition (iv) now follows. Using (3.72) to calculate the left-hand side of the equation $U_{A^{-1}}(U_A \xi)(v) = \xi(v)$ for a.e. $v \in r(A)$ as well as (3.75), we obtain (iii). We now turn to (ii).

We first show that for $A, B \in G^{op}$,

$$L^{AB}(xy) = L^A(x) L^B(y) \tag{3.77}$$

for ν^2-a.e. $(x, y) \in A \dot\times B$. Using (3.72), (3.74) and the uniqueness part of Theorem E.0.4, and equating the resulting expressions for $U_A(U_B \xi)(v)$ and $U_{AB} \xi(v)$ a.e., we obtain that for $v \in r(AB) \sim E$, where E is a μ-null set in $r(AB)$,

$$u^{AB}(v) = u^A(v) u^B(\alpha_A(v)). \tag{3.78}$$

Let $N = \{a \in A : r(a) \in E\}$ and $W = \{(a, b) \in A \dot\times B : a \in N\}$. It is sufficient to show that W is ν^2-null since if $(x, y) \in (A \dot\times B) \sim W$, use of (3.78) gives : $L^{AB}(xy) = u^{AB}(r(x)) = u^A(r(x)) u^B(\alpha_A(r(x))) = u^A(r(x)) u^B(d(x)) = L^A(x) L^B(y)$. Now by (3.49), $\nu^2(W) \leq \mu(d(N))$. But $\mu(d(N)) = 0$ since $\mu(r(N)) \leq \mu(E) = 0$ and μ is quasi-invariant. So (3.77) holds except on the ν^2-null set W. Applying a similar argument to deal with the μ-null sets where L^{AB}, L^A, L^B differ from L, we obtain (ii).

So L is a representation of G. We now show that π_L coincides with π on S. Let $A \in S$. Then using (3.57), (i) of Proposition 3.2.1, (3.72), (3.53),

(3.73) and the fact that $L(x) = u^A(r(x))$ ν-a.e. on A, we have

$$
\begin{aligned}
\langle \pi_L(A)\xi, \eta \rangle \\
&= \int_A \langle L(x)(\xi(d(x))), \eta(r(x)) \rangle D^{-1/2}(x)\, d\nu(x). \\
&= \int_{r(A)} \langle L(r_A^{-1}(u))(\xi(\alpha_A(u))), \eta(u) \rangle D^{-1/2}(r_A^{-1}(u))\, d\mu(u) \\
&= \int_{r(A)} \langle u^A(u)(\xi(\alpha_A(u))), \eta(u) \rangle \delta_A^{1/2}(u)\, d\mu(u) \\
&= \langle U_A\xi, \eta \rangle.
\end{aligned}
$$

So $\pi_L = \pi$ on S. $\square$

Corollary 3.2.1   *Let G be, in addition, an ample groupoid. If L is a representation of G, then the representation π_L of G^a, where π_L is defined by (3.57), is an additive representation of G^a. Conversely, every additive representation of G^a is of the form π_L for some representation L of G.*

Proof.   By Theorem 3.2.1, we just have to show that every additive representation π of G^a is both regular and subadditive. Regularity is immediate since every member of G^a is compact. To prove subadditivity, let U_i $(1 \le i \le n)$ be compact open subsets of G^0 and $U = \cup_{i=1}^n U_i$. Now the family of compact open subsets of G^0 is closed under finite unions, intersections and differences. Hence we can express U as a disjoint finite union of compact open sets B_j with each B_j contained in some U_i. Then $\pi(U) = \sum_j \pi(B_j) \le \sum_i \pi(U_i)$ by the additivity of π. $\square$

We note that the map $L \to \pi_L$ of Theorem 3.2.1 also respects the natural equivalence relations on the classes of G-representations and regular, additive, subadditive representations of S (cf. [230, p.52]). A similar comment applies in the ample case of the above corollary.

Theorem 3.2.1 relates representations of an additive inverse subsemigroup S of G^{op} to G-representations. Renault's original argument related $C_c(G)$-representations to G-representations. We now discuss how $C_c(G)$-representations relate to G-representations through the mediation of S-representations. This will be needed later for proving Theorem 3.3.1 and gives a proof of Theorem 3.1.1 in the r-discrete case.

Let Π be a representation of $C_c(G)$ on a Hilbert space $\mathcal{H}$. From the discussion following Theorem 3.1.1, the representation Π extends to a representation Π' of $B_c(G)$, and this determines a representation π_1 of G^{op}. It is easy to check that π_1 is regular, additive and subadditive. Further, if $g \in C_c(A)$, where $A \in G^{op}$, then by (3.56), we have $g = (g \circ r_A^{-1}) * \chi_A$, and so

$$\Pi(g) = \phi(g \circ r_A^{-1})\pi_1(A). \tag{3.79}$$

It is left to the reader to check that if we start with $\pi = \pi_1$ in the proof of Theorem 3.2.1, then the integrated form π_L of the G-representation L in that proof coincides with Π. This gives a proof in the r-discrete case of Theorem 3.1.1.

We conclude this section by showing that for an ample groupoid G, the C^*-algebra $C^*(G)$ is related to G^a in a particularly elegant way inasmuch as it can be intrinsically expressed in terms of G^a (with its additive structure) alone using the semigroup algebra $\ell^1(G^a)$.

Indeed, let G be an r-discrete groupoid and S be an additive inverse subsemigroup of G^{op}. We call the representations of $\ell^1(S)$ associated with additive representations of S *additive*. Just as the representations of S are associated with the enveloping C^*-algebra of $\ell^1(S)$ so also the additive representations of S are associated with the enveloping C^*-algebra of a quotient B of $\ell^1(S)$ which is described below.

Recall that a sequence $A_1, \ldots, A_n$ in S is called *orthogonal* if $A_i \perp A_j$ whenever $i \neq j$.

Proposition 3.2.9   *Let I be the closure of the subspace of $\ell^1(S)$ spanned by elements of the form $\cup_{i=1}^n A_i - \sum_{i=1}^n A_i$ where $\{A_i\}$ is an orthogonal sequence in S. Then I is a closed *-ideal in $\ell^1(S)$. Let $\ell^1_{add}(S)$ be the Banach *-algebra $\ell^1(S)/I$. Then the *-representations of $\ell^1_{add}(S)$ on a Hilbert space can be canonically identified with the additive representations of S.*

Proof.   Let $\{A_i\}$ be an orthogonal sequence in S and let $A = \cup_{i=1}^n A_i$. Let $B \in S$. We claim that $\{BA_i\}$ is orthogonal. For suppose that $i \neq j$. Then using (3.59), $BA_i(BA_j)^{-1} = BA_i A_j^{-1} B^{-1} = 0$. On the other hand, since B is a G-set, $B^{-1}B \subset G^0$ and so

$$(BA_i)^{-1}BA_j = A_i^{-1}B^{-1}BA_j \subset A_i^{-1}A_j = 0.$$

So $(BA_i)^{-1}BA_j = 0$. It follows that $\{BA_i\}$ is an orthogonal sequence in S and that $B(A - \sum_{i=1}^n A_i) = BA - \sum_{i=1}^n BA_i \in I$. Similarly, $(A - \sum_{i=1}^n A_i)B \in I$. It follows that I is a (closed) ideal in $\ell^1(S)$. Since $\{A_i^{-1}\}$ is (by the symmetry of (3.59)) orthogonal, we have that I is a *-ideal in $\ell^1(S)$. Finally, the *-representations of $\ell^1_{add}(S)$ can be identified with those which factor through I, and these are precisely the representations π of $\ell^1(S)$ for which $\pi(A) = \sum_{i=1}^n \pi(A_i)$ for all orthogonal sequences $\{A_i\}$ with $A = \cup_{i=1}^n A_i$. But these representations are the additive ones from (3.62). $\square$

Theorem 3.2.2   *Let G be an ample groupoid. Then $C^*(G)$ is canonically isomorphic to the enveloping C^*-algebra $C^*_{add}(G^a)$ of the Banach *-algebra $\ell^1_{add}(G^a)$.*

Proof. Let $S = G^a$. The map $A \to \chi_A$ is a multiplicative *-homomorphism from S into $C_c(G) \subset C^*(G)$ (Proposition 2.2.6) and so extends to a *-homomorphism Γ from $\mathbf{C}(S)$ into $C^*(G)$. This in turn extends to a norm decreasing *-homomorphism from $\ell^1(S)$ into $C^*(G)$ (since each $\Gamma(A)$, for $A \in S$, is a partial isometry and so has norm ≤ 1.)

From Proposition 2.2.7, $V = \Gamma(\mathbf{C}(S))$ is an I-norm dense *-subalgebra of $C_c(G)$ and hence is also a dense *-subalgebra of $C^*(G)$. If $\{A_1, \ldots, A_n\}$ is an orthogonal sequence in S with union A, then trivially, $\Gamma(A - \cup_{i=1}^n A_i) = 0$. So Γ factors through the closed ideal I of Proposition 3.2.9 and therefore induces a *-homomorphism $\Gamma' : \ell^1_{add}(S) \to C^*(G)$ whose range contains V.

Next, if π is a representation of $C_c(G)$, then the map $\pi \circ \Gamma'$ defines an additive representation of S and hence (Proposition 3.2.9) a representation of $\ell^1_{add}(S)$. Conversely, given a representation Θ of $\ell^1_{add}(S)$, it defines an additive representation Θ' of S, which in turn (Corollary 3.2.1) defines a representation L of G which in turn integrates up to give the representation π_L of $C_c(G)$ and hence of $C^*(G)$. By Theorem 3.2.1, the map Θ' extends by linearity to give the representation π_L on the dense subalgebra V of $C^*(G)$. So $\Theta = \pi_L \circ \Gamma'$, and it follows that, under Γ', the representations of $\ell^1_{add}(S)$ correspond to the representations of $C^*(G)$. So Γ' is an isometric *-isomorphism from a dense subalgebra of $C^*_{add}(S)$ onto a dense subalgebra of $C^*(G)$. Hence $C^*_{add}(S) \cong C^*(G)$, the isomorphism being canonically implemented by the map $A \to \chi_A$. $\square$

3.3 Groupoid and covariance C^*-algebras

In this section, we will show (Theorem 3.3.1) that if G is an r-discrete groupoid, then $C^*(G)$ is isomorphic to the crossed product C^*-algebra $C_0(G^0) \times_\beta S$ for any additive (countable) inverse subsemigroup S of G^{op} with its natural localization action ((3.51)) on G^0, the crossed product being taken in the sense of Sieben ([256]).[7] The converse is also true under very general conditions (Corollary 3.3.2). Indeed, for any given localization (X, S), we will explicitly construct a natural r-discrete groupoid $G(X, S)$ (Theorem 3.3.2). Every localization is effectively equivalent to a localization with a certain *additive* property, and in the additive case, $C_0(X) \times_\beta S \cong C^*(G(X, S))$. The Sieben theory, generalized to the non-unital case, is covered in the early part of the section.

We recall the notion of a *localization* from **3.2**. Let X be a locally compact Hausdorff space and S have a right action $x \to x.s$ of S on X. We assume that the domain D_s of each map α_s, where $\alpha_s(x) = x.s$, is open, and

---

[7] The reader is referred to the work of Nandor Sieben and John Quigg (e.g. [257, 258]) for further advances in the theory of inverse semigroup covariant systems.

that α_s is a homeomorphism from D_s onto its range $R_s = D_{s^*}$. The family of such domains D_s is further assumed to form a basis for the topology of X. The pair (X, S) is then called a *localization*. Since $D_s = D_{ss^*}$, it follows that the preceding basis condition is equivalent to the family $\{D_e : e \in E(S)\}$ forming a basis for the topology of X.

In the above definition of a localization, we have extended the terminology of A. Kumjian ([148]). Kumjian applied the notion in the case where S is actually given as a semigroup of partial homeomorphisms on X, i.e. the case where the map $s \to \alpha_s$ is an anti-isomorphism (rather than just an anti-homomorphism). However, since we need the localization theory to apply to pairs such as (G^0, S) above and such pairs do not usually give localizations in the sense of Kumjian (see, for example, **4.3**, Example 4), it seems preferable to use the extended notion of a localization given above.

As in the case of an inverse subsemigroup of G^{op} which is a basis for the topology of the r-discrete groupoid G (**3.2**), we shall avoid the trivial case where X is a singleton so that for some $s \in S$, we have $\alpha_s = \emptyset$. We now briefly discuss two kinds of localization of particular importance.

For the first, let G be an r-discrete groupoid and S be an inverse subsemigroup of G^{op} which is a basis for the topology of G. As discussed in **3.2**, there is a canonical right action $x \to x.s = s^{-1}xs$ of S on G^0. (See (3.51).) Then $D_s = r(s) = ss^* \in E(S)$ and is open in G^0. And of course each map $x \to x.s$ is a homeomorphism from D_s onto D_{s^*}. Lastly, since $S^0 = \{D_e : e \in E(S)\}$ is a basis for G^0, it follows that (G^0, S) is a localization. This localization will feature prominently in this section.

The second example of a localization (X, S) is that associated with the r-discrete holonomy groupoid G_T discussed in Example 2 of **2.3**. In that case, the space X was what we called T, an appropriate union of local transverse sections T_r of the foliation. Associated with T was a pseudogroup S generated by partial diffeomorphisms T_γ between open subsets of T. Since the set of domains of the elements of S is a basis for the topology of T, it follows that the pair (T, S) is indeed a localization. We also noted that G_T can be obtained from the pair (T, S) as the sheaf of S-germs. These germs are of the form $\overline{(x, s)}$ for $s \in S$ and x in the domain of s. This gives a strong clue about how localizations in general should give rise to r-discrete groupoids, and that theme is developed in this section and in Chapter 4.

In **3.2**, we discussed the notion of an *additive* inverse subsemigroup of G^{op}. The following formulates an analogous (though weaker) notion for localizations. (A stronger version will be considered later.) We shall say that a localization (X, S) is *extendible* if whenever $e_1, e_2 \in E(S)$ then there exists $e_3 \in E(S)$ such that $D_{e_3} = D_{e_1} \cup D_{e_2}$. It is obvious that if G is r-discrete, $X = G^0$ and S is a (countable) additive subsemigroup of G^{op} acting canonically on X, then the localization (X, S) is extendible.

Let (X, S) be a localization. If U is an open subset of X, then we regard $C_0(U)$ as a closed ideal in $C_0(X)$ (extending a $C_0(U)$ function to X by making it zero outside U). In particular, since D_s is open in X, we have that $C_0(D_s)$ is a closed ideal in $C_0(X)$. For each $s \in S$, define a map $\beta_s : C_0(D_{s^*}) \to C_0(D_s)$ by dualizing the right S-action on X:

$$\beta_s(F)(x) = F(x.s) \quad (x \in D_s). \tag{3.80}$$

It is obvious that β_s is an isomorphism between closed ideals of $C_0(X)$. Now $C_0(U) \cap C_0(V) = C_0(U \cap V)$ for open subsets U, V of X. Using this and the fact that $s \to \alpha_s$ is an antihomomorphism, we see that the dual map $s \to \beta_s$ is a homomorphism from S into $\mathcal{I}(C_0(X))$, the inverse semigroup of partial one-to-one maps on $C_0(X)$. We will sometimes write $\beta_s(F)$ as sF. Since $x.e = x$ for all $e \in E(S)$ and all $x \in D_e$, it follows that for $e \in E(S)$, we have $eF = F$ for all $F \in C_0(D_e)$.

In fact, the β_s's and their domains, the ideals $C_0(D_{s^*})$, define a situation close to what Sieben ([256]) calls an *action* of S on $C_0(X)$.[8] However, we do not have here the unital requirements of the Sieben theory. For this reason, we now adapt the theory of Sieben to the non-unital case. This will be applied later in the localization context in which the C^*-algebra A will be $C_0(X)$. Since it is not much more difficult to present the theory in the general case, the non-unital version will be given for general actions on a C^*-algebra. Let A be a C^*-algebra and S be an inverse semigroup.

Definition 3.3.1 *An* action *of S on A is defined to be a homomorphism β from S into the inverse semigroup of partial one-to-one maps on subsets of A such that:*

(i) *the domain E_{s^*} of every β_s ($s \in S$) is a closed ideal in A, and β_s is an isomorphism from the ideal E_{s^*} onto the ideal E_s;*

(ii) *if $s, t \in S$ then there exists $w \in S$ such that $E_s \cup E_t \subset E_w$;*

(iii) *the set $B = \cup_{s \in S} E_s$ is a dense subalgebra of A.*

We shall sometimes say that the triple (A, β, S) *gives* or even *is* an (inverse semigroup) *covariant system*. As in the localization case, we will sometimes write sa for $\beta_s(a)$ ($s \in S, a \in E_{s^*}$).

The reason for using the notation E_{s^*} in this definition rather than E_s for the domain of β_s is illustrated by the localization case (X, S) discussed above where A is $C_0(X)$ and the domain of β_s is $C_0(D_{s^*})$.

A useful fact ([256]) is that for $s, t \in S$,

$$s(E_{s^*} E_t) = E_{st}. \tag{3.81}$$

---

[8]Inverse semigroup actions are closely related to the theory of partial actions on C^*-algebras developed by Exel and Maclanahan ([94, 171]).

To prove this, observe that if I, J are closed ideals in a C^*-algebra, then $I \cap J = IJ$. (Trivially, IJ is contained in the closed ideal $I \cap J$ and the reverse inclusion follows since any closed ideal K in a C^*-algebra has a bounded approximate identity and so $K^2 = K$ by Cohen's theorem ([121, p.270]).) So $E_{st} = s(tE_{t^*}) = s(E_{s^*} \cap E_t) = s(E_{s^*}E_t)$.

The condition (ii) just says that the ideals E_s are directed upwards. From (ii), the set B of (iii) is a $*$-subalgebra of A. The density requirement of (iii) follows from the need to be able to capture a dense subalgebra of A from the E_s's in order to have a chance of obtaining a representation of A in the development of the theory of covariant representations below. Note also that for each $s \in S$, the ideal E_s equals E_{ss^*}. So $B = \cup_{e \in E(S)} E_e$ and (ii) and (iii) above can be formulated in terms of E_e's rather than in terms of E_s's. The conditions (ii), (iii) follow immediately if it is assumed that S is unital and $E_1 = A$. This is the situation of [256].

When S is a group acting on A, then every $E_s = A$, and, of course, the triple (A, β, S) is a covariant system in the usual sense.

Of particular interest to us is the covariant system $(C_0(X), \beta, S)$ given by (3.80), where (X, S) is an extendible localization. To check that this is a covariant system, we have to show that the conditions (i),(ii) and (iii) of Definition 3.3.1 hold. In this case, $E_s = C_0(D_s) = C_0(D_{ss^*})$ and $\beta_s(f)(x) = f(x.s)$ for $x \in D_{s^*}$. Now given $e_1, e_2 \in E(S)$, we have, by the extendibility of (X, S), that there exists $e_3 \in E(S)$ such that $D_{e_3} = D_{e_1} \cup D_{e_2} \in E(S)$. Then $E_{e_1} \cup E_{e_2} \subset E_{e_3}$, giving (ii). Since every compact subset of X is covered by a finite number of D_e's ($e \in E(S)$) (as the family of D_e's is a basis for X), it follows that $B \supset C_c(X)$ so that condition (iii) also holds. We therefore have the following proposition.

Proposition 3.3.1 *If (X, S) is an extendible localization, then the triple $(C_0(X), \beta, S)$ is a covariant system.*

Turning to the general case again, let (A, β, S) be a covariant system. Let $\mathbf{C}(A, S)$ be the space of functions $\theta : S \to A$ such that $\theta(s) \in E_s \subset A$ for all $s \in S$ and θ vanishes off a finite subset of S. Then $\mathbf{C}(A, S)$ is a vector space under pointwise operations.

Let $V(A, S)$ be the set of elements $(a, s) \in \mathbf{C}(A, S)$ where $a \in E_s$ and $(a, s)(t) = \delta_{s,t} a$ for $t \in S$. Note that the map $a \to (a, s)$ is linear from E_s into $\mathbf{C}(A, S)$. Clearly, $\mathbf{C}(A, S)$ is spanned by $V(A, S)$, and indeed is the vector space over $\mathbf{C}$ generated by $V(A, S)$ subject to the relations determined by the linearity of the maps $a \to (a, s)$. One then shows, as in the standard case for discrete group actions, that $\mathbf{C}(A, S)$ is a $*$-algebra with product and involution determined by the following product and involution on $V(A, S)$:

$$(a, s)(b, t) = (s[(s^*a)b], st), \qquad (a, s)^* = (s^*a^*, s^*). \qquad (3.82)$$

Note above that $(s^*a)b \in E_{s^*}E_t = E_{s^*} \cap E_t$, and since every element of

E_t is in tE_{t^*}, we have that $s[(s^*a)b] \in (st)E_{(st)^*} = E_{st}$. We would like to write "$(a, s)(b, t) = (a(sb), st)$" but unfortunately sb does not usually make sense! However, as we will see below, in the case of the covariant system associated with an extendible localization, we can use a formula like that. In general, the inverse semigroup S is not contained as a subsemigroup of $V(A, S)$ in any natural way.

The slightly involved formula for the product in $V(A, S)$ necessitates a little care in dealing with this product. To illustrate this, let us prove that this product is associative. Let $(a, s), (b, t), (c, u) \in V(A, S)$. By multiplying out both sides, we see that $((a, s)(b, t))(c, u) = (a, s)((b, t)(c, u))$ if and only if

$$(st)\{(st)^*[s[(s^*a)b]]c\} = s\{(s^*a)t[(t^*b)c]\}.$$

Let $\{e_\delta\}$ be a bounded approximate identity for E_{t^*}. We then argue:

$$
\begin{aligned}
(st)\{(st)^*[s[(s^*a)b]]c\} &= (st)\{t^*(s^*s)[(s^*a)b]c \\
&= (st)\{t^*[(s^*a)b]c\} \\
&= \lim_\delta(st)\{t^*[(s^*a)b]e_\delta c\} \\
&= \lim_\delta s\{tt^*[(s^*a)bt(e_\delta c)]\} \\
&= \lim_\delta s[(s^*a)bt(e_\delta c)] \\
&= \lim_\delta s[(s^*a)tt^*(b)t(e_\delta c)] \\
&= \lim_\delta s[(s^*a)t[(t^*b)e_\delta c]] \\
&= s\{(s^*a)t[(t^*b)c]\}
\end{aligned}
$$

as required.

Now define $\|.\|_1 : \mathbf{C}(A, S) \to [0, \infty)$ by: $\|\theta\|_1 = \sum_{s \in S} \|\theta(s)\|$. It is obvious that, as for the ℓ^1-norm on a group algebra, the algebra $\mathbf{C}(A, S)$ is a normed algebra under $\|.\|_1$. Note that for fixed s, the linear map $a \to (a, s)$ from E_s into $\mathbf{C}(A, S)$ is continuous.

When (X, S) is an extendible localization, the definition of the product above for the associated covariant system can be simplified. For if $(f, s), (g, t) \in V(X, S)$ (i.e. $f \in C_0(D_s), g \in C_0(D_t)$), we can define

$$(f, s)(g, t) = (f(sg), st), \tag{3.83}$$

where we define the function sg on X by setting $sg(x) = g(x.s)$ if $x \in D_s$ and zero otherwise. (It is easy to check that $s[(s^*f)g] = f(sg)$.) Of course, from (3.82), the involution on $\mathbf{C}(C_0(X), S)$ is given by:

$$(f, s)^* = (s^*\overline{f}, s^*). \tag{3.84}$$

As for the case of group actions, we want to relate the representations of $\mathbf{C}(A, S)$ to "covariant" pairs of representations of A and S. Following Sieben ([256, Definition 4.5]), we define a *covariant representation* for a covariant system (A, β, S) to be a pair of representations ϕ of A and π of S on a Hilbert space $\mathcal{H}$ such that for all $s \in S$, the initial subspace H_s of $\pi(s)$ is $\phi(E_{s^*})\mathcal{H}$, and for all $a \in E_{s^*}$,

$$\pi(s)\phi(a)\pi(s^*) = \phi(sa). \tag{3.85}$$

In connection with the above definition, the set $\phi(E_{s^*})\mathcal{H} = \{\phi(a)\xi : a \in E_{s^*}, \xi \in \mathcal{H}\}$ is, by Cohen's theorem ([121, p.268]) already a closed linear subspace of $\mathcal{H}$.

Using (3.85) and the fact that β_{s^*s} is the identity map on E_{s^*}, we have

$$\pi(s^*s)\phi(a)\pi(s^*s) = \phi(s^*sa) = \phi(a),$$

and it follows that

$$\pi(s^*s)\phi(a) = \phi(a) = \phi(a)\pi(s^*s). \tag{3.86}$$

(This can also be proved using the fact that $H_s = \phi(E_{s^*})\mathcal{H}$.)

A *representation* of $\mathbf{C}(A, S)$ on a Hilbert space $\mathcal{H}$ is a norm-continuous, *-homomorphism $\Phi : \mathbf{C}(A, S) \rightarrow B(\mathcal{H})$ which is non-degenerate and satisfies the following property: for all $(a, e_1), (a, e_2) \in V(A, S)$ with $e_1, e_2 \in E(S)$, we have

$$\Phi((a, e_1)) = \Phi((a, e_2)). \tag{3.87}$$

The significance of (3.87) is that if a is in two E_e's then Φ does not distinguish between the two (a, e)'s. In particular, it enables us to associate with Φ a well-defined map on A given by $a \rightarrow \Phi((a, e))$. This will give the ϕ when Φ is realized as a covariant pair (ϕ, π) in Proposition 3.3.3.

Proposition 3.3.2   *Let (ϕ, π) be a covariant representation for the covariant system (A, β, S) on a Hilbert space $\mathcal{H}$. Then the map*

$$\Phi : \mathbf{C}(A, S) \rightarrow B(\mathcal{H})$$

given by

$$\Phi((b, s)) = \phi(b)\pi(s) \tag{3.88}$$

for $(b, s) \in V(A, S)$, is a representation of $\mathbf{C}(A, S)$ on $\mathcal{H}$.

Proof.   Let $(b, s), (c, t) \in V(A, S)$. Then $ss^*b = b$, and using (3.86) and (3.85), we have that:

$$
\begin{aligned}
\Phi((b, s))\Phi((c, t)) &= \phi(b)\pi(s)\phi(c)\pi(t) \\
&= \phi(s(s^*b))\pi(s)\phi(c)\pi(t)
\end{aligned}
$$

$$
\begin{aligned}
&= \pi(s)\phi(s^*b)\pi(s^*s)\phi(c)\pi(t) \\
&= \pi(s)\phi(s^*b)\phi(c)\pi(t) \\
&= \pi(s)\phi([(s^*b)c])\pi(s^*s)\pi(t) \\
&= (\pi(s)\phi([(s^*b)c])\pi(s^*))\pi(st) \\
&= \Phi((b,s)(c,t)).
\end{aligned}
$$

We now show that Φ is a *-mapping. Using (3.85) and (3.86) (with s^*, b^* in place of s, a), we then have $\Phi((b,s)^*) = \Phi((s^*b^*, s^*)) = \phi(s^*b^*)\pi(s^*) = \pi(s^*)\phi(b^*)\pi(ss^*) = \pi(s^*)\phi(b^*) = (\phi(b)\pi(s))^* = \Phi((b,s))^*$. So Φ is *-homomorphism.

That Φ is non-degenerate follows since $\Phi((b, ss^*)) = \phi(b)$ (using (3.86)) and ϕ is non-degenerate. Next, since $\|\Phi((b,s))\| = \|\phi(b)\pi(s)\| \leq \|b\| = \|((b,s))\|_1$, it follows that Φ is norm continuous. Lastly, let $(b,e) \in V(A,S)$ with $e \in E(S)$. Using (3.86), we have $\Phi((b,e)) = \phi(b)\pi(e) = \phi(b)$ and this gives (3.87). $\square$

The next result is the converse to Proposition 3.3.2. The interesting proof is due to Sieben ([256, Proposition 5.6]), with some minor modifications needed for the non-unital situation.

Proposition 3.3.3   *Let (A, β, S) be a covariant system. Let Φ be a representation of $\mathbf{C}(A, S)$ on the Hilbert space $\mathcal{H}$. Then there exists a covariant representation (ϕ, π) for the system such that*

$$
\Phi((b,s)) = \phi(b)\pi(s) \tag{3.89}
$$

for $(b, s) \in V(A, S)$.

Proof.   For $e \in E(S)$ and $a \in E_e$, define

$$
\phi(a) = \Phi((a, e)). \tag{3.90}
$$

The map ϕ is well-defined by (3.87). Then $\phi(a)^* = \Phi((a,e))^* = \Phi((a,e)^*) = \Phi((ea^*, e)) = \phi(a^*)$ and $\phi(\lambda a) = \lambda\phi(a)$ for $\lambda \in \mathbf{C}$. Let $f \in E(S)$ and $b \in E_f$. By condition (ii) of Definition 3.3.1, there exists $e_1 \in E(S)$ such that $a, b \in E_{e_1}$. Since $(ab, e_1) = (a, e_1)(b, e_1)$ and $(a+b, e_1) = (a, e_1)+(b, e_1)$ in $\mathbf{C}(A, S)$, we have $\Phi((ab, e_1)) = \Phi((a, e_1))\Phi((b, e_1))$ and $\Phi((a+b, e_1)) = \Phi((a, e_1)) + \Phi((b, e_1))$. Use of (3.90) then gives $\phi(ab) = \phi(a)\phi(b)$ and $\phi(a+b) = \phi(a) + \phi(b)$. So ϕ is a *-homomorphism from A into $B(\mathcal{H})$. Since $\|\phi(a)\| \leq \|\Phi\|\|a\|$, it follows that ϕ is continuous. Using (iii) of Definition 3.3.1, the map ϕ extends to a homomorphism, also denoted by ϕ, from A into $B(\mathcal{H})$.

For $s \in S$ recall that $H_s = \phi(E_{s^*})\mathcal{H}$. Then $H_s = H_{ss^*}$. Let $s \in S$ and $\{p_\delta\}$ be a positive bounded approximate identity for E_s. We claim that the strong operator limit $\lim_\delta \Phi((p_\delta, s))$ exists and is a partial isometry with initial subspace H_s and final subspace H_{s^*}. We will take

$$\pi(s) = \lim_\delta \Phi((p_\delta, s)). \tag{3.91}$$

To this end, let $\xi \in H_s$. Then for some $\eta \in \mathcal{H}$ and some $a \in E_{s^*}$, we have $\xi = \phi(a)\eta \ (= \Phi((a, s^*s))\eta)$. Then

$$\Phi((p_\delta, s))\xi = \Phi((s[(s^*p_\delta)a], ss^*s))\eta \rightarrow \Phi((sa, s))\eta$$

since $\{s^*p_\delta\}$ is a bounded approximate identity for $s^*(E_s) = E_{s^*}$. Now suppose that $\xi_1 \in H_s^\perp$. Then for all $b \in E_{s^*}$, $\phi(b)\xi_1 = 0$ since $\langle \xi_1, \phi(b^*)\mathcal{H} \rangle = \{0\}$. Using the former equality with $b = s^*p_\delta^{1/2} \in E_{s^*}$, as well as the homomorphism property of Φ, we have

$$
\begin{aligned}
\Phi((p_\delta, s))\xi_1 &= \Phi(s[(s^*p_\delta^{1/2})(s^*p_\delta^{1/2})], s(s^*s))\xi_1 \\
&= \Phi(((s^*p_\delta^{1/2}), s)((s^*p_\delta^{1/2}), s^*s))\xi_1 \\
&= \Phi(((s^*p_\delta^{1/2}), s))\phi(s^*p_\delta^{1/2})\xi_1 \\
&= 0.
\end{aligned}
$$

So the strong operator limit of (3.91) exists and $\pi(s)$ is defined. Note that

$$\pi(s)(\phi(a)\eta) = \Phi((sa, s))\eta \tag{3.92}$$

for all $a \in E_{s^*}, \eta \in \mathcal{H}$. Then (3.92) defines $\pi(s)$ on H_s, and $\pi(s)$ so defined is independent of the choice of $\{p_\delta\}$ and vanishes on $H_s^\perp$. It follows from the above proof using (3.92) that if $e \in E(S)$, then $\pi(e)$ is the orthogonal projection of $\mathcal{H}$ onto H_e.

We now show that $\pi(s)$ is a partial isometry with initial subspace H_s and that π is $*$-preserving. In the earlier notation, $\pi(s)(\phi(a)\eta) = \Phi((sa, s))\eta = \lim_\delta \Phi(((sa)p_\delta, s))\eta = \lim_\delta \Phi((sa, ss^*)(p_\delta, s))\eta$. Since Φ is a homomorphism and $\Phi((sa, ss^*)) = \phi(sa)$ with $sa \in E_s$, it follows that $\pi(s)(H_s) \subset H_{s^*}$. Also, with $\{q_\sigma\}$ a positive bounded approximate identity for E_{s^*}, we have $\Phi((q_\sigma, s^*))\pi(s)(\phi(a)\eta) = \Phi((q_\sigma, s^*))\Phi((sa, s))\eta = \Phi((s^*[(sq_\sigma)(sa)], s^*s))\eta \rightarrow \Phi((a, s^*s))\eta = \phi(a)\eta$ since $\{sq_\sigma\}$ is a bounded approximate identity for $s(E_{s^*}) = E_s$. Using the $\{q_\sigma\}$ to define $\pi(s^*)$, we obtain

$$\pi(s^*)\pi(s) = \pi(ss^*). \tag{3.93}$$

Next,

$$\Phi((p_\delta, s))^* = \Phi((s^*p_\delta, s^*)). \tag{3.94}$$

Using $\{s^*p_\delta\}$ to define $\pi(s^*)$ and the facts that the weak operator topology is weaker than the strong operator topology and that the $*$-operation is continuous in the weak operator topology, we obtain from (3.94) that $\pi(s^*) = \pi(s)^*$. So π is a *-map and, using (3.93), $\pi(s)$ is a partial isometry with initial subspace H_s.

We now show that π is a homomorphism. Let $s, t \in S$. We first show that $\pi(s)\pi(t)$ and $\pi(st)$ coincide on H_{st}. We will then show that $\pi(s)\pi(t)$ vanishes on $H_{st}^\perp$ and this will give $\pi(s)\pi(t) = \pi(st)$.

Let $b \in E_{(st)^*}, \eta \in \mathcal{H}$. Then by (3.81), $tb \in E_{s^*} = s^* E_s$. So $stb \in E_s$ and $tb = s^*(stb)$. Then using the definition of π given by (3.91), we have

$$\begin{aligned}
\pi(s)\pi(t)(\phi(b)\eta) &= \pi(s)\Phi((tb,t))\eta \\
&= \lim_\delta \Phi((p_\delta, s)(tb,t))\eta \\
&= \lim_\delta \Phi((s[(s^*p_\delta)tb], st))\eta \\
&= \lim_\delta \Phi((s[(s^*p_\delta)(s^*(stb))], st))\eta \\
&= \lim_\delta \Phi((p_\delta st(b), st))\eta \\
&= \pi(st)(\phi(b)\eta).
\end{aligned}$$

So $\pi(s)\pi(t)$ and $\pi(st)$ coincide on H_{st}.

We now show that $\pi(s)\pi(t)$ vanishes on $H_{st}^\perp$. Let $\xi_1 \in H_{st}^\perp$. As earlier, $\pi(st)\xi_1 = 0$. Let $\{w_\mu\}$ be a bounded approximate identity for E_t. Then using the joint continuity of multiplication on bounded sets for the strong operator topology, we have

$$\begin{aligned}
\pi(s)\pi(t)\xi &= \lim_{\delta,\mu} \Phi((p_\delta, s)(w_\mu, t))\xi_1 \\
&= \lim_{\delta,\mu} \Phi((s[(s^*p_\delta)w_\mu], st))\xi_1. \qquad (3.95)
\end{aligned}$$

Fix δ, μ for the present. Now by (3.81), $s[(s^*p_\delta)w_\mu] \in s(E_{s^*}E_t) = E_{st}$. Since E_{st} has a bounded approximate identity, Cohen's factorization theorem applies to give that there exist c, d (depending on δ, μ) in E_{st}, such that $s[(s^*p_\delta)w_\mu] = cd$. Then $\Phi((s[(s^*p_\delta)w_\mu], st))\xi_1 = \Phi((cd, st))\xi_1 = \Phi((c, st))((st)^*(d), (st)^*(st))\xi_1 = \{0\}$ since the fact that $H_{st} = \phi(E_{(st)^*})\mathcal{H}$ gives that for $\xi_1 \in H_{st}^\perp$, we have $\phi(d')\xi_1 = 0$ for all $d' \in \phi(E_{(st)^*})$. (Take $d' = (st)^*(d)$ above.) So π is a homomorphism.

Next, we need to show that for $b \in E_s$, we have

$$\Phi((b, s)) = \phi(b)\pi(s).$$

Taking $e = ss^*$, we have

$$\begin{aligned}
\phi(b)\pi(s) &= \lim_\delta \Phi((b, ss^*))\Phi((p_\delta, s)) \\
&= \lim_\delta \Phi((e[(eb)p_\delta], ss^*s)) \\
&= \Phi((b, s))
\end{aligned}$$

as required. The covariant condition (3.85) for the pair (ϕ, π) proceeeds as follows:

$$
\begin{aligned}
\pi(s)\phi(a)\pi(s^*) &= \lim_\delta \Phi((p_\delta, s))\Phi((a, s^*)) \\
&= \lim_\delta \Phi((s[(s^* p_\delta)a], ss^*)) \\
&= \Phi((sa, ss^*)) = \phi(sa).
\end{aligned}
$$

It remains finally to show that both ϕ and π are non-degenerate. Let $\xi \in \mathcal{H}$ be such that $\pi(S)\xi = \{0\}$. By (3.89), $\Phi((b, s))\xi = 0$ for all $(b, s) \in V(A, S)$. Since Φ is non-degenerate, it follows that $\xi = 0$. So π is non-degenerate. Next suppose that $\phi(A)\eta = 0$ for some $\eta \in \mathcal{H}$. Let $(b, s) \in V(A, S)$ and $a = s^*b, e = s^*s$. Now $\phi(a) = \Phi((a, e))$, and arguing as above, we have $\Phi((b, s))\eta = \Phi((sa, s))\eta = \pi(s)(\phi(a)\eta) = 0$. So $\eta = 0$ and the non-degeneracy of ϕ now follows.   □

Corollary 3.3.1   *Let (A, β, S) be a covariant system. Then there is a one-to-one correspondence, given by (3.89), between the representations Φ of $\mathbf{C}(A, S)$ and the covariant pairs (ϕ, π).*

We now discuss some simple but useful facts about the action associated with a localization (X, S). (These facts, presented in this paragraph, do not depend on the set of D_e's being a basis for X and will be used in the non-localization context of **4.3**.) Recall that $x \leq s$ means that $x \in D_s$. Recall also that if $e \in E(S)$, then (Proposition 2.1.1) $ses^* \in E(S)$. Note that $x \leq s$ if and only if $x \leq ss^*$. If $x \leq s$ and $e \in E(S)$, then $x.s \leq e \Leftrightarrow x.se = x.s \Leftrightarrow x.(ses^*) = x.ss^* = x$. So

$$x.s \leq e \Leftrightarrow x \leq ses^*. \tag{3.96}$$

Putting $f = ses^*$ in (3.96) gives that for $x \leq s$,

$$x \leq f \Leftrightarrow x.s \leq s^* f s. \tag{3.97}$$

The next proposition shows that in the case of the inverse semigroup covariant system associated with an extendible localization (X, S) ($A = C_0(X)$), the covariant condition on a pair of representations (ϕ, π) assumes a more manageable form that relates readily, as we shall see, to the corresponding notion of covariance for r-discrete groupoids. This proposition is the localization parallel to Proposition 3.2.8.

Proposition 3.3.4   *Let (X, S) be an extendible localization. Then the covariant pairs (ϕ, π) on a Hilbert space $\mathcal{H}$ are canonically given by pairs (P, π) where P is a resolution of the identity on $\mathcal{B}(X)$ for the Hilbert space $\mathcal{H}$, π is a representation of S on $\mathcal{H}$ and for all $e \in E(S)$, we have $P(D_e) = \pi(e)$.*

Proof. Suppose that the pair (P, π) is as in the above statement. Define $\phi : B(X) \to B(\mathcal{H})$ by:

$$\phi(f) = \int_X f \, dP. \tag{3.98}$$

Then ϕ, and its restriction (also denoted by ϕ) to $C_0(X)$, are representations. To prove (3.85), we have, for $s \in S$ and $e \in E(S)$ with $e \leq s^*s$, that

$$\pi(s)\phi(\chi_{D_e})\pi(s^*) = \pi(ses^*) = \phi(\chi_{D_{ses^*}})$$

so that

$$\pi(s)\phi(\chi_{D_e})\pi(s^*) = \phi(s\chi_{D_e}), \tag{3.99}$$

where we have used (3.96) and where the action of s on $C_0(D_{s^*})$ is extended to $B(D_{s^*})$ in the obvious way. Now the set of such D_e's is a basis for X. By using (3.98) and the result on monotone classes, it follows that (3.99) holds with D_e replaced by any Borel subset W of D_{s^*}. Approximating uniformly any $a \in C_0(D_{s^*})$ by linear combinations of such functions χ_W, we obtain that (ϕ, π) satisfies (3.85). Now $\phi(\chi_{D_{s^*s}})$ is the strong operator limit of a sequence in $\phi(C_0(D_{s^*}))$, and a simple argument gives that the space $\phi(C_0(D_{s^*}))\mathcal{H}$ is the initial subspace of the orthogonal projection $\phi(\chi_{D_{s^*s}}) = \pi(s^*s)$. So the pair (ϕ, π) is covariant.

For the converse, let the pair (ϕ, π) be covariant on a Hilbert space $\mathcal{H}$. Then ϕ determines a canonical resolution of the identity P on $\mathcal{B}(X)$ satisfying (3.98). Let $e \in E(S)$. Extend ϕ canonically to $B(D_e)$. As above, $P(D_e) = \phi(\chi_{D_e})$ is the orthogonal projection onto $\phi(C_0(D_e))\mathcal{H}$ which (by the definition of "covariant pair") is the projection onto H_e, i.e. is $\pi(e)$. $\square$

Let (A, β, S) be a covariant system. We now define a seminorm on $\mathbf{C}(A, S)$ by setting, for $\theta \in \mathbf{C}(A, S)$,

$$\|\theta\| = \sup_\Phi \|\Phi(\theta)\|$$

where Φ ranges over the representations of $\mathbf{C}(A, S)$ (which, by the above, are the same as the covariant pairs (ϕ, π)). Then $\|.\|$ is a C^*-seminorm on $\mathbf{C}(A, S)$ and so gives a C^*-norm on the quotient of $\mathbf{C}(A, S)$ by the ideal $\{\theta : \|\theta\| = 0\}$. The completion of this quotient is by definition the enveloping C^*-algebra $A \times_\beta S$, and is called the *covariance C^*-algebra* (for (A, β, S)). It follows from Proposition 3.3.5 later that, for groupoid reasons, in the case of an additive localization, $\|.\|$ is actually a norm on $\mathbf{C}(C_0(X), S)$ so that we just need to complete the latter to get $A \times_\beta S$.

The next result implies that if G is an r-discrete groupoid and S is an additive inverse subsemigroup of G^{op} (**3.2**), then $C^*(G)$ is isomorphic to the covariance C^*-algebra $C_0(G^0) \times_\beta S$ for the canonical localization (G^0, S).

Theorem 3.3.1   *Let G be an r-discrete groupoid and S be an additive inverse subsemigroup of G^{op}. Then the triple $(C_0(G^0), \beta, S)$ is a covariant system, and $C^*(G)$ is canonically isomorphic to $C_0(G^0) \times_\beta S$.*

Proof.   Let $X = G^0$ and consider the localization (X, S). Since S is additive as an inverse subsemigroup of G, it follows that (X, S) is an extendible localization, and hence by Proposition 3.3.1, the triple $(C_0(X), \beta, S)$ is a covariant system. Recall that the canonical action of S on X is given by $s \rightarrow \alpha_s$ where $\alpha_s(u) = s^{-1}us$. Set $V(X, S) = V(C_0(X), S)$.

We now relate $V(X, S)$ to $C_c(G)$. For $(f, s) \in V(X, S)$, define $f_s : G \rightarrow \mathbf{C}$ by setting $f_s(x) = f(r(x))$ $(x \in s)$ and $f_s(w) = 0$ for all $w \in G \sim s$. Then $f_s \in C_c(s)$. Let $(f, s), (g, t) \in V(X, S)$. Then by (2.30), $f_s * g_t \in C_c(st)$ and for $y = ab, a \in s, b \in t$, we have $f_s * g_t(y) = f_s(a)g_t(b) = f(r(a))g(r(b))$. Now $(f(sg)), st) \in V(X, S)$, and using (3.52), we have $(f(sg))_{st}(y) = f(r(ab))(sg)(r(ab)) = f(r(a))g(\alpha_s(r(a))) = f(r(a))g(d(a)) = f(r(a))g(r(b))$, and it follows that $f_s * g_t = (f(sg))_{st}$. Similarly, $(s^*\bar{f})_{s^*} = f_s^*$. Since the map $f \rightarrow f_s$ for $f \in C_0(D_s)$ is also linear and $\mathbf{C}(C_0(X), S)$ is generated by $V(X, S)$ subject to the relations given by the linearity of the maps $f \rightarrow (f, s)$, it follows using (3.83) and (3.84) that the map $(f, s) \rightarrow f_s$ extends to a *-homomorphism $\Delta : \mathbf{C}(C_0(X), S) \rightarrow C_c(G)$.

The map Δ is onto $C_c(G)$. Indeed, suppose that U is an open Hausdorff subset of G and $F \in C_c(U) \subset C_c(G)$. If $U = s \in S$, then $F = f_s = \Delta((f, s))$, where $f(u) = F(r_s^{-1}(u))$ for $u \in r(U)$. Since S is a basis for the topology of G, a partition of unity argument gives for general U that F is a finite sum of elements of the form $\Delta((g_i, s_i))$. The isomorphism $C_0(X) \times_\beta S \cong C^*(G)$ will follow once we have shown that Δ is isometric for the C^*-norms on $\mathbf{C}(C_0(X), S)$ and $C_c(G)$ respectively. (As defined earlier, we only had a C^*-seminorm for $\mathbf{C}(C_0(X), S)$; however, once we have shown that Δ is isometric, it will follow that this seminorm is actually a norm.)

Let Φ be a representation of $\mathbf{C}(C_0(X), S)$ on a Hilbert space $\mathcal{H}$. By Proposition 3.3.3 and Proposition 3.3.4, Φ is given by a covariant pair (ϕ, π) through (3.88), and ϕ is determined by integrating a resolution of the identity P on X such that $P(D_e) = \pi(e)$ for all $e \in E(S)$. By Proposition 3.2.8 and Theorem 3.2.1, the pair (P, π) determines (**3.2**) a representation π' of the multiplier algebra $M(B_c(G)) \supset C_c(G)$. Further, π' restricted to $C_0(X)$ coincides with ϕ, and when restricted to $S = \{\chi_s : s \in S\}$, coincides with π. Also, for $x \in s, f \in C_c(X)$, we have, by (3.56), $f * \chi_s(x) = f(xs^{-1}) = f(r(x))$ and $f * \chi_s(x) = 0$ if $x \in G \sim s$. It follows that $f * \chi_s = f_s$. Applying π', we have for $(f, s) \in V(X, S)$ that

$$\pi'(f_s) = \pi'(f)\pi'(\chi_s) = \phi(f)\pi(s) = \Phi((f, s)).$$

So π' restricted to $C_c(G)$, composed with Δ, equals Φ.

Conversely, if Π is a representation of $C_c(G)$, then it defines, again by Proposition 3.2.8 and Theorem 3.2.1, a pair (P, π_1) where P is a resolution of the identity on X, and π_1 is a representation of S coinciding with P on $E(S)$. Let ϕ be the integrated form of P on $C_0(X)$. By Proposition 3.3.4, the pair (ϕ, π_1) is covariant. By (3.79), $\Pi(f_s) = \phi(f)\pi_1(s) = \Phi((f,s))$ where Φ is the representation of $\mathbf{C}(C_0(X), S)$ given by (3.88). So $\Pi \circ \Delta = \Phi$. It follows that Δ is an isomorphism from $C^*(G)$ onto $C_0(G^0) \times_\beta S$. $\square$

Our next objective is to establish a converse to Theorem 3.3.1, i.e. to show that, under appropriate conditions, the covariance C^*-algebra associated with a localization (X, S) is the C^*-algebra of a certain r-discrete groupoid. In the case of a foliation localization, this groupoid will be the holonomy groupoid for the foliation, i.e. the sheaf groupoid of germs of the pseudogroup maps.

In general, it could happen for an extendible localization (X, S) that different e's in $E(S)$ have the same domains. This cannot happen for an inverse subsemigroup of G^{op} acting on G^0. In order to relate localizations to r-discrete groupoids, we thus replace (X, S) by an extendible localization (X, S_1) which has the same covariance C^*-algebra as (X, S) but for which the map $e \to D_e$ on $E(S_1)$ *is* one-to-one.

More precisely, let R be the congruence (**2.1**) on S generated by the following relation ρ on S: for $s, t \in S$, we write $s\rho t$ if $s, t \in E(S)$ and $D_s = D_t$. Let S_1 be the inverse semigroup S/R. Since the anti-homomorphism $s \to \beta_s$ factors through R, we have that S_1 also has the natural right action on X with a corresponding left action β_1 on $C_0(X)$. Also, (X, S_1) is a localization, which is extendible if (X, S) is extendible. Further, from Proposition 3.3.4, if (ϕ, π) is a covariant representation for $(C_0(X), S, \beta)$ on the Hilbert space $\mathcal{H}$, then, since the initial subspace H_e of $\pi(e)$ is $\phi(C_0(D_e))\mathcal{H}$ for $e \in E(S)$, we have $\pi(e) = \pi(f)$ if eRf in $E(S)$, and so π also factors through R and thus defines a representation π_1 of S_1 that factors through the natural homomorphism from $\mathbf{C}(C_0(X), S)$ onto $\mathbf{C}(C_0(X), S_1)$. It follows that this homomorphism gives an isomorphism from $C_0(X) \times_\beta S$ onto $C_0(X) \times_{\beta_1} S_1$. The upshot is that for the consideration of covariance C^*-algebras associated with extendible localizations, we can suppose that the map $e \to D_e$ is an isomorphism on $E(S)$, i.e. that we can identify $e \in E(S)$ with D_e. *We will assume that this map is an isomorphism for all future localizations.* Extendibility for S then assumes the form that if $e, f \in E(S)$, then $e \cup f \in E(S)$.

Let (X, S) be a localization. We will write $E = E(S)$. We want to relate the pair (X, S) to an r-discrete groupoid $G(X, S)$. We construct a "first approximation" Ξ to $G(X, S)$ motivated by how one constructs the transformation groupoid for a group acting on a locally compact space.

Let Ξ be the subset of $X \times S$ given by:

$$\Xi = \{(x,s) : x \in D_s, s \in S\}.$$

The set Ξ^2 of "composable pairs" in Ξ is defined as the set of pairs

$$((x,s),(x.s,t))$$

in $\Xi \times \Xi$, and for such a pair, we define the product :

$$(x,s)(x.s,t) = (x,st). \tag{3.100}$$

The set $\{(x,e) : x \in D_e, e \in E\}$ is denoted by Ξ^0. An inverse map $a \to a^{-1}$ on Ξ is defined by

$$(x,s)^{-1} = (x.s, s^*). \tag{3.101}$$

As discussed in Chapter 1, the groupoid axioms do not hold in general for Ξ, and it is necessary to take the quotient of Ξ by an equivalence relation $\sim$ which will identify $(x.s, s^*st)$ with $(x.s, t)$ and (x, stt^*) with (x, s). (See (1.14).) To this end, we define $(x,s) \sim (y,t)$ to mean that $x = y$ and there exists $e \in E$ such that $x \le e$ (i.e. $x \in D_e$) and $es = et$. By multiplying e on the left by $(ss^*)(tt^*)$, we can suppose whenever we want that $e \le (ss^*)(tt^*)$. The following lemma shows that $\sim$ is indeed an equivalence relation.

Lemma 3.3.1 *The relation $\sim$ is an equivalence relation on Ξ.*

Proof. Let $(x,s) \in \Xi$. Then using $e = ss^*$ we obtain $(x,s) \sim (x,s)$. So $\sim$ is reflexive. The symmetry of $\sim$ is obvious. Now suppose that $(y,s) \sim (y,t)$ and $(y,t) \sim (y,w)$ in Ξ. Let $e_1, e_2 \in E$ be such that $y \le e_1, e_2$ and $e_1s = e_1t$, $e_2t = e_2w$. Let $f = e_1e_2 \ (= e_2e_1)$. Then $y \le f$, and $fs = e_1(e_2t) = fw$. Thus $\sim$ is an equivalence relation on Ξ. $\square$

Let $G(X,S) = \Xi/\sim$ and $\psi : \Xi \to G(X,S)$ be the quotient map. For each $s \in S$, let $\psi_X(s) = \{\psi((x,s)) : x \in D_s\} \subset G(X,S)$. The equivalence class $\psi((x,s))$ of $(x,s) \in \Xi$ will also be denoted by $\overline{(x,s)}$.

The product and inversion operations in $G(X,S)$ are those inherited from Ξ. We will show below that these are well-defined. We note here the following useful result: if $(y,s_1) \sim (y,s_2)$ in Ξ, then $y.s_1 = y.s_2$. Indeed, there exists $e_1 \in E(S)$ such that $y \le e_1$ and $e_1s_1 = e_1s_2$. Then

$$y.s_1 = y.e_1s_1 = y.e_1s_2 = y.s_2. \tag{3.102}$$

The next theorem is fundamental for this work: it shows that $G(X,S)$ is an r-discrete groupoid with ψ_X a homomorphism from S into G^{op}. We will use it again in Chapter 4 (Theorem 4.3.1) to establish the existence of the universal groupoid for an inverse semigroup. Another interesting application of the theorem is in constructing a C^*-algebra of observables for quantum mechanical systems on a tiling (**4.2**, Example 5).

Theorem 3.3.2 *Let (X, S) be a localization. Then $G(X, S)$ is an r-discrete groupoid with the following operations and topology. The composable pairs are those of the form $(\overline{(x, s)}, \overline{(x.s, t)})$, and the product and involution on $G(X, S)$ are given by:*

$$(\overline{(x, s)}, \overline{(x.s, t)}) \to \overline{(x, st)}, \quad \overline{(x, s)} \to \overline{(x.s, s^*)}.$$

A basis for the topology on $G(X, S)$ is given by the family of sets of the form $D(U, s)$, where $s \in S$, U is an open subset of D_s, and

$$D(U, s) = \{\overline{(x, s)} : x \in U\}.$$

The unit space $G(X, S)^0$ is canonically identified with X, the map ψ_X, where $\psi_X(s) = \{\overline{(x, s)} : x \in D_s\}$, is a homomorphism from S into G^{op}, and $\psi_X(S)$ is a basis for $G(X, S)$.

Proof. We first prove that $G = G(X, S)$ is a groupoid under the structure inherited from Ξ.

The composable pairs in G will be those of the form $(\overline{z}, \overline{w})$ with $(z, w) \in \Xi^2$ and the product of such a pair will be $\overline{zw}$. The inverse will be the map $\overline{z} \to \overline{z^{-1}}$. Note that once we have made sense of these operations, then, since $(z^{-1})^{-1} = z$ and $(z, z^{-1}) \in \Xi^2$, we will automatically have $(\overline{z^{-1}})^{-1} = \overline{z}$ and $(\overline{z}, \overline{z^{-1}}) \in G^2$.

To establish that the product and inversion make sense and give a groupoid structure on G, it is sufficient, from the axioms of a groupoid ((1.10), (1.11)), to show that in Ξ:

(α) if $z_1 \sim z_2$, $w_1 \sim w_2$ and $(z_1, w_1) \in \Xi^2$, then $(z_2, w_2) \in \Xi^2$ and $z_1 w_1 \sim z_2 w_2$;

(β) if $z_1 \sim z_2$ then $z_1^{-1} \sim z_2^{-1}$;

(γ) if $(\overline{z}, \overline{w})$, $(\overline{w}, \overline{t}) \in G^2$, then both $(\overline{zw}, \overline{t})$, $(\overline{z}, \overline{wt}) \in G^2$ and $(\overline{zw})\overline{t} = \overline{z}(\overline{wt})$.

(δ) if (v, w) belongs to Ξ^2, then $v^{-1}(vw) \sim w$ and $(vw)w^{-1} \sim v$.

Starting with (α), let $z_1 = (y, s_1)$, $z_2 = (y, s_2)$, $w_1 = (y.s_1, t_1)$ and $w_2 = (y.s_1, t_2)$ where $z_1 \sim z_2$ and $w_1 \sim w_2$. Since $z_1 \sim z_2$, we have by (3.102) that $y.s_1 = y.s_2$ so that $(z_2, w_2) \in \Xi^2$. By definition of $\sim$, there also exist $e_1, e_2 \in E$ such that $y \leq e_1$, $y.s_1 = y.s_2 \leq e_2$ and $e_1 s_1 = e_1 s_2$, $e_2 t_1 = e_2 t_2$. Using (3.96), we have $y \leq (s_1 e_2 s_1^*)(s_2 e_2 s_2^*)$. We can also take $e_1 \leq (s_1 e_2 s_1^*)(s_2 e_2 s_2^*)$ and (since $y.s_2 \leq e_2$) take $e_2 \leq s_2^* s_2$. Then $e_1 s_1 t_1 = (e_1 s_2) t_1 = e_1 s_2 e_2 s_2^* s_2 t_1 = e_1 s_2 e_2 t_1 = e_1 s_2 e_2 t_2 = e_1 s_2 e_2 s_2^*(s_2 t_2) = e_1 s_2 t_2$. Since each $z_i w_i = (y, s_i t_i)$, it follows that $z_1 w_1 \sim z_2 w_2$ and (α) is proved. We thus have a well-defined product $(\overline{z}\,\overline{w}) \to \overline{zw}$ on $G^2 = \{(\overline{z}, \overline{w}) : (z, w) \in \Xi^2\}$.

Turning to (β), let z_1, z_2, and e_1 be as above. Then $z_i^{-1} = (y.s_i, s_i^*)$. We can take $e_1 \leq (s_1 s_1^*)(s_2 s_2^*)$. Let $g = (s_1^* e_1 s_1)(s_2^* e_1 s_2) \in E(S)$. Then using (3.97), $y.s_1 = y.s_2 \leq g$, and $gs_1^* = (s_2^* e_1 s_2)(s_1^* e_1 s_1)s_1^* = (s_2^* e_1 s_2)(s_1^* e_1) = (s_2^* e_1 s_2)(e_1 s_1)^* = (s_2^* e_1 s_2)(e_1 s_2)^* = s_2^* e_1 s_2 s_2^* e_1 = s_2^* e_1$. Similarly, $gs_2^* = s_1^* e_1$ and so $gs_1^* = gs_2^*$ since $s_2^* e_1 = (e_1 s_2)^* = (e_1 s_1)^* = s_1^* e_1$. So $z_1^{-1} \sim z_2^{-1}$.

(γ) and (δ) follow using (α), (β) and the corresponding "groupoid" conditions (1.10) and (1.14) for Ξ.

Let $\mathcal{B} = \{D(U, s) : U \text{ open in } D_s, s \in S\}$. We first show that $\mathcal{B}$ is a basis for a topology on G. Obviously, $\bigcup \mathcal{B} = G$. Let $h \in D(U, s) \cap D(V, t)$. Then for some $y \in U \cap V$, we have $h = \overline{(y, s)} = \overline{(y, t)}$. Further, there exists $e \in E$ with $y \leq e$ and $es = et$. Let $Z = U \cap V \cap D_e$. Then

$$h \in D(Z, s) = D(Z, t) \subset D(U, s) \cap D(V, t). \tag{3.103}$$

So $\mathcal{B}$ is a basis for a topology $\mathcal{T}$ on G.

Next we show that $(G, \mathcal{T})$ is a topological groupoid (**2.2**). Suppose that $\overline{z_\delta} \to \overline{z}$, $\overline{w_\delta} \to \overline{w}$ in G, where $(\overline{z_\delta}, \overline{w_\delta})$, $(\overline{z}, \overline{w}) \in G^2$. We can write $z_\delta = (y_\delta, s_\delta)$, $w_\delta = (y_\delta.s_\delta, t_\delta)$, $z = (y, s)$ and $w = (y.s, t)$. By considering neighborhoods $D(U, s)$, $D(V, t)$ of $\overline{z}$, $\overline{w}$, we see that $y_\delta \to y$, $y_\delta.s_\delta \to y.s$ and eventually, $(y_\delta, s_\delta) \sim (y_\delta, s)$ and $(y_\delta.s_\delta, t_\delta) \sim (y_\delta.s_\delta, t)$. By (3.102), we have $y_\delta.s_\delta = y_\delta.s$, and we can take $z_\delta = (y_\delta, s), w_\delta = (y_\delta.s, t)$. Then $z_\delta w_\delta = (y_\delta, st)$ and $zw = (y, st)$. Let W be an open neighborhood of $\overline{zw}$ in G. Then there exists an open neighborhood Z of y in Y such that $D(Z, st) \subset W$. So $\overline{z_\delta w_\delta} \in D(Z, st)$ eventually, and $\overline{z_\delta w_\delta} \to \overline{zw}$. Noting that $z_\delta^{-1} = (y_\delta.s, s^*)$ and $z^{-1} = (y.s, s^*)$, a similar argument gives $\overline{z_\delta^{-1}} \to \overline{z^{-1}}$. So $(G, \mathcal{T})$ is a topological groupoid.

The unit space of G is $r(G)$, the set of elements $\overline{(x, e)}$ where $e \in E$ and $x \in D_e$. If we also have $x \in D_f$ for some $f \in E(S)$, then since $(ef)e = (ef)f$, we have $\overline{(x, e)} = \overline{(x, f)}$, and so the map $x \to \overline{(x, e)}$ is well-defined and bijective. Using this map, we identify X with G^0. Now if $\overline{(x, e)} \in D(U, s)$ for some U and s, then for some $f \in E$, we have $x \leq f$ and $fe = fs$. Replacing U by a smaller open neighborhood of x, we can suppose that $f = s$ and then the identification of X with G^0 identifies U with $D(U, s)$, i.e. $G^0 = X$ as a topological space as well. In particular, the unit space of G is locally compact Hausdorff (and is open in G since it is a union of open sets $D(U, e)$).

A similar argument to that which identified X with G^0 topologically shows that if U is an open subset of D_s, then the map $x \to \overline{(x, s)}$ is a homeomorphism from U onto $D(U, s)$. Since S is countable and X is second countable, we obtain a countable basis for the topology of G consisting of sets of the form $D(U, s)$. Also by requiring U to be the interior of a compact subset of X, we can take each $D(U, s)$ to be the interior of a compact subset of G. Next, by identifying $D(D_s, s)$ with D_s, U with $D(U, ss^*)$ and $U.s$

with $D(U.s, s^*s)$, we see that r and d, restricted to $D(U, s)$, are respectively the identity map and the map $u \rightarrow u.s$, each of which is a homeomorphism. Hence $D(U, s) \in G^{op}$. Let $\mathcal{C}$ be the family of these sets $D(U, S)$. From the discussion preceding Proposition 2.2.4, the topological groupoid G is r-discrete.

Lastly, easy calculations show that $D(D_s, s)D(D_t, t) = D(D_{st}, st)$ so that the map $\psi_X : S \rightarrow G^{op}$ is a homomorphism. Since the family of sets D_s is a basis for X, it follows that $\psi_X(S)$ is a basis for the topology of G.
□

Example

Of special interest is the case of a localization (X, S) in the original sense of Kumjian, i.e. the case where S is given as an inverse semigroup of partial homeomorphisms on X. In that case, in the above notation, the pairs $(x, s), (x, t)$ are identified if and only if s and t coincide in a neighborhood of x, so that $G(X, S) = \Xi/\sim$ is just the set of germs of the maps s ($s \in S$). The topology on $G(X, S)$, given by the $D(U, s)$, is just the germ topology on $G(X, S)$ so that $G(X, S)$ *is the sheaf of germs defined by the maps $s \in S$.* This sheaf is an r-discrete groupoid with product and involution given in the statement of Theorem 3.3.2. In particular, when S is a pseudogroup for a foliated manifold, then the groupoid $G(X, S)$ is the r-discrete holonomy groupoid in the form of G_S^{hol}, discussed earlier in **2.3**, Example 2.

We are particularly interested in a condition which will ensure that the homomorphism ψ_X of Theorem 3.3.2 is an isomorphism into $G(X, S)^{op}$ and that $\psi_X(S)$ is additive in $G(X, S)^{op}$. This will enable us to relate the representation theory of additive inverse semigroups associated with an r-discrete groupoid, developed in **3.2**, to localization covariance C^*-algebras.

This condition, which ensures additivity at the groupoid level, will appropriately be called additivity at the localization level. We first define the notion of *compatibility* for partial homeomorphisms on X.

Let (X, S) be a localization and $s, t \in S$. Recall that each $e \in E(S)$ is identified with D_e. Compatibility is a condition which will enable us to "combine together" s and t rather as we did earlier with the additive condition in the groupoid context. More precisely, let us say that the pair (s, t) is *r-compatible* if for all $x \in ss^* \cap tt^*$, there exists $e \in E(S)$ with $x \leq e$ and $es = et$. We say that the pair (s, t) is *compatible* if both pairs $(s, t), (s^*, t^*)$ are r-compatible.[9] Thinking of compatibility for G^{op}-sets in the groupoid context, what it means is that the range and source maps, r, d, are one-to-one on $s \cup t$ so that $s \cup t \in G^{op}$. When S is given as an

---

[9]Compatibility is effectively the same as Kumjian's notion of *coherence* in [148].

inverse semigroup of partial homeomorphisms on X, then the pair (s,t) is compatible if and only if there exists a partial homeomorphism $w : D_s \cup D_t \to D_{s^*} \cup D_{t^*}$ such that $w_{|D_s} = s$ and $w_{|D_t} = t$.

To capture additivity for a localization, we still need to formulate the existence and uniqueness of "$s \cup t$" in S. To this end, define a relation $\sim$ on S by setting $s \sim t$ in S if $ss^* = tt^*$ and for every $x \in ss^*$, there exists $e \in E(S)$ such that $x \in e(= D_e)$ and $es = et$. The relation $\sim$ on S is easily checked to be a congruence on S. (The reader is invited to show that $S/\sim$ has a natural right action on X inherited from S, and that the localization $(X, S/\sim)$ satisfies (i) below.) In the foliation context, the additive requirement below is Plante's fourth condition for the definition of a *pseudogroup* in [207, p.329].

We define the localization (X, S) to be *additive* when (i) and (ii) hold:

(i) if $s \sim t$ then $s = t$ (i.e. $\sim = \{(s,s) : s \in S\}$);

(ii) if the pair (s,t) is compatible and $f = ss^* \cup tt^* \subset X$, then there exists $w \in S$ such that $f = ww^*$ and if $x \leq ss^*$ (or $x \leq tt^*$), then there exists $e \in E(S)$ such that $x \leq e$ and $es = ew$ (or $et = ew$).

What (ii) above is expressing in the groupoid context is that there is a candidate w for $s \cup t$. What (i) then expresses effectively is that this candidate is *unique*. (For if w' was another candidate, we would have $w \sim w'$ whence $w = w'$.) All this will be made more precise in the next result. Note for the present however that an additive localization is always extendible since (e, f) is compatible for any $e, f \in E(S)$.

When a localization (X, S) is such that S is given as an inverse semigroup of partial homeomorphisms on X, then the inverse semigroup S can always be extended to a larger inverse semigroup T of partial homeomorphisms on X such that (X, T) is an additive localization. Indeed, in that case, (i) is automatically satisfied. Also, if (s,t) is a compatible pair, then the map w of (ii) is the homeomorphism on $D_s \cup D_t$ extending both s, t. The set of such w's is called S_2 and $S \subset S_2$. The notion of compatibility extends in the obvious way to the case of i-tuples $(s_1, \ldots, s_i)$ $(s_r \in S)$, and one constructs S_i $(i \geq 3)$ using compatible i-tuples in the same way as S_2 was constructed using compatible pairs. Then $S_i^* = S_i$, $S_i \subset S_{i+1}$ and $S_i S_j \subset S_{ij}$. The inverse semigroup T is defined to be the union of the S_i's and it is readily checked that T is a countable inverse semigroup and that (X, T) is an additive localization. Note that if $t \in T$ and $x \leq tt^*$, then there exists $s \in S$ and $e \in E(S)$ such that $x \leq e$ and $es = et$. In particular, it follows that $G(X, T) = G(X, S)$, so that we have not affected the groupoid $G(X, S)$ by enlarging S to T.

Note also that a groupoid localization (G^0, S) is additive if and only if $S \subset G^{op}$ is additive.

We now show that if (X, S) is an additive localization, then the map ψ_X of Theorem 3.3.2 is an isomorphism onto an additive subsemigroup of $G(X, S)^{op}$ and that if G is an r-discrete groupoid with S an inverse subsemigroup of G^{op} which is a basis for the topology of the r-discrete groupoid G, then G is identifiable with $G(X, S)$ where $X = G^0$.

Proposition 3.3.5

(i) *If the localization (X, S) is additive, then the map ψ_X of Theorem 3.3.2 is an isomorphism onto an additive inverse subsemigroup of $G(X, S)$.*

(ii) *Let G be an r-discrete groupoid with $X = G^0$ and S be an inverse subsemigroup of G^{op} which is a basis for the topology of G. Then the map $xs \rightarrow \overline{(x, s)}$ is an isomorphism of r-discrete groupoids from G onto $G(X, S)$, and this isomorphism takes $s \in S$ to $\psi_X(s)$ and the action of S on X to the corresponding action of $\psi_X(S)$ on X.*

Proof. (i) Suppose that (X, S) is additive. Suppose also that $s, t \in S$ and $\psi_X(s) = \psi_X(t)$. Then $ss^* = tt^* = D \subset X$, let us say. For each $x \in D$, we have $\overline{(x, s)} = \overline{(x, t)}$, and so there exists $e \in E(S)$ with $x \leq e$ and $es = et$. By (i) of the definition of *additive localization* above, we have $s = t$. So ψ_X is an isomorphism. Of course, $T = \psi_X(S)$ is a basis for the topology of $G(X, S)$ (Theorem 3.3.2).

Next we show that T is additive. Let $t, t' \in T$ be such that $t \cup t' \in G(X, S)^{op}$. Let $s, s' \in S$ be such that $\psi_X(s) = t, \psi_X(s') = t'$. We claim that the pair (s, s') is compatible. Indeed, let $x \in ss^* \cap s'(s')^*$. Then $\overline{(x, s)} \in t$ and $\overline{(x, s')} \in t'$, and since r is one-to-one on the G^{op} set $t \cup t'$, it follows that $\overline{(x, s)} = \overline{(x, s')}$. So there exists $e \in E(S)$ such that $x \leq e$ and $es = et$. So the pair (s, s') is r-compatible. Similarly, the pair $(s^*, (s')^*)$ is r-compatible (using the fact that d is one-to-one on $t \cup t'$) and so the pair (s, s') is compatible. Since the localization (X, S) is assumed to be additive, there exists (by (ii) of the above definition) $w \in S$ such that $ww^* = ss^* \cup tt^*$ and if $x \leq ss^*$ (or $x \leq tt^*$), then there exists $e \in E(S)$ such that $x \leq e$ and $es = ew$ (or $et = ew$). It follows that if $x \in w$, then $\overline{(x, w)} = \overline{(x, s)}$ if $x \leq ss^* = tt^*$ or $\overline{(x, w)} = \overline{(x, s')}$ if $x \leq s'(s')^* = t'(t')^*$. Hence $\psi_X(w) = t \cup t' \in T$. So T is additive.

(ii) Define $\Psi : G \rightarrow G(X, S)$ by: $\Psi(xs) = \overline{(x, s)}$ for $x \in D_s = r(s), s \in S$. We need to show that Ψ is well-defined. Note first that since S is a basis for the topology of G, every element of G is of the form xs. Suppose that for some $s, t \in S$ and $x \in D_s \cap D_t = r(s) \cap r(t)$, we have $xs = xt = y$ for some y. Then $s \cap t$ is an open neighborhood of y and again by the basis property of S, there exists $e \in E(S)$ with $x \in e \subset r(s \cap t)$ such that $es = et \subset s \cap t$. So $\overline{(x, s)} = \overline{(x, t)}$ and Ψ is well-defined. The preceding argument reverses to give that $xs = xt$ if $\overline{(x, s)} = \overline{(x, t)}$.

It follows that Ψ is a bijection from G onto $G(X, S)$. Next, the map Ψ takes a basis set s in G to a basis set $\psi_X(s)$ in $G(X, S)$, a product $(xs)((x.s)t)$ to the product $\overline{(x, s)}\,\overline{(x.s, t)}$ and the inverse $(xs)^{-1} = (x.s)s^*$ to the inverse $\overline{(x, s)^{-1}} = \overline{(x.s, s^*)}$. So Ψ is an isomorphism of r-discrete groupoids. The other claims of (ii) are then obvious. $\square$

Corollary 3.3.2   *If G is an r-discrete groupoid and S is an additive inverse subsemigroup of G^{op}, then $C^*(G)$ is canonically isomorphic to the covariance C^*-algebra $C_0(G^0) \times_\beta S$ for the localization (G^0, S). Also, if (X, S) is an additive localization then $C_0(X) \times_\beta S$ is canonically isomorphic to $C^*(G(X, S))$.*

Proof.   The first assertion of the corollary is just Theorem 3.3.1. The last assertion of the Corollary follows from the first part using (i) of Proposition 3.3.5 to regard S as an additive inverse subsemigroup of $G(X, S)^{op}$. $\square$

To illustrate the second part of the above corollary, let $(M, \mathcal{F})$ be a foliated manifold and (X, S) be a pseudogroup localization giving the r-discrete holonomy groupoid G_S^{hol} as in **2.3**. By enlarging S as described in the discussion preceding Proposition 3.3.5, we can suppose that (X, S) is an additive localization. It then follows that $G_S^{hol} = G(X, S)$ and

$$C_0(X) \times_\beta S \cong C^*(G(X, S)).$$

So the r-discrete version of the C^*-algebra of a foliation is a covariance C^*-algebra, which perhaps helps explain why discrete group covariance C^*-algebras and foliation C^*-algebras have similar properties. Indeed, both group orbit spaces and the leaf spaces of foliations are included in the list of badly behaved topological spaces whose problems are resolved by such C^*-algebras in the context of noncommutative geometry ([56, Chapter 2]). The foliation C^*-algebra is a covariance C^*-algebra *as well*, not for a discrete group but for an *inverse semigroup*.

CHAPTER 4

The Groupoid C^*-Algebras of Inverse Semigroups

4.1 Introduction

In Chapter 3, we investigated the relationship between r-discrete groupoids and inverse semigroup actions (in the form of localizations (X, S)). We showed in particular (Theorem 3.3.1) that for every r-discrete groupoid G, there is an inverse subsemigroup S of G^{op} such that $C^*(G)$ is isomorphic to the covariance algebra $C_0(G^0) \times_\beta S$. Conversely, under reasonable conditions, covariance C^*-algebras are r-discrete groupoid C^*-algebras (Corollary 3.3.2). So r-discrete groupoids are essentially the same as localizations.

In this chapter, we turn to the natural question: given an abstract inverse semigroup S, are groupoids also involved in the representation theory of S? Could, for example, $C^*(S)$ be isomorphic to $C^*(G)$ for some r-discrete groupoid G? We will show that the answer is "yes".

In the context of an abstract inverse semigroup, i.e. one not given as acting on a locally compact Hausdorff space, the groupoids that are relevant are the special r-discrete groupoids that we have called *ample* (**2.2**). The ample groupoids G naturally associated with S are what we will call the *S-groupoids*. An S-groupoid is roughly an ample groupoid whose structure (both topological and algebraic) is determined by a homomorphism ψ from S into G^a.

The precise definition of *S-groupoid* is postponed to **4.3**. The reason for this is that it is important to motivate and justify the development of the theory relating inverse semigroups to ample groupoids by discussing first a number of examples, many of which have appeared in the literature, in which particular inverse semigroups are in fact related to concrete ample groupoids, the relating extending to the level of their C^*-algebras. This also gives a source of examples that will be used to illustrate the later theory.

The whole of **4.2** is devoted to providing such examples. The first example is that of the free monogenic inverse semigroup whose C^*-algebra was investigated by Conway, Duncan and Paterson [62] and Hancock and Raeburn [118]. The second example discusses the Wiener-Hopf algebras. The groupoid aspect of this was developed by Muhly and Renault in [180]. In order to bring in (discrete) inverse semigroups, it is natural to require that the group involved be discrete. In understanding the relations between inverse semigroups and groupoids in this context, important progress has been made in the work of Nica ([187, 188, 189]). The third example covers the inverse semigroups and groupoids associated with the Cuntz, Cuntz-Krieger and vertex/path algebras (in increasing order of generality). The theory of Cuntz semigroups and groupoids, developed by Renault ([230]), is particularly transparent and has motivated much of the progress in the area of this book. The Cuntz groupoid involves pairs of infinite strings on the generators together with an integer. The inverse semigroup aspect of the Cuntz-Krieger case was investigated by Hancock and Raeburn ([118]). We give a description, in terms of generators and relations of what can reasonably be called the Cuntz-Krieger inverse semigroup.

A Cuntz-Krieger C^*-algebra is determined by a matrix A each of whose entries is 0 or 1. The Cuntz-Krieger groupoid is best thought of within the more general context of the *directed graph* associated with the matrix A. Along similar lines to the Cuntz groupoid, pairs of infinite vertex paths together with an integer can be used to obtain a groupoid and hence its associated C^*-algebras. We will show that there is a natural inverse semigroup involving pairs of *finite* paths associated with such a graph. In this work, the paths are thought of as *vertex* paths and produce *vertex* algebras. From another point of view, work of Ocneanu ([192, 193]) led to the notion of paths in terms of *edges* rather than vertices. Pask and Sutherland ([194]) then developed a theory of *path* algebras for pointed graphs. The theory covers a number of C^*-algebras including the Doplicher-Roberts algebras ([74, 75]). We will show that if the pointed condition is modified, then we get a wider class of algebras. This gives a theory of path algebras which effectively coincides with that of vertex algebras.

Next, we discuss briefly the groupoids associated with Clifford semigroups. The inverse semigroup S is called a Clifford inverse semigroup if it is a union of groups. The C^*-algebras of Clifford inverse semigroups were studied by Duncan and Paterson ([81]). Not surprisingly, Clifford semigroups are associated with groupoids which are unions of groups.

The final example of **4.2** deals with applications of groupoids arising out of the modelling of quasicrystals using *tilings*. After a brief discussion of crystals and quasicrystals, we examine some of the theory of Penrose tilings, the best known tilings used to model quasicrystals. This leads to and gives motivation for the space X_p with its equivalence relation groupoid R_p discussed earlier in the book. We then examine some of the ideas of J. Kellendonk in which he obtains a C^*-algebra of observables for certain quantum mechanical particle systems on tilings. This algebra is the C^*-algebra of a certain r-discrete groupoid, and we obtain a version of this groupoid using the theory of vertex groupoids above. The groupoid obtained is an S-groupoid for a natural inverse semigroup S.

In **4.3**, we begin the development of the theory of the chapter proper. We define what is meant by an S-*groupoid* and show (Theorem 4.3.1) that there is a *universal* S-groupoid $G_{\mathbf{u}}$. The proof of this theorem relies on the localization result Theorem 3.3.2. While the natural action of S on the unit space of an S-groupoid is not normally a localization, we can, nevertheless, obtain a larger inverse semigroup S' within the semigroup algebra $\mathbf{C}(S)$ which has a natural localization action on the unit space and extends the S-action. It is to this localization that Theorem 3.3.2 is applied.

The universal groupoid $G_{\mathbf{u}}$ is, in the appropriate sense, the "largest" groupoid with which S is naturally associated, and every other S-groupoid is obtained from the universal one by taking a certain homomorphic image of a reduction of $G_{\mathbf{u}}$ (Proposition 4.3.5). Intuitively, the inverse semigroup S can be regarded as the integrated form of the universal groupoid. The rest of **4.3** interprets the examples of **4.2** in terms of the universal groupoid. For example, since $G_{\mathbf{u}}$ is explicitly constructed, we are able to calculate it for the Cuntz inverse semigroup in Example 3 of **4.3**. This groupoid is effectively the Cuntz-Toeplitz groupoid. In particular, we calculate the universal groupoid for the Cuntz inverse semigroup with infinitely many generators and obtain the Cuntz groupoid in this case. This solves a problem raised by its treatment in [230].

The objective of **4.4** is to show that $C^*(S) \cong C^*(G_{\mathbf{u}})$ and that the reduced C^*-algebras of S and $G_{\mathbf{u}}$ are also isomorphic.

Turning to the first of these, we embed S inside a larger inverse semigroup S'' in the semigroup algebra $\mathbf{C}(S)$. (The inverse semigroup S'' contains the semigroup S' alluded to above.) The significance of S'' in this context is that it is isomorphic (in an "additive" sense) to the ample semigroup $(G_{\mathbf{u}})^a$ of $G_{\mathbf{u}}$. We can then use Corollary 3.2.1 to identify the representations of $G_{\mathbf{u}}$ with the additive representations of S'' and we will show that these can be identified with the representations of S. This gives the isomorphism $C^*(S) \cong C^*(G_{\mathbf{u}})$.

We then turn to the case of the reduced C^*-algebra $C^*_{red}(S)$ of S. This was discussed in **2.1**. Renault's theory ([230, p.82f.]) of the reduced C^*-algebra $C^*_{red}(G)$ of a locally compact groupoid G was discussed in **3.1**. A simple, but important, observation for relating $C^*_{red}(S)$ to $C^*_{red}(G_{\mathbf{u}})$ is Proposition 1.0.1 which shows that the inverse semigroup S itself can be

identified with a (discrete) groupoid G_S with unit space $E = E(S)$ by suit-
ably restricting the multiplication. Now E can be identified with a subset of
the unit space X of $G_\mathbf{u}$ and G_S can be identified with a dense subgroupoid
of $G_\mathbf{u}$. Take any probability measure μ equivalent to counting measure on
$E \subset X$. One then shows that the regular representation of μ generates
a C^*-algebra isomorphic to $C^*_{red}(G_\mathbf{u})$. Finally this is linked up with the
regular representation of S through the G_S connection.

Now it is a well-known fact that for a locally compact group, its universal
and reduced C^*-algebras coincide if and only if the group G is amenable. In
the discrete case, the amenability of the group is equivalent to the amenability (=injectivity) of the von Neumann algebra $VN(G)$ generated by the
reduced C^*-algebra of the group sitting on its ℓ^2-space. The final section
4.5 is concerned with a condition on S which will ensure the amenability of
the von Neumann algebra $VN(S)$ generated by $C^*_{red}(S)$. The condition is
the very accessible one that *all maximal subgroups of S be amenable*. The
proof goes via the identification of the reduced C^*-algebras of S and $G_\mathbf{u}$ in
4.5 and uses the theory of amenable groupoids developed by Renault ([230,
pp.86f.]).

4.2 Examples of inverse semigroups and their associated groupoids

In this section, we will briefly discuss some examples of inverse semigroups
and related groupoids and C^*-algebras. These will give a stock of examples
to illustrate the theory developed in **4.3**.

Example 1. Monogenic inverse semigroups.

An inverse semigroup S is called *monogenic* if it is generated by $\{u, u^*\}$ for
some $u \in S$. The representation theory of monogenic inverse semigroups
is determined canonically by that of the free monogenic inverse semigroup
I_1. (For the existence of free inverse semigroups, see, for example, [202,
Chapter 8].)

We will need later an explicit model for I_1. This is due to L. M. Gluskin
and accounts of it are given in the paper by G. B. Preston [217] (who calls
it GB) and in the book by Petrich ([202, Chapter 9]). Let $\mathbf{N}$ and $\mathbf{Z}$ be the
set of non-negative integers and the set of all integers respectively. Then
define

$$
\begin{aligned}
GB \quad = \quad & \{(k, l, m) \in \mathbf{N} \times \mathbf{Z} \times \mathbf{N} : (k + l) \in \mathbf{N}, \\
& (l + m) \in \mathbf{N}, (k + l + m) > 0\}
\end{aligned} \tag{4.1}
$$

with the multiplication and involution given by:

$$(k, l, m)(k', l', m') = (k \vee (k' - l), l + l', m' \vee (m - l')), \qquad (4.2)$$

$$(k, l, m)^* = (k + l, -l, l + m). \qquad (4.3)$$

Then GB is isomorphic to I_1, the generating element u for I_1 corresponding to $(0, 1, 0)$. We will identify I_1 with GB. It is easily checked that the idempotents of I_1 are the elements of the form $(k, 0, m)$ and that for any $s = (k, l, m) \in I_1$, we have

$$ss^* = (k, 0, l + m) \qquad s^*s = (k + l, 0, m).$$

The algebra $C^*(I_1)$ was determined in [62, 118]. For each $n \geq 2$, let $J_n : \mathbf{C}^n \to \mathbf{C}^n$ be the truncated shift : $J_n(z_1, \ldots, z_n) = (0, z_1, \ldots, z_{n-1})$ and let $J = \bigoplus_{n=2}^{\infty} J_n \in B(\bigoplus_{n=2}^{\infty} \mathbf{C}^n)$. Then $C^*(I_1)$ *is canonically isomorphic to* $C^*(J, J^*)$, *the* C^*-*subalgebra of* $B(\bigoplus_{n=2}^{\infty} \mathbf{C}^n)$ *generated by* J *and* J^*. (There is a gap in the proof given in [62]. A complete proof is given in [118].)

The proof uses the Halmos-Wallen theorem ([117]) which decomposes a power partial isometry in terms of a unitary, the unilateral shift and its adjoint, and the J_n's. We will compute the universal groupoid for I_1 in **4.3**.

Example 2.([180, 187, 188, 189]) **Wiener-Hopf algebras.**

For Wiener-Hopf algebras, consideration of the groupoid aspect preceded that of the inverse semigroup. Classically, the Wiener-Hopf algebra is the C^*-subalgebra $\mathcal{W}$ of $B(L^2([0, \infty)))$ generated by the operators $W(f)$ ($f \in L^1(\mathbf{R})$) in $B(L^2([0, \infty)))$ where for $\xi \in L^2([0, \infty))$ and $s \geq 0$,

$$W(f)\xi(s) = \int_0^{\infty} f(s - t)\xi(t) \, dt.$$

In their paper [180], Muhly and Renault showed that $\mathcal{W}$ (and a much wider class of C^*-algebras) are actually reduced groupoid C^*-algebras. More precisely, for a closed subsemigroup P of a second countable, locally compact group G satisfying certain conditions, the groupoid in question is a reduction $\mathcal{G}_{G,P}$ of a certain transformation groupoid $Y \times G$. The Wiener-Hopf algebra $\mathcal{W}(G, P)$ is just $C^*_{red}(\mathcal{G}_{G,P})$. The groupoid $\mathcal{G}_{G,P}$ is then called the *Wiener-Hopf groupoid.*

As far as the author can ascertain, there is little known about topological inverse semigroups and in this work only the discrete case will be discussed. Hence we concentrate on the case where the group G is discrete. This has been investigated in detail in the context of inverse semigroups by A. Nica ([188]) and is briefly discussed below. (In [188], Nica, in fact, considerably extends the theory to cover the case of $\tilde{F}$-inverse semigroups but we will not consider this extension here. We should also note that the semigroups of isometries considered by Salas ([244]) are related to the work of Nica.)

Let G be a discrete group and P be a subsemigroup of G such that $e \in P$ and $PP^{-1} = G$. For each $x \in G$, let β_x be the partial one-to-one map on P defined as follows: $t \in P$ is in the domain of β_x if and only if $xt \in P$, and for such a t, $\beta_x(t) = xt$. In terms of the regular representation π_2 of G on $\ell^2(G)$, β_x is the compression of $\pi_2(x)$ to $\ell^2(P)$. The inverse subsemigroup of $\mathcal{I}(P)$ generated by $\{\beta_x : x \in G\}$ is called $S_{G,P}$. This is the natural inverse semigroup associated with $\mathcal{W}(G, P)$ and is appropriately called by Nica the *Toeplitz* inverse semigroup.

It is easily checked that $\beta_x^{-1} = \beta_{x^{-1}}$ so that a typical element of $S_{G,P}$ is of the form

$$\alpha = \beta_{x_1} \cdots \beta_{x_n}$$

where $x_i \in G$. Now α is never empty. Indeed ([188, p.9]) since $PP^{-1} = G$, we can write $x_n = s_n t_n^{-1}, x_{n-1} s_n = s_{n-1} t_{n-1}^{-1}, \ldots, x_1 s_2 = s_1 t_1^{-1}$ where $s_i, t_i \in P$. Then $t = t_n t_{n-1} \cdots t_1 \in P$ and for all j, $x_j x_{j+1} \cdots x_n t = s_j t_{j-1} \cdots t_1 \in P$. So t belongs to the domain $d\alpha$ of α. Further, $x_\alpha = x_1 x_2 \cdots x_n$ is uniquely defined. For if $t \in d\alpha$, then $x_1 x_2 \cdots x_n = \alpha(t) t^{-1}$. So α is just the restriction of β_{x_α} to $d\alpha$. Further, $d\alpha$ is the set

$$\cap_{i=1}^{n+1} (x_i \cdots x_n)^{-1} P \tag{4.4}$$

where the $(n+1)$th term is defined by *fiat* to be P. In particular, if $x \in G$, then $d_{\beta_x} = x^{-1} P \cap P \ (= \{p \in P : xp \in P\})$. Replacing α by α^{-1} in (4.4) gives that the range of α is

$$\cap_{i=1}^{n+1} (x_1 x_2 \cdots x_i) P. \tag{4.5}$$

Now $(x_1 x_2 \cdots x_i) P \cap P$ is the range of $\beta_{x_1 x_2 \cdots x_i}$, and it follows from (4.5) that if $e = \alpha \alpha^{-1}$ and

$$e_i = \beta_{x_1 x_2 \cdots x_i} \beta_{x_1 x_2 \cdots x_i}^{-1}$$

then

$$e = e_1 \cdots e_n \tag{4.6}$$

in the semilattice E of idempotents of $S_{G,P}$.

We now describe Nica's construction of the Wiener-Hopf groupoid $\mathcal{G}_{G,P}$. Define $\tau : P \to \{0,1\}^G$ as follows: $\tau(p)(x) = 1$ if and only if $p \in xP$. Let Ω be the closure of $\tau(P)$ in the compact space $\{0,1\}^G$. The unit space of $\mathcal{G}_{G,P}$ will be Ω, and each $\alpha = \beta_{x_1} \cdots \beta_{x_n} \in S_{G,P}$ acts on Ω by defining $\alpha(\tau(p)) = \tau(x_\alpha p)$ for p in the domain $d\alpha$ of α and extending by continuity to the closure D_α of $\tau(d\alpha)$ in Ω. (The set D_α is an open compact subset of Ω.)

The groupoid $\mathcal{G}_{G,P}$ then consists of pairs (x,ω) $(x \in G, \ \omega \in D_{\beta_x})$ with the relative topology inherited from $G \times \Omega$, and product given by:

$$(x,\omega)(x',\omega') = (xx',\omega') \tag{4.7}$$

the latter being only defined when $\beta_{x'}(\omega') = \omega$. The inverse of (x,ω) is defined to be $(x^{-1}, \beta_x(\omega))$. Then $\mathcal{G}_{G,P}$ is easily checked to be an ample Hausdorff groupoid.

In **4.3**, when we construct the universal groupoid of an inverse semigroup, we will have to consider *all* of the elements s in the semigroup. In the above case of (4.7) above, only $x \in G$ needs to be used in the first component, rather than a typical semigroup element $\alpha = \beta_{x_1} \cdots \beta_{x_n}$. The reason for this is that, as noted above, α is just a restriction of β_{x_α}. The fact that this does not hold for general inverse semigroups is one of the reasons why the Hausdorff property fails in general for the groupoids associated with inverse semigroups.

Nica shows that under certain conditions, $C^*(\mathcal{S}_{G,P})$ is isomorphic to $C^*(\mathcal{G}_{G,P})$.

Example 3.([230, 66, 67, 118])   **Cuntz, Cuntz-Krieger and vertex/path algebras.**

The groupoid perspective explains some of the properties of the Cuntz algebra from a dynamical perspective. Indeed, in the more general Cuntz-Krieger situation (considered below) Cuntz ([67, p.25]) comments that the algebra $K \otimes \mathcal{O}_A$ reflects very precisely the topological properties of a certain flow space F_A and the dynamics of a topological Markov chain σ_A because it can be represented as the convolution algebra of a certain topological groupoid. (This groupoid and the inverse semigroups associated with it will be discussed below.)

Although the Cuntz algebra is a special case of the Cuntz-Krieger algebra, it is convenient to treat it independently first, as it illustrates very clearly the interaction between inverse semigroups, groupoids and C^*-algebras.

The *Cuntz semigroup* S_n $(1 < n \leq \infty)$, due to Renault, was discussed in Chapter 1. We will continue to assume for the time being that $n < \infty$. The case $n = \infty$ requires care and will be discussed in **4.3**, Example 3 after the theory of the universal groupoid has been developed.

Recall that S_n is generated by the unit 1, the zero z_0 and the elements s_i, t_i $(1 \leq i \leq n)$ where (1.3) $t_i s_j = \delta_{ij} 1$. Apart from 1 and z_0, the elements of S_n are uniquely of the form $s_\alpha t_\beta$ where α, β are finite strings of integers i $(1 \leq i \leq n)$ (one of which could be empty). The product in S_n is readily checked (c.f. (4.15), (4.16)) to be given by :

$$(s_\alpha t_{\beta\gamma})(s_\beta t_{\beta'}) = s_\alpha t_{\beta'\gamma} \tag{4.8}$$

$$(s_\alpha t_\beta)(s_{\beta\gamma} t_{\beta'}) \quad = \quad s_{\alpha\gamma} t_{\beta'}, \tag{4.9}$$

with all other products, apart from the obvious ones in which a 1 is present, being z_0. The involution on S_n is determined by $s_i^* = t_i$ and is given by

$$(s_\alpha t_\beta)^* = s_\beta t_\alpha. \tag{4.10}$$

An obvious representation of S_n generates the Cuntz algebra O_n.

In Chapter 1, we also referred to the *Cuntz groupoid* G_n associated with S_n, this groupoid being also due to Renault. This groupoid is actually a localization groupoid $G(Z, S_n)$ as in **3.3** (see Example 4 below) but we will give the direct description.

Recall that $Z = \{1, \cdots, n\}^{\mathbf{P}}$ where $\mathbf{P}$ is the set of positive integers and Z has the product topology. So the elements of Z are just sequences in $1, 2, \cdots n$. The elements of G_n are triples

$$(\alpha\gamma, \ell(\alpha) - \ell(\beta), \beta\gamma) \tag{4.11}$$

where $\gamma \in Z$, α, β are finite strings as above and $\ell(\alpha)$ is the length of α. (Note that different choices of α, β, γ can give the same triple.) Multiplication and inversion in G_n are given by :

$$(z, k, z')(z', k', z'') = (z, k + k', z'') \quad (z, k, z')^{-1} = (z', -k, z). \tag{4.12}$$

Then $r((z, k, z')) = z, d((z, k, z')) = z'$ where we have identified $z \in Z$ with $(z, 0, z) \in G_n^0$.

A basis for the topology of G_n is given by sets of the form $U_{\alpha, \beta, V'}$, where V' is open in Z and

$$U_{\alpha, \beta, V'} = \{(\alpha\gamma, \ell(\alpha) - \ell(\beta), \beta\gamma) : \gamma \in V'\}.$$

(Note that, as for the Penrose tiling equivalence relation (**2.2**) and the Kronecker foliation equivalence relation (**2.3**), the topology on the Cuntz groupoid is not the product topology.) It is straightforward to check that the groupoid G_n is a Hausdorff ample groupoid (cf. the proof of Theorem 3.3.2). The sets $U_{\alpha, \beta, V'}$ belong to G_n^a.

Set $U_{\alpha, \beta, z} = U_{\alpha, \beta}$. Then

$$U_{\alpha, \beta, V'} = (\alpha V') U_{\alpha, \beta}. \tag{4.13}$$

(Note also for later use that the family of sets $\alpha V'$ is a basis for Z.)

In fact, the family of sets αZ (α ranging over all finite strings) is a basis for the topology of $Z - \alpha Z$ is just the set of sequences $w \in Z$ with $w_i = \alpha_i$ for $1 \leq i \leq \ell(\alpha)$ and we have the product topology on Z. It follows that the family of sets $U_{\alpha, \beta}$ is a basis for the topology of G_n.

It is easy to check that the map $\psi : S_n \to G_n^a$, where $\psi(s_\alpha t_\beta) = U_{\alpha,\beta}$, is an isomorphism. So $T_n = \psi(S_n)$ is an inverse subsemigroup of G_n^a which is a basis for the topology of G_n. However, if $n \geq 2$, then T_n is not additive. For example, $U_{1,1} \cup U_{2,2} \in G_n^a \sim T_n$. So the theory developed earlier for the additive case, including Theorem 3.2.1, does not apply directly.

Accordingly, in the discussion below, we will work with the ample semigroup G_n^a rather than T_n. However, T_n is large enough to determine G_n^a in the sense that *every $C \in G_n^a$ is the disjoint union of a finite number of members of T_n*. Indeed, since T_n is a basis for G_n and C is both open and compact, we have that C is the union of a finite collection $\mathcal{X}$ of sets $U_{\alpha,\beta}$. Since each $U_{\alpha,\beta} = \cup_{i=1}^n U_{\alpha i, \beta i}$ we can arrange that all of the sets $U_{\alpha,\beta} \in \mathcal{X}$ are such that the α's have the same length. Then $U_{\alpha,\beta} \cap U_{\alpha',\beta'} = \emptyset$ if $\alpha \neq \alpha'$. If $U_{\alpha,\beta}, U_{\alpha,\beta'} \in \mathcal{X}$, then $U_{\alpha,\beta} \cap U_{\alpha,\beta'} = \emptyset$ if $\ell(\beta) \neq \ell(\beta')$, while if $\ell(\beta) = \ell(\beta')$, then $U_{\alpha,\beta} \cap U_{\alpha,\beta'} = \emptyset$ if $\beta \neq \beta'$. Removing repeated $U_{\alpha,\beta}$'s, we have C as the disjoint union of members of T_n.

We now prove Renault's result ([230, pp.140-145]) that $C^*(G_n)$ *is isomorphic to the Cuntz algebra O_n*. Let Φ be a representation of $C_c(G_n)$ generating $C^*(G_n)$. By Theorem 3.2.2, $C^*(G_n)$ is generated by the restriction Φ' of Φ to G_n^a and Φ' is an additive representation of G_n^a. Since, by the above paragraph, every $C \in G_n^a$ is a disjoint union of members of T_n and Φ' is additive, it follows that the vector space spanned by $\Phi'(T_n)$, where T_n is regarded as contained in $C_c(G_n)$, itself contains $\Phi'(G_n^a)$. So $\pi(S_n)$, where $\pi = \Phi' \circ \psi$, generates $C^*(G_n)$. The representation π is faithful since ψ and Φ are faithful (**3.1**). Note that

$$\psi(s_i) = \{(i\gamma, 1, \gamma) : \gamma \in Z\} \quad \psi(s_i^*) = \{(\gamma, -1, i\gamma) : \gamma \in Z\}.$$

So $\psi(s_i s_i^*)$ is the set of products $(i\gamma, 1, \gamma)(\gamma, -1, i\gamma)$ and so equals $U_{i,i}$, the subset of Z whose members begin with i. Similarly, $\psi(s_i^* s_i) = Z$. Now the $U_{i,i}$ are disjoint and non-empty, and their union is the identity G_n^0 of G_n^a. Using the additivity of Φ', we get

$$\sum_{i=1}^n \pi(s_i s_i^*) = \Phi'(\cup_{i=1}^n U_{i,i}) = I.$$

Since $\psi(s_i^* s_i) = Z$, we get $\pi(s_i^* s_i) = I$ and each $\pi(s_i)$ is an isometry. So $C^*(G_n)$ is generated by the isometries $\pi(s_i)$ and these satisfy the Cuntz condition (1.1). It follows that $C^*(G_n) = O_n$.

We now discuss the Cuntz-Kreiger algebras. These are generalizations of the Cuntz algebras. We first recall how the Cuntz-Krieger algebras $\mathcal{O}_A$ are defined ([66]). Let A be an $n \times n$ matrix such that each $A(i,j) = A_{ij}$ is either 1 or 0. Every row and every column of A is assumed to be non-zero. A C^*-algebra $\mathcal{O}_A$ is then a non-degenerate C^*-algebra on a Hilbert space, generated by partial isometries s_i $(1 \leq i \leq n)$ such that if $p_i = s_i s_i^*$ and $q_i = s_i^* s_i$, then

$$p_i p_j = 0 \quad (i \neq j) \qquad q_i = \sum_{j=1}^{n} A(i,j) p_j. \qquad (4.14)$$

In general, different choices of s_i's will generate non-isomorphic C^*-algebras. Cuntz and Krieger show that if A satisfies a certain property called *condition (I)* then $\mathcal{O}_A$ is independent of the Hilbert space on which it is realized. To obtain the original Cuntz algebras, we take $A_{ij} = 1$ for all i, j. (Note that since we are presupposing non-degeneracy for $\mathcal{O}_A$, (4.14) entails that when every $A_{ij} = 1$, then every $q_i = 1$.)

The discussion of the relation between Cuntz-Krieger algebras, groupoids and inverse semigroups is best carried on in the more general context of directed graphs to which we now turn. (A description of an inverse semigroup, associated with the Cuntz-Krieger algebras, in terms of generators and relations, will be postponed until later.)

The groupoid C^*-algebras associated with the vertices (and edges) of directed graphs have been considered by a number of people, although the present writer does not know of a convenient reference. The elements of the unit space of the associated groupoid are the sequences of vertices of infinite paths in the graph, and for that reason, we shall call such groupoids *vertex groupoids*. The large class of vertex groupoids includes, in particular, the natural groupoids giving the Cuntz-Krieger algebras. Another interesting example of a vertex groupoid, arising in the context of tilings, is given in Example 5 below.

Path groupoids and path algebras, in which sequences of edges of infinite paths in the graph are taken in place of vertex sequences, have also been considered. We shall call the groupoids associated with such path edges *edge groupoids*.

Starting first with vertex groupoids, we consider a directed graph $\mathcal{G}$ with vertex set V and edge set E. The initial vertex of an edge e is denoted by $d(e)$ and the end vertex of e by $r(e)$. The set X_0 is defined to be the set of v's in V such that $v \neq r(e)$ for all $e \in E$ if there is such a v, and is defined otherwise to be V. It is assumed that for every $v \in V$, there exists $e \in E$ such that $v = d(e)$. The sets X_i are defined recursively as follows: a vertex v belongs to X_i if and only if there is a path $(v_0, v_1, \ldots, v_i)$ of vertices where $v_j \in X_j$, $v_i = v$ and for each j $(0 \leq j \leq (i-1))$, there is an edge e_j such that $d(e_j) = v_j, r(e_j) = v_{j+1}$. Such a path will be called a *path of length i* in $\mathcal{G}$. Note that $X_i \neq \emptyset$ for $i > 0$. Of course, the X_j's can overlap. We assume that $\cup_{j=0}^{\infty} X_j = V$. Such a directed graph $\mathcal{G}$ will be called *admissible*.

If $X_0 = V$ then every vertex v is the end of an edge and so is the end of a path of vertices of arbitrary length. So in that case, $X_i = V$ for all $i \geq 0$, and admissibility is automatic. The directed graphs associated with the Cuntz-Krieger cases all have $X_0 = V$ as we will see below.

Let $\mathcal{G}$ be an admissible directed graph. Associated with $\mathcal{G}$ is the space X of infinite paths $(v_0, v_1, \ldots)$ such that for each i, $(v_0, v_1, \ldots, v_i)$ is a path of length i in $\mathcal{G}$. The space X will be the unit space for the groupoid $G_{\mathcal{G}}$ associated with the graph. (If $X_0 = V$, then we take $X = V^{\mathbf{P}}$.) If the X_i's are finite then X can be regarded as a closed subset of the infinite product of the X_i's and as such is compact Hausdorff in the product topology. When there are infinitely many infinite X_i then it is not clear how the topology should be defined. (However, as we will see when discussing the groupoid associated with O_∞ (**4.3**, Example 3), the inverse semigroup approach gives a general procedure for constructing the appropriate groupoid and its unit space. This procedure shows that we need to consider *finite* sequences for X as well as infinite ones.)

Assuming the finiteness of the X_i's, we then obtain a locally compact, Hausdorff ample groupoid $G_{\mathcal{G}}$ and associated C^*-algebra $C^*(G_{\mathcal{G}})$ in a manner entirely analogous to that of the Cuntz groupoid G_n. The elements of $G_{\mathcal{G}}$ are triples $(\alpha\gamma, \ell(\alpha) - \ell(\beta), \beta\gamma)$ where $\alpha\gamma, \beta\gamma \in X$ with α, β initial segments of finite length. The topology is given by the $U_{\alpha,\beta,V'}$ just as in the Cuntz case $((4.13))$, as are products and inverses in $G_{\mathcal{G}}$.

As far as V' above is concerned, we only consider pairs α, β for which there is a $\gamma = (\gamma_1, \gamma_2, \ldots)$ such that $\alpha\gamma, \beta\gamma \in X$. So if $\ell(\alpha) = r$, $\ell(\beta) = s$, then $\gamma_1 \in X_{r+1} \cap X_{s+1}$ and there are edges from α_r, β_s to γ_1. Let W be the set of such paths γ. Then W is compact Hausdorff under the relative product topology .

One easily checks that the sets of such $\alpha\gamma$'s and $\beta\gamma$'s ($\gamma \in W$) are open in X, and we take V' to be open in W. In the Cuntz case earlier, the family of sets αZ was a basis for the topology of X and the family of sets $U_{\alpha,\beta} = U_{\alpha,\beta,Z}$ was a basis for the topology of G_n. In the present more general situation, the family of sets of elements $\alpha\gamma$ (fixed α, all possible γ's allowed) is a basis for the topology of X, while the family of sets $U_{\alpha,\beta}$ of elements $(\alpha\gamma, \ell(\alpha) - \ell(\beta), \beta\gamma)$ (fixed α, β, all possible γ's allowed) is a basis for the topology of $G_{\mathcal{G}}$.

In particular, in the case of a Cuntz-Krieger algebra specified by A, we have the graph $\mathcal{G}$ whose vertices are $s_1, \ldots, s_n$ (corresponding to the canonical generators of $\mathcal{O}_A$) with an edge going from s_i to s_j if and only if $A(i,j) = 1$. In this case, $\mathcal{G}$ is admissible. (The reason for this is that the columns and rows of A are non-zero.) Also $X_0 = V$ and $X_i = \{s_1, \ldots, s_n\}$ for all $i \geq 0$. It can be shown as for the Cuntz case earlier that (under *condition (I)*) $\mathcal{O}_A \cong C^*(G_{\mathcal{G}})$.

For a general admissible directed graph with the X_i's finite, we can associate an inverse semigroup $S_{\mathcal{G}}$ with the graph $\mathcal{G}$ in a manner analogous to that of the Cuntz case. (We will treat the case $X_0 \neq V$, the case where $X_0 = V$ being similar.) For a path of finite length $\alpha = (v_0, \ldots, v_i)$ in $\mathcal{G}$, define $r(\alpha) = v_i$. Then $S_{\mathcal{G}}$ is the semigroup of pairs of finite vertex paths (α, α') such that $r(\alpha) = r(\alpha')$, with zero adjoined. (A significant difference from the S_n-case is that neither of the paths α, α' is allowed to be empty.) The only non-zero products (where μ is allowed to be empty) are given by:

$$(\alpha, \alpha'\mu)(\alpha', \beta) \quad = \quad (\alpha, \beta\mu) \tag{4.15}$$

$$(\alpha, \alpha')(\alpha'\mu, \beta') \quad = \quad (\alpha\mu, \beta') \tag{4.16}$$

and $(\alpha, \alpha')^* = (\alpha', \alpha)$. Geometrically, the meaning of this product is very simple – as illustrated in Fig. 4.2.1 below, we just delete the dotted path α' when forming products.

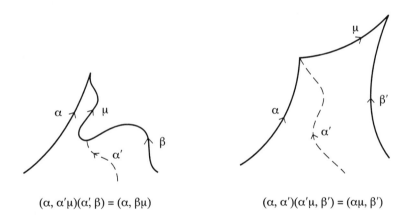

$$(\alpha, \alpha'\mu)(\alpha', \beta) = (\alpha, \beta\mu) \qquad\qquad (\alpha, \alpha')(\alpha'\mu, \beta') = (\alpha\mu, \beta')$$

Figure 4.2.1

It is routine to check that $S_{\mathcal{G}}$ is an inverse semigroup. Note that when $\mathcal{G}$ is the directed graph associated with the Cuntz case, then $S_{\mathcal{G}}$ can be identified with a proper subsemigroup of S_n. The Cuntz inverse semigroup seems rather exceptional in the class of $S_{\mathcal{G}}$'s. For in that case, we can adjoin generators s_i in $S_{\mathcal{G}}$ and the unit 1 to get another inverse semigroup S_n. We can also remove the condition $r(\alpha) = r(\alpha')$ for pairs $(\alpha, \alpha') \in S_{\mathcal{G}}$. (As we will see below, however, in the Cuntz-Krieger case something similar can be done.)

Edge groupoids, where for the units, we take sequences of edges in a graph rather than vertices, are associated with a class of algebras called *path algebras.* These produce a number of interesting C^*-algebras such as the AF-algebras, the Doplicher-Roberts algebras and the Cuntz-Krieger algebras ([74, 75, 192, 193, 194, 173]).

In the treatment of path algebras in [194], Pask and Sutherland require their graphs to be pointed. A pointed graph is one for which there is a distinguished vertex $*$ in the graph from which all edge paths to be considered start. However, such graphs are admissible in the sense above, and working in the more general admissible context, their construction of a unit space consisting of infinite paths of edges and its associated groupoid goes through. In fact, the vertex and edge groupoids come to the same thing using the natural bijection between vertex paths $\{v_i\}$ and edge paths $\{e_j\}$ implemented by taking e_j to be the edge from v_j to v_{j+1}. So in the context of admissible graphs, we get the same theory whether we deal with vertex paths or edge paths.

The notion of *path algebra* is due to Ocneanu ([192, 193]). The theory is further developed in [194]. We now present a brief alternative description of such algebras in the vertex context and will call the algebra the *vertex algebra*. Let $\mathcal{G}$ be an admissible directed graph. We can interpret the vertex algebra in a very natural way as the quotient of the semigroup algebra of $S_{\mathcal{G}}$.

The *vertex algebra* associated with $G_{\mathcal{G}}$ is defined as $\mathbf{C}(S_{\mathcal{G}})/I$ where I is the ideal in $\mathbf{C}(S_{\mathcal{G}})$ generated by elements of the form

$$(\alpha, \beta) - \sum_v (\alpha v, \beta v) \tag{4.17}$$

the sum being taken over the vertices v for which there is an edge e with $d(e) = r(\alpha) = r(\beta)$ and $r(e) = v$, i.e. over all vertices that can be added on to α (and β) to give a vertex path. (When translated into the edge context, the inverse semigroup element (α, β) corresponds, in the notation of Pask and Sutherland ([194]) to $e_{\alpha,\beta}$ and it is easy to see that the vertex algebra as defined above corresponds to their path algebra.)

Note that $U_{\alpha,\beta}$ is the disjoint union of the finite number of sets $U_{\alpha v,\beta v}$ so that ((4.17)) the ideal I is contained in the kernel of the map ψ where $\psi((\alpha, \beta)) = \chi_{U_{\alpha,\beta}}$. It is routine to check that ψ is a homomorphism. (We met this homomorphism ψ above in the Cuntz case except that there we regarded ψ as G_n^a-valued.) The homomorphism ψ thus extends canonically to a homomorphism from the vertex algebra $\mathbf{C}(S_{\mathcal{G}})$ into $C_c(G_{\mathcal{G}})$. A result communicated to the author by D. Pask implies that this algebra homomorphism ψ is an isomorphism. (This is also related to the theorem proved later (Theorem 4.4.1) that for *any* inverse semigroup S, ψ_u is an isomorphism on $\mathbf{C}(S)$, where $(\psi_u, G_{\mathbf{u}})$ is the universal groupoid of S.)

We finally discuss another inverse semigroup associated with the Cuntz-Krieger algebra $\mathcal{O}_A$. Of course we always have the appropriate $S_\mathcal{G}$ with $\mathcal{G}$ the directed graph defined by A. However, this is not given in terms of generators and relations as is S_n in the Cuntz case.

However, Hancock and Raeburn ([118]) showed that $\mathcal{O}_A$ is generated by a naturally associated inverse semigroup C_A generated by s_i, t_i satisfying explicit relations. To motivate this semigroup, if the s_i generate $\mathcal{O}_A$ on a Hilbert space and if $t_i = s_i^*$, then by the partial isometry property we have firstly $t_i s_i t_i = t_i$ and $s_i t_i s_i = s_i$. Next, recalling that $p_i = s_i t_i, q_i = t_i s_i$ and using (4.14), we have $t_j s_i = t_j p_j p_i s_i$ so that secondly $t_j s_i = 0$ if $i \neq j$. Next, substituting in the right-hand side of (4.14) for q_i, we obtain thirdly $(t_i s_i)(s_j t_j) = A(i,j)(s_j t_j) = (s_j t_j)(t_i s_i)$. Similar substitution for q_i, q_j gives fourthly that $(t_i s_i)(t_j s_j) = (t_j s_j)(t_i s_i)$.

Hancock and Raeburn then defined the *Cuntz-Krieger semigroup* C_A to be the semigroup with a zero 0 and generated by elements s_i, t_i for $1 \leq i \leq n$ subject to the four conditions of the preceding paragraph:

1. $t_i s_i t_i = t_i$ and $s_i t_i s_i = s_i$;

2. $t_j s_i = 0$ if $i \neq j$;

3. $(t_i s_i)(s_j t_j) = A(i,j)(s_j t_j) = (s_j t_j)(t_i s_i)$ for all i, j;

4. $(t_i s_i)(t_j s_j) = (t_j s_j)(t_i s_i)$ for all i, j.

(Of course we interpret $0s = 0$ and $1s = s$ for all $s \in C_A$.) These four conditions easily imply other relations among the s_i, t_j such as the following ([118, p.343]): $t_j q_i = A(i,j)t_j$ and $q_i s_j = A(i,j)s_j$.

It is of interest to have available a model S_A for C_A analogous to the $s_\alpha t_\beta$ formulation for the elements of the Cuntz inverse semigroup. We will not have occasion to use it in this work, but since it may be useful for determining the range of groupoids associated with Cuntz-Krieger algebras (we will do this for the Cuntz case in **4.3**, Example 3) we give the following model. (The proof is omitted.)

Consider triples of the form (μ, B, ν) where $\mu = \mu_1 \mu_2 \cdots \mu_r$ and $\nu = \nu_1 \nu_2 \cdots \nu_l$ are vertex paths in the Cuntz-Krieger graph and $B \subset \{1, 2, \ldots, n\}$. We think of a triple $(\mu, \{b_1, \ldots, b_k\}, \nu)$ as the product

$$s_{\mu_1} s_{\mu_2} \cdots s_{\mu_r} q_{b_1} q_{b_2} \cdots q_{b_k} t_{\nu_l} t_{\nu_{l-1}} \cdots t_{\nu_1}.$$

Such a triple (μ, B, ν) is called *reduced* if μ_r, ν_l do not belong to B. The set of reduced triples is denoted by S_A. The *reduced form* of a triple (μ, B, ν) is (μ, B', ν) where $B' = B \sim \{\mu_r, \nu_l\}$. Let e be the empty vertex path. For $B \subset \{1, 2, \cdots, n\}$ and $1 \leq i \leq n$, let $A(B, i) = \prod_{b \in B} A(b, i)$.

One can show that S_A is an inverse semigroup with $(e, \emptyset, e)$ as zero under the following product. Here products are put into reduced form if necessary, and any product with any undefined terms as well as all unlisted products are defined to be zero:

(i) $(\mu, B, \mu'\nu)(\mu', B', \nu') = A(B', \nu_1)(\mu, B, \nu'\nu);$

(ii) $(\mu, B, \nu)(\nu\mu'', B', \rho) = A(B, \mu_1'')(\mu\mu'', B', \rho);$

(iii) $(\mu, B, \nu)(\nu, B', \rho) = (\mu, B \cup \{\nu_l\} \cup B', \rho).$

The involution on S_A is given by: $(\mu, B, \nu)^* = (\nu, B, \mu)$. Further, there is an isomorphism $\phi : C_A \rightarrow S_A$ defined by: $\phi(s_i) = (i, \emptyset, e)$, $\phi(t_i) = (e, \emptyset, i)$ and $\phi(0) = (e, \emptyset, e)$.

Example 4.([148]) Localizations and Clifford semigroups.

Localizations were discussed in detail in **3.3** where we saw that if (X, S) is a localization, then it is associated with an r-discrete groupoid $G(X, S)$ (Theorem 3.3.2). In the case where the elements of S are given as partial homeomorphisms on X, this groupoid is just the sheaf of germs of the elements of S as maps on X. The covariance and groupoid C^*-algebras associated with (X, S) were, in the additive case, shown to be the same (Corollary 3.3.2).

Sometimes a groupoid naturally associated with an abstract inverse semigroup S is given by a localization (X, S). An example of this is given by the Cuntz case in Example 3 above. However, in general, while S always acts on the unit space of any of its associated groupoids, the domains will not usually give a basis for the topology of the unit space. (This is, in general, the case even when S is a semilattice.) As we shall see in **4.3**, we need to imbed S in a larger inverse semigroup S' to achieve the required basis condition, and then the theory of **3.3** applies to the localization (X, S').

Finally, we briefly discuss *Clifford semigroups* ([50, (4.2)]). An inverse semigroup S is called a *Clifford semigroup* if it is a union of groups.[1] Thinking of such an inverse semigroup in terms of partial one-to-one maps (as in the Vagner-Preston theorem of Proposition 2.1.3) and noting that the domain and range of any element of S have to coincide, it is easy to see that this implies that E *is contained in the center of* S. Then $S = \cup_{e \in E(S)} H_e$ where H_e is the maximal subgroup of S with e as identity. Note that $H_e H_f \subset H_{ef}$.

The representation theory of Clifford semigroups is discussed in [79] although there, the groupoid aspect is only implicit. This aspect will be briefly discussed later in **4.3**, Example 4, the groupoids in question being group bundles.

---

[1] In the algebraic theory of semigroups, one (of many) characterizations of a Clifford semigroup is that it is a semigroup which is a *semilattice of groups* ([133, p.94]).

Example 5. Quasicrystals and the non-commutative geometry of tilings

In this example we will discuss, together with other related topics, a remarkable construction by J. Kellendonk in which he associates with a tiling an inverse semigroup and r-discrete groupoid. The motivation and background for this lies in the physics of quasicrystals. These are modelled on tilings – the Penrose tilings are good examples of the kinds of tilings that need to be considered – and for quantum mechanical purposes, one needs a C^*-algebra of observables. In the work of Kellendonk, this is achieved by constructing an r-discrete groupoid G whose C^*-algebra is this algebra of observables, and he uses its K_0-group to investigate the labelling of the gaps in the spectrum of discrete Schrödinger operators of quantum mechanical systems on the tiling. (The use of K-theory for gap labelling purposes was originally proposed by Belissard ([14]).) We will not discuss that part of the theory, but will focus on the construction of the groupoid, which illustrates very well the theory developed in Chapters 3 and 4 of the book. However, before discussing the construction, we will sketch very briefly some of the background in the theory of crystals, quasicrystals, Penrose tilings and the noncommutative geometric aspect of the mathematics involved.

In classical physics, tilings arise in connection with *crystals* (e.g. [9, 41, 136]). A crystalline solid – such as a metal – is distinguished from non-crystalline solids – such as wood – by the fact that the crystal is built up out of identical crystalline *units*, e.g. an atom, atomic cluster, molecule, ion, etc., fitting together to form a tiling, a 3-dimensional array in which the unit corresponds to a tile which is repeated to cover all of space. (Of course, in practice there are finitely many units making a particular finite crystalline solid, but it is convenient for the model to extend the configuration to all of space.)

In the classical theory of crystals, each tile is replaced by its center of gravity, so that the tiling is replaced by a *lattice* of points in $\mathbf{R}^3$. (This simplified matters since one no longer had to decide what the building blocks of real crystals were.) The simplest kind of lattice is the *point lattice* in which any two points of the lattice are equivalent under a translation that preserves the lattice. Such a lattice is called a *Bravais* lattice. For such a lattice L, we can chose a point in L as the origin. There are then three linearly independent vectors $a_1, a_2, a_3 \in L$ such that every element of L is an *integer* linear combination of the a_i's. Following the approach of Klein's Erlangen program, in which a geometrical structure is to be classified by the group of symmetries which preserve that structure, it was natural to classify the geometry of a point lattice L by considering the group of isometries of $\mathbf{R}^3$ that preserve the lattice. The translations in this group identify with L, and since every isometry is a sum of a translation and a linear isometry, one looks instead at the group G_L of *linear* isometries of $\mathbf{R}^3$ that preserve L. Bravais showed in 1845 that there were 14 possibilities for G_L, so that there are 14 symmetry classes of Bravais lattices. This had applications to crystallography that made the lattice approach to the subject attractive. In two dimensions, there are five possibilities for G_L (Fig. 4.2.2).

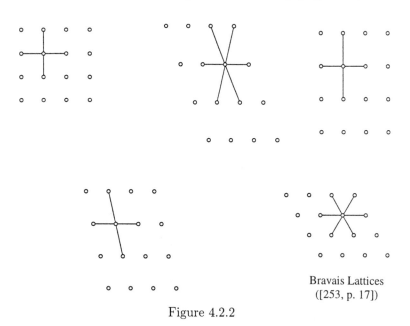

Bravais Lattices
([253, p. 17])

Figure 4.2.2

An important (and indeed elementary) restriction on a (non-trivial) element R of G_L (in two or three dimensions) is that *its order is either 2,3,4 or 6*. (See, for example, [253, p.50].) This restriction on the orders of elements of G_L is called the *crystallographic restriction*. We give the delightful argument for the case of two dimensions. In that case, there is a basis $\{a_1, a_2\}$ for $\mathbf{R}^2$ such that L is the set of integer linear combinations of a_1 and a_2. The matrix of R for that basis has integer entries (since it preserves the lattice) and at the same time is conjugate either to a reflection (when the result is trivial) or to a rotation about the origin through some angle θ. In that case, it follows by considering the trace of R that $2\cos\theta \in \mathbf{Z}$, and as $|\cos\theta| \le 1$, the only possible values for $\cos\theta$ are $0, \pm 1/2, \pm 1$. The result restricting the order of R to 2,3,4 or 6 now follows. In particular, R cannot have order 5, so that *neither pentagonal nor icosahedral symmetry is allowed in crystallography*.

As observed above, in the classical theory, the tiles were replaced by lattice points. The group G of isometries leaving the Bravais lattice L invariant acts transitively on L, so that L is a homogeneous space for G, equivalently an orbit for G, equivalently identifiable with G/G_L (since G_L is the stabilizer subgroup at the point 0). So group theory completely suffices to determine the lattice.[2]

However, in the 20th century, further investigations into crystalline structures showed that crystal structures in general could no longer be incorporated within the lattice framework. Fundamentally important information about the crystal units – which in the lattice approach, had been replaced by points – was obtained through the use of X-ray diffraction patterns in the work of Max von Laue in 1912. Indeed, as Marjorie Senechal points out ([253, p.24]), the experiments of Laue "settled three controversial questions simultaneously: it proved that X-rays are a form of light, that atoms exist and that the atoms in crystals are arranged in an orderly way". The latter is shown by the fact that crystals function as diffraction gratings for X-rays. To explain these, Fourier techniques had to be used, in particular, Bohr's theory of almost periodic functions.

In 1984, an alloy of aluminum and manganese was produced whose diffraction pattern showed the icosahedral (pentagonal) symmetry forbidden (as we saw above) in the lattice approach to crystallography. This challenged many widely held assumptions about crystallography. It led to the introduction of the concept of a *quasicrystal* applied to such crystalline structures which exhibit symmetry forbidden by the crystallographic restriction ([253, p.31]). In this context, groups acting on a homogeneous lattice space are no longer adequate to describe the symmetry present. Instead, groupoids are the appropriate mechanism for analyzing symmetry in the quasicrystal context, and groupoids arise naturally, as we shall see, in the mathematical modelling of quasicrystals.

The inadequate lattice approach is replaced by what are called *quasilattices* ([260, p.4]) which effectively are tilings of $\mathbf{R}^2$ (resp. $\mathbf{R}^3$) by a finite number of polygonal (resp. polyhedral) types. These types represent the "tiles" of the quasicrystal. In the crystal context, there is only one such type. For quasicrystals, there is more than one. In their beautiful book [105], Grünbaum and Shephard develop a coherent theory of tilings in general. In particular, they give a discussion ([105, Ch.10]) of the most celebrated examples of the tilings relevant to quasicrystals. These are the 2-dimensional *Penrose tilings* initially developed by Roger Penrose in his paper [201]. This tiling is essentially a completion of a partial tiling of Kepler in the early 1600's ([253, pp.14, 171]).

---

[2]However, as noted in the Preface, groupoids enter naturally even in the classical context when screw and glide motions are to be considered.

Penrose tilings can be constructed using "kites and darts" ([108]), which in turn ([105, p.540]) can be regarded ([240]) as tilings by *two triangles*, a large one L_A and a small one S_A. The triangles have "decorations" – each has exactly one edge with an arrow and the vertices are colored either black or white as shown. The number τ is the *golden number* $(1 + \sqrt{5})/2$. The triangles L_A, S_A are shown in Fig. 4.2.3 below, and tilings are built up out of them by fitting copies of the triangles together so that vertex colors and edge arrows match. Fig. 4.2.3 also shows part of a Penrose tiling made up out of these two triangles. (A few of the triangles are explicitly given with their vertices colored and arrows on some edges.) For the present purposes, a Penrose tiling is any tiling of $\mathbf{R}^2$ by these two triangles.

R. M. Robinson ([240]) investigated the connection between Penrose tilings. His work is described by Grünbaum and Shephard in [105, p.540, p.568]. (Grünbaum and Shephard also use in their account an article by Gardner ([101]) and private communications from R. Penrose, R. Ammann and J. Conway. See also [56, p.175f.].) Such a tiling is determined by an *index* sequence. This is a sequence $\{x_n\}$ where x_n is either 0 or 1 and such that if $x_n = 1$ then $x_{n+1} = 0$. So the set of index sequences is just the set X_p that we met in **2.2** and **3.1**. So every index sequence determines (non-uniquely) a tiling. The converse also holds true so that each tiling determines (non-uniquely) an index sequence.

In more detail, let T be a Penrose tiling. An index sequence $\{x_n\}$ for T is obtained by producing a sequence T_n of tilings of $\mathbf{R}^2$, where the tiles of T_n are copies of the (decorated) triangles L_n, S_n. We start with $T_1 = T$. We take $L_1 = L_A, S_A = S_1$. The triangle L_n is obtained by deleting the common edge of an adjacent pair L_{n-1}, S_{n-1} in T_{n-1} and amalgamating. For example, L_2 is obtained from L_1 and S_1 by deleting the dotted edge between them in Fig. 4.2.3. We take $S_2 = L_1$. Performing these deletions gives a Penrose tiling T_2 of $\mathbf{R}^2$ by copies of L_2 and S_2. Repeating the procedure gives for any n a tiling T_n of $\mathbf{R}^2$ by copies of L_n and S_n. The triangle S_n is always the same as L_{n-1}, and L_{n+2}, S_{n+2} are respectively similar to L_n, S_n but enlarged in the ratio of τ to 1. In going from L_n, S_n to L_{n+2}, S_{n+2}, the colors of the vertices are interchanged. One fixes a point P of $\mathbf{R}^2$ interior to one of the triangles of T and defines the sequence $\{x_n\} \in X_p$ by requiring $x_n = 0$ if P lies in an L_n triangle of T_n, and $x_n = 1$ if P lies in an S_n triangle of T_n. The fact that $x_n = 1$ implies that $x_{n+1} = 0$ follows since $L_n = S_{n-1}$. Conversely, given $\{x_n\}$ in X_p, the x_n's tell us how to construct a tiling T given that sequence. For example, if $x_1 = 0$, then we start with an L_1 triangle containing P. If $x_2 = 0$, then that L_1 triangle is amalgamated with an S_1 to give an L_2 triangle. If $x_2 = 1$, then regard the L_1 triangle as an S_2 triangle. Continuing in this way, we get an increasing sequence of triangles which can be extended to give a Penrose tiling.

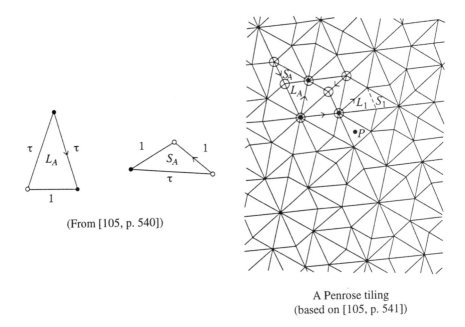

(From [105, p. 540])

A Penrose tiling
(based on [105, p. 541])

Figure 4.2.3

However, clearly, the index sequence constructed above for the tiling T depends on the choice of P and rotated tilings yield the same index sequence, and it is here that the equivalence relation R_p, discussed in **2.2** and **3.1**, arises. Identifying two tilings if they differ by a rotation, it can be shown that a change in the choice of P only changes the sequence $\{x_n\}$ in a finite number of places, and conversely, if two such sequences differ in only a finite number of places, then we get the same tiling. It follows, as Connes points out in his book, that the set of Penrose tilings is a *noncommutative space*. (The leaf space of the Kronecker foliation gives another example of a noncommutative space. (See Appendix F.)) For, the set of Penrose tilings is then identified with the quotient space X_p/R_p (recall that $\{x_n\}R_p\{y_n\}$ if and only if $x_n = y_n$ eventually) and *the quotient topology on X_p/R_p is trivial*. It also follows that there are uncountably many distinct Penrose tilings. We note the remarkable result (cf. [105, p.562]) which asserts that *every finite pattern in a given Penrose tiling is congruent under translation to infinitely many patterns in any Penrose tiling.*

The equivalence relation groupoid R_p is, as we saw in **2.2**, an r-discrete groupoid in the inductive limit topology. In **3.1**, we discussed the C^*-algebra of R_p and and showed (following Connes) that it was an AF-algebra. Connes goes on ([56, p.93]) to interpret the occurrence of the golden number in results on the density of tiles or patterns in Penrose tilings in terms of the unique trace on that algebra and the latter's K_0-group.

It follows from the above that given any pattern of a Penrose tiling, it does not make sense to ask *which* tiling the pattern is in – it is in *every* Penrose tiling. The empirical evidence of a quasicrystal will only give such a pattern, so that if we want to model the quasicrystal using such a tiling, we have to consider the *set* of Penrose tilings, none being privileged. It is in this way that a quasicrystal so modelled is the noncommutative space X_p/R_p which (as we have seen) is studied through the r-discrete equivalence relation groupoid R_p and its associated C^*-algebra.

With this background, we can now consider a version of some of the remarkable results of J. Kellendonk ([144, 141, 142, 143]).[3] His results operate within the context of a very general class of tilings (which includes the Penrose tilings). Kellendonk produces a compact space Ω which is the unit space of a certain r-discrete groupoid, the construction not relying on index sequences (the latter being special to the Penrose case). Instead, the key idea is to use equivalence classes of *pointed* tiles, patterns and tilings under translation in $\mathbf{R}^d$. The K_0-groups of the C^*-algebra of this r-discrete groupoid give gap labelling information for the spectrum of discrete Schrödinger operators on the tiling.

Each of the sets X_p in the Penrose case and the set Ω in the general case is associated with a set of tilings, the latter being obtained from the former as the set of equivalence classes under a natural equivalence relation. In the case when T_0 is a Penrose tiling, the space Ω is not the same as the space X_p. In addition, unlike that X_p-case, the r-discrete groupoid that will be considered for the Ω case is not an equivalence relation. Also, as Kellendonk points out ([141, p.39]), the coding of the Penrose tilings by $0, 1$ sequences preserves equivalence classes of tilings not just under translations but *also* under rotations and reflections. So in the Penrose case, X_p is actually a quotient of Ω by "orientational symmetry".

After discussing some preliminaries on tilings and pointed classes, we will construct a variant of Kellendonk's r-discrete groupoid. The finite type condition that we require is slightly weaker than that of [143], and an advantage for the purposes of the present work is that the groupoid $G(T_0)$ that we construct below fits in well with Example 4. In fact, it is a reduction of a vertex groupoid. The author is uncertain about the precise relationship between the r-discrete groupoid $G(T_0)$ below and that given in [142, 143], but he believes that the latter is a homomorphic image of the former (and that the same holds also at the inverse semigroup level).[4]

(In [143], Kellendonk compares his construction with the construction of the universal groupoid in [197]. The universal groupoid will be consid-

---

[3]The author is grateful to Johannes Kellendonk for very helpful correspondence.

[4]Kellendonk comments ([142, p.1173]) that under certain special circumstances, the hull can be identified with a space of paths on a graph.

ered in detail in the next section. The relation between universal groupoids and localization groupoids (a reduction of one of which is the groupoid constructed in this example) is considered in **4.3**, Example 4.) In his construction, Kellendonk uses a certain "almost groupoid", which, when a 0 is adjoined, gives an inverse semigroup. There is also, as we saw in Example 3, an inverse semigroup associated with an admissible directed graph with finite X_i's. Inverse semigroups are not used in the construction of the r-discrete groupoid below, but, as we will see at the end, there is a natural inverse semigroup S such that that groupoid is an S-groupoid.

We start by defining what we mean by a *tiling*. Precisely (cf. [105, p.16ff.]), a *tiling* of $\mathbf{R}^d$ is defined to be a cover T of $\mathbf{R}^d$ by compact sets A, where each A is homeomorphic to the closed ball $\{x \in \mathbf{R}^d : \|x\|_2 \leq 1\}$, and if $A, B \in T$, then $A \cap B$ is contained in the boundary of A (and B). The elements of T are called *tiles*. Each tile is the closure of its interior and its boundary is a null set. The tiles need not be polyhedra. Since each tile interior A^0 is non-empty and the set $\{A^0 : A \in T\}$ is pairwise disjoint, it follows that T is countable. As in the case of the Penrose tilings, we allow possible decorations of the tiles. A *pattern* is defined to be a finite subset P of the tiling T. In what follows, T_0 is a fixed tiling of $\mathbf{R}^d$.

An interesting feature of Kellendonk's theory is that it uses *pointed* pattern classes of T_0. A *pointed pattern* of T_0 is a pair (P, x), which we will usually abbreviate to Px, where P is a pattern and $x \in P$. (So x is a tile of P.) The notion extends in the obvious way to *pointed subsets* of T_0.

We now consider the equivalence relation associated with the (partial) action of the translation group of $\mathbf{R}^d$ on the pointed subsets of T_0. Firstly, for $a \in \mathbf{R}^d$ and a tile $A \in T_0$, define $A + a \subset \mathbf{R}^d$ by: $A + a = \{z + a : z \in A\}$. If P is a pattern, then we define $P + a = \{A + a : A \in P\}$. If Px is a pointed pattern of T_0, then define $Px + a$ to be the pair $(P + a)(x + a)$. The relation $\sim$ on the set of pointed patterns of T_0 is defined by specifying that $Px \sim P'x'$ if, for some a, $Px + a = P'x'$ and the decorations on any tile of P are preserved under the a translation. (Equality between pointed subsets will always be understood to include the decorations being preserved, as well as their identity as sets.) Trivially, $\sim$ is an equivalence relation. The equivalence class of Px is denoted by $\overline{Px}$. We define

$$\mathcal{P}_p = \{\overline{Px} : Px \text{ is a pointed pattern of } T_0\}.$$

For a tile x, we will write $\overline{x}$ in place of $\overline{\{x\}x}$.

Of course in general, $\mathcal{P}_p$ could be very large, so that the pointed pattern equivalence classes would be small, and taking them would not help much. However, for the tilings that we will be considering, those satisfying the finite type condition defined below, the equivalence classes *are* very large. One can see this in the case of a Penrose tiling. In that case, each pattern class $\overline{Px}$ is infinite, since, as observed earlier, there are infinitely many translations that can take a given pattern of such a tiling onto another pattern of the tiling.

A physical reason for considering equivalence classes is that in terms of modelling a quasicrystal, what we observe is obviously a *pattern* (a finite part of the tiling), and the pattern stays the same if the quasicrystal is translated around.[5]

For any pattern P in T_0, let $d(P)$ be the diameter of $\cup P$. Obviously, if P, Q are equivalent patterns, then $d(P) = d(Q)$. For any pointed pattern class $\overline{Px}$, we can thus define $d(\overline{Px}) = d(P)$. We will assume the following finite type condition (cf. [143, p.6]) on the tiles:

$d_{max} = \sup_{A \in T_0} d(A) < \infty$, *and for any $r > 0$, the set $\{\overline{Px} : d(\overline{Px}) < r\}$ is finite.*

Very many tilings in the literature satisfy this condition.[6]

This condition ensures some kind of translational symmetry for T_0 in a "bounded" sense, i.e. up to translation equivalence, only finitely many pointed pattern classes of diameter $< r$ are possible. Here are some useful consequences of the finite type condition. The existence of $d_{\max}$ implies that the total set of classes of *tiles* is finite. Since Lebesgue measure λ on $\mathbf{R}^d$ is preserved under translation, it also follows that $\min\{\lambda(A) : A$ is a tile $\} > 0$. Next, using the definition of *tiling* given earlier, we see that if $A \in T_0$, then $\lambda(A) = \lambda(A^0) > 0$, and that if $B \in T_0$ and $A \neq B$, then $A^0 \cap B^0 = \emptyset$. It follows that any bounded region of $\mathbf{R}^d$ contains only finitely many tiles.

Using the terminology of Example 4, we now construct a directed graph $V = \cup_{r=0}^{\infty} X_r$ associated with T_0. Without loss of generality, we can suppose that $d_{\max} < 1$. Define $X_0 = \{\overline{x} : x$ is a tile $\}$. For $r \geq 1$, let

$$X_r = \{\overline{Px} \in \mathcal{P}_p : d(\overline{Px}) < r\}.$$

Obviously, $X_r \subset X_{r+1}$ and $V = \cup_{r=1}^{\infty} X_r = \mathcal{P}_p$. Each X_r is finite by the finite type condition. We now need to specify when there is an arrow from $\overline{Px} \in X_r$ to $\overline{Qy} \in X_{r+1}$. We say that $\overline{Px} \leq \overline{Qy}$ if there is a subset P' of Q such that $\overline{Px} = \overline{P'y}$. It is easy to check that the partial ordering $\leq$ is well

---

[5]Of course, this also holds when a general isometry is applied, not just a translation. Indeed for the pointed equivalence classes above, one could have used the full isometry group instead of just its translation subgroup. Also, as was mentioned earlier, the full isometry group has to be considered when relating the coding of Penrose tilings by 0, 1-sequences to the approach of Kellendonk. The use of the translation subgroup in that approach is due to the fact that the groupoid resulting gives the correct C^*-algebra for the purposes of K-theoretic gap labelling.

[6]I am grateful to Johannes Kellendonk for pointing out that Penrose tilings, Amman-Beenker tilings and Socolar tilings do satisfy the finite type condition, but that the pinwheel tilings ([253, p.220]) are not of finite type.

defined on V. We specify that there is an edge from $\overline{Px}$ to $\overline{Qy}$ whenever $\overline{Px} \leq \overline{Qy}$. Then V with these edges is a directed graph $\mathcal{G}$.

We now claim that $\mathcal{G}$ is admissible. For if $v = \overline{Px} \in X_r$, then there exists an edge from v to some element w of X_{r+1}. For let $a, b \in \cup P$ be such that $\|a - b\|_2 = d(P)$. Then a is not in the interior of any tile of P, and there is a tile A of T_0 not in P such that a lies on the boundary of A. Let $Q = P \cup \{A\}$. Since $d(A) < 1$ and $d(P) < r$, we have $d(Q) < (r+1)$. So $\overline{Px} \leq \overline{Qx}$ and $\overline{Qx} \in X_{r+1}$. Hence there exists an edge e such that $d(e) = \overline{Px}$. If $v = \overline{Px} \in X_r \sim X_0$ then, by taking away tiles from P, we obtain an edge e' such that $v = r(e')$. If $v \in X_0$ then there is no edge whose range is v. So X_0, as we have defined it, is the set of initial vertices (in the terminology of Example 4).

So $\mathcal{G}$ is an admissible directed graph with finite X_i's, and the results of Example 4 apply. In particular, there exists an r-discrete Hausdorff groupoid $G_{\mathcal{G}}$ whose elements are of the form $(\alpha\gamma, \ell(\alpha) - \ell(\beta), \beta\gamma)$ and an inverse semigroup $S_{\mathcal{G}}$ whose non-zero elements are pairs of finite vertex paths (α, α') with $r(\alpha) = r(\alpha')$. The unit space X of $G_{\mathcal{G}}$ is the set of infinite paths $(v_0, v_1, \ldots)$ in $\mathcal{G}$.

We now need to interpret the elements of X as pointed *subsets* of T_0. The idea (cf. [142, p.1141]) is that we translate the components $\{v_r\}$ back to the initial tile class and take the resultant union. This gives the pointed subset that we require. Precisely, let $x = (v_0, v_1, \ldots) \in X$ and write $v_r = \overline{P_r x_r} \in X_r$. Since $v_0 \leq v_1 \leq \ldots$, there exists, for each r, an element $b_r \in \mathbf{R}^d$ such that $P_r + b_r \subset P_{r+1}, x_r + b_r = x_{r+1}$. Let $a_r = \sum_{i=0}^{r-1} b_i$, $a_0 = 0$. Note that $x_r = x_0 + a_r$. Then for all r,

$$\{x_0\} = P_0 - a_0 \subset P_r - a_r \subset P_{r+1} - a_{r+1}.$$

Let $M = \cup(P_r - a_r)$. Then M is a "tiling" of a region in $\mathbf{R}^d$. The dependence of M on the choice of the $P_r x_r$'s can be removed by associating x with the pointed subset class $\overline{Mx_0}$ of T_0. Let $\alpha(x) = \overline{Mx_0}$. Note that every pattern of M is equivalent to a pattern of T_0 (since it is equivalent to a pattern contained in some P_r).

Of course, we are really interested in tilings of the *whole* of $\mathbf{R}^d$ in the sense defined earlier. Note first that $d(x_r) = d(x_0)$ since the x_i's are equivalent. To characterize the x for which the M is a tiling, let $d = d(x_0)$, and for each $N > 0$, let U_N^r be an open ball in $\mathbf{R}^d$ of radius N such that every point in x_r is distance $\geq (N - d)$ from the boundary of the ball. Then M is a tiling if and only if for each N, there exists an r such that $P_r \supset U_N^r$. Indeed, after translating, M contains every ball U_N^0, and the union of these balls is $\mathbf{R}^d$.

Elements x for which M is a tiling can be explicitly constructed as follows for a given sequence of equivalent tiles $\{x_r\}$. Let $a_r \in \mathbf{R}^d$ be such that $x_r - a_r = x_0$. We can construct recursively a sequence of pointed patterns $P_r x_r$ such that P_r is a *maximal pattern* in $U_{r/2}^r$ and $P_r - a_r \subset P_{r+1} - a_{r+1}$ for each r. Note that $d(U_{r/2}^r) < r$ so that $\overline{P_r x_r} \in X_r$. Then set $x = (v_0, v_1, \ldots)$ where $v_r = \overline{P_r x_r}$.

We now show that $M = \cup(P_r - a_r)$ is a tiling for $\mathbf{R}^d$. To this end, if there is a point of $U_{r/2}^r$ not in $\cup P_r$ whose distance from the boundary of $U_{r/2}^r$ is $> d_{\max}$, then, since T_0 covers $\mathbf{R}^d$ and every tile has diameter $\leq d_{\max}$, we contradict the maximality of P_r. So $\cup P_r$ contains an open ball B with $x_r \subset B$ and such that the boundary of B is of distance $\geq (r/2 - d - d_{\max})$ from every point of x_r. In particular, the radius of B is $\geq (r/2 - d - d_{\max})$. Then $P_r - a_r$ contains x_0 and $\cup(P_r - a_r)$ contains a ball B_r whose boundary is of distance $\geq (r/2 - d - d_{\max})$ from every point of x_0. So M covers $\mathbf{R}^d$ as required.

We write $x = \beta(\{x_r\})$. The sequence $\{x_r\}$ can be regarded as a generalized "index sequence" for the tiling M. An element x of X of the type just constructed will be called *maximal*.

The simplest example of this construction is in the case where $x_r = x_0$ for all r, so that all of the a_r's are 0. In that case, M is just T_0 so that $x = \alpha(\overline{T_0 x_0})$. In the general case, however, tilings which are different from T_0 arise through the construction since pattern classes based at other tiles are being translated and piled up around the one fixed tile x_0.

Different x's can give the same tiling $\alpha(x)$. However, for any $y \in X$ giving a tiling, we can find a maximal x so that $\alpha(x) = \alpha(y)$. In this regard, it is left to the reader to check that if $y = \{\overline{Q_r y_r}\}$ gives a tiling, and if $x = \beta(\{y_r\}) = \{\overline{P_r y_r}\}$, then for each r, there is an s such that $\overline{Q_r y_r} \leq \overline{P_s y_s}$ and conversely, so that $\alpha(y) = \alpha(x)$.

So we get all of the tilings associated with elements of X by considering only the maximal ones. Let Ω be the set of maximal x's in X. Recalling that the topology on Ω is the product topology, it is easy to show that the set of elements of the form $\alpha(\overline{T_0 x_0})$ are dense in Ω. (This is effectively how the "hull" is defined in [142]).

We now claim that Ω is a closed (and therefore compact) subset of X. Indeed, any cluster point x of Ω in X is, as above, such that $P_r - a_r$ contains B_r and so $x \in \Omega$. So (cf. the comments before Lemma 2.3.1) the reduction of the vertex groupoid $G_{\mathcal{G}}$ to Ω is an r-discrete, Hausdorff groupoid. (Note, however, that Ω is not an invariant subset of X.) This is the groupoid $G(T_0)$ whose construction was our objective in this part of the example.

It remains to consider the inverse semigroup aspect of $G(T_0)$. In [141], Kellendonk constructs his r-discrete groupoid using a right action of the inverse semigroup of *doubly pointed* pattern classes (with zero adjoined) on its unit space. This action is a localization and Theorem 3.3.2 gives that groupoid. In the approach adopted above, there is a natural inverse semigroup $S(T_0)$ associated with $G(T_0)$. Here, $S(T_0)$ is the inverse subsemigroup of elements $(\alpha, \alpha') \in S_{\mathcal{G}}$ for which α, α' are *maximal*, i.e. for some γ, both $\alpha\gamma, \alpha'\gamma$ belong to Ω. The action of $S(T_0)$ on Ω is just that obtained by restricting to Ω the action of $S(T_0) \subset S_{\mathcal{G}}$ on X, and the pair $(\Omega, S(T_0))$ is a localization. Straightforward calculation shows that $G(T_0)$ is the localization groupoid associated with $(\Omega, S(T_0))$ through Theorem 3.3.2.

4.3 The universal groupoid of an inverse semigroup

Let S be an inverse semigroup and let $E = E(S)$. In this section, we will construct an ample groupoid $G_{\mathbf{u}}$ associated with S. This groupoid, which will be called (for categorical reasons) the *universal groupoid* for S, determines all of the representation theory of S. Recall (**2.2**) that a locally compact groupoid G is defined to be *ample* when the family G^a of compact, open, Hausdorff G-sets is a basis for the topology of G.

Let S be an inverse semigroup. We now want to specify the class of ample groupoids G that are naturally associated with S. It is reasonable to restrict attention to those ample groupoids G which are determined by a homomorphic image of S in G^a. Such groupoids, which will be called *S-groupoids*, are defined precisely below. The appropriate formulation of an *S-groupoid* for unital S was given in [197]. The definition below extends this formulation to the nonunital case.

It may be wondered why, in this definition of S-groupoid, we should allow only ample rather than the more general, r-discrete, groupoids with which we were concerned in Chapter 3. Examples do indicate that the ample class is the correct class when starting from an abstract inverse semigroup. For example, if S is a semilattice then the natural groupoid associated with S is the groupoid of units given by the space of non-zero semicharacters of S, and this groupoid is totally disconnected (i.e. ample). Further, in the case of the Cuntz inverse semigroup S_n, the Cuntz groupoid G_n of **4.2**, Example 3 is ample.

The reason why the localization context of Chapter 3 does not apply directly here is, as noted earlier, that there are ample groupoids G associated with S, groupoids that we want to count as S-groupoids, such that the natural right action of S on G^0 fails to satisfy the basis condition of a localization. However, in the ample context, the domains of the transformations associated with the elements of S are *compact* as well as open, so that their complements in G^0 are also open. (The use of complements of domains to "cut down" the size of open sets is reflected in the presence of the sets $\psi(e_i)^c$ in the following definition of an S-groupoid.) Then by taking finite intersections of domains D_s with such complements in G^0, we actually *do* obtain a basis, and this allows the localization theory of Chapter 3 to be applied, the semigroup of the localization being an inverse semigroup S' containing S.

Definition 4.3.1. Let G be an ample groupoid and $\psi : S \to G^a$ be a homomorphism. Then the pair (G, ψ) is called an S-*groupoid* if:

(i) $\cup_{s \in S} \psi(s) = G$;

(ii) the family of sets of the form

$$U_{e,e_1,\cdots,e_n}$$

where, for $e, e_1, \cdots, e_n \in E$,

$$U_{e,e_1,\ldots,e_n} = [\psi(e) \cap \psi(e_1)^c \cap \cdots \cap \psi(e_n)^c], \qquad (4.18)$$

is a basis for the topology of G^0. (Here, $A^c = G \sim A$ for $A \subset G$.)

It follows from (i), (ii) and the fact that the $\psi(s)$'s are in G^a that a basis for the topology of G is given by sets of the form $U_{e,e_1,\cdots,e_n}\psi(s)$. These sets are also in G^a.

The Cuntz groupoid G_n of **4.1**, Example 3 is an example of an S_n-groupoid where $\psi(s_\alpha t_\beta) = U_{\alpha,\beta}$. We note that in this example, condition (ii) of the above definition follows since the family of sets $U_{\alpha,\alpha} = \psi(s_\alpha t_\alpha)$ is a basis for the topology on G_n^0. The equation (4.13) effectively expresses the fact that the sets $U_{e,e_1,\ldots,e_n}\psi(s)$ form a basis for the topology of G_n. In that case, each $U_{\alpha,\alpha}$, as we have seen, is, in the notation of (ii), just a $U_e = \psi(e)$ and we do not need to consider complements $\psi(e_i)^c$. (The pair (G_n^0, S_n) is a localization.)

The main objective of this section is to show that for general S, there exists an explicitly constructible S-groupoid $G_\mathbf{u}$ which is, in a sense to be described, *universal*. (We will see later that G_n is not the universal groupoid for S_n.)

We first introduce the space X which will be the unit space of $G_\mathbf{u}$. (We met X earlier in the course of proving Wordingham's theorem (Theorem 2.1.1).) It is simply the maximal ideal space of the (commutative) convolution Banach algebra $\ell^1(E)$, X being given the Gelfand topology. (Recall that $E = E(S)$.) Equivalently, since every multiplicative linear functional on $\ell^1(E)$ is determined by its values on the elements of the commutative semigroup E and these are all idempotent, it follows that X is the set of non-zero semicharacters $x : E \to \{0, 1\}$ with the topology of pointwise convergence. Then X is a locally compact, totally disconnected Hausdorff space. For computational purposes, it is sometimes useful to regard the elements of X as the *filters* of E. (The definition of *filter* is given below.) The space X is then called the *filter completion* of E ([103, 78, 81]).

A *filter* in the semilattice E is a subsemigroup A of E with the property that if $f \in E$, $e \in A$ and $f \geq e$, then $f \in A$. (So a filter is a subsemigroup of E such that any element of E that is bigger than an element of the subsemigroup is in the subsemigroup.) For $x \in X$, define $A_x \subset E$ by :

$$A_x = \{e \in E : x(e) = 1\}.$$

Since x is a homomorphism, it follows that A_x is a subsemigroup of E. Also if $e \in A_x$, $f \in E$ and $f \geq e$, then $fe = e$, and so $x(f)x(e) = x(e)$ giving that $f \in A_x$. So A_x is a filter.

Conversely, as is easily checked, every filter A in E equals A_x for some $x \in X$ – take x to be the characteristic function of A. Of importance for this chapter is that the elements of E can themselves be regarded as filters: we associate with each $e \in E$ the filter $\bar{e} = \{f \in E : f \geq e\}$. As a semicharacter on E, $\bar{e}$ is given by: $\bar{e}(f) = 1$ if $f \geq e$ and is 0 otherwise. (We also met $\bar{e}$ earlier in **2.1**.) Of course, $\overline{E}$ is the set of $\bar{e}$'s ($e \in E$). We now discuss the topology on X.

Recall that the topology of X is that of pointwise convergence on E. Sets of the form

$$\{y \in X : |\, x(e_i) - y(e_i)\,| < \epsilon\} \tag{4.19}$$

where $x \in X$, $0 < \epsilon < 1$ and $e_1, \ldots, e_n \in E$, thus give a basis of compact open sets for the topology of X. Reordering the e_i's if necessary, we can suppose that for some r, $x(e_i) = 0$ for $1 \leq i \leq r$ and $x(e_i) = 1$ for $r + 1 \leq i \leq n$. Setting $e = e_{r+1} \cdots e_n$, the set of (4.19) will be denoted by $D_{e,e_1,\ldots,e_r}$, and we have

$$D_{e,e_1,\ldots,e_r} = \{y \in X : y(e) = 1, y(e_i) = 0, 1 \leq i \leq r\}. \tag{4.20}$$

As commented above, the set $D_{e,e_1,\ldots,e_r}$ is a compact open subset of X. By replacing e_i by ee_i, we can take $e_i \leq e$. Since we are only interested in non-empty basis sets $D_{e,e_1,\ldots,e_r}$, we can in fact suppose that each $e_i < e$. For $e \in E$, the set D_e is defined by:

$$D_e = \{x \in X : x(e) = 1\}. \tag{4.21}$$

Note that

$$D_{e,e_1,\ldots,e_r} = D_e \cap D_{e_1}^c \cap \cdots \cap D_{e_r}^c, \tag{4.22}$$

complementation being taken in X. (The connection between the sets of (4.22) and the $U_{e,e_1,\ldots,e_n}$ in Definition 4.3.1 is apparent.)

Since $\bar{e} \in D_{e,e_1,\ldots,e_r}$ and the latter sets form a basis for X, we get the following easy, well-known result:

Proposition 4.3.1 $\overline{E} = \{\bar{e} : e \in E\}$ *is a dense subset of* X.

We now describe a natural right action of S on X. For $s \in S$, let

$$D_s = \{x \in X : x(ss^*) = 1\} = D_{ss^*} \qquad (4.23)$$

and $R_s = D_{s^*}$. Of course, if $e \in E$, then the D_e defined in (4.23) coincides with the earlier D_e in (4.21). Each D_s is a compact open subset of X, and from (4.22), the family $\{D_e : e \in E\} \cup \{D_e^c : e \in E\}$ is a sub-basis for the topology of X. For $x \in X$, $s \in S$ we will sometimes write $x \leq s$ if $x \in D_s$.

Define a map $\beta(s) : D_s \to \ell^\infty(E)$ by:

$$\beta(s)(x)(e) = x(ses^*). \qquad (4.24)$$

We will usually write $x.s$ in place of $\beta(s)(x)$. One readily checks that if $e \in E$ where $e \leq ss^*$, then

$$\bar{e}.s = \overline{s^*es}. \qquad (4.25)$$

(The action of S on $\overline{E}$ given in (4.25) is effectively that of (3.50).)

Proposition 4.3.2   The map $x \to x.s = \beta(s)(x)$ is a homeomorphism from D_s onto R_s and the map $s \to \beta(s)$ defines a right action of S on X.

Proof.   Let $s \in S$, $x \in D_s$ and $e, f \in E$. Then $x.s(ef) = x(sefs^*) = x(sefs^*ss^*) = x(ses^*sfs^*) = [x.s(e)][x.s(f)]$. Since $x.s(s^*s) = x(ss^*) = 1$, it follows that $x.s \in X$. It also follows that $D_s.s \subset D_{s^*} = R_s$. Further, $D_{s^*}.s^* \subset D_s$. Since $(x.s).s^*(e) = x(ss^*ess^*) = x(e)x(ss^*) = x(e)$, it follows that $(x.s).s^* = x$. Clearly, the map $x \to x.s$ is a homeomorphism from D_s onto R_s with inverse $y \to y.s^*$. Next, the domain of $x \to (x.s).t$ is $D_t.s^* = \{x \in X : x.s(tt^*) = 1\} = \{x \in X : x(st(st)^*) = 1\} = D_{st}$, and it is immediate that $(x.s).t = x.st$.   □

From (3.97), we see that if $f \leq ss^*$ in E, then

$$y \leq f \Leftrightarrow y.s \leq s^*fs. \qquad (4.26)$$

To deal with the fact that, as discussed above, the pair (X, S) is usually not a localization, we extend the action of S on X to a larger inverse semigroup S' whose action *will* give a localization. The semigroup S' is located in the semigroup algebra $\mathbf{C}(S)$ of S.

Let E' be the subset of $\mathbf{C}(S)$ whose elements are of the form

$$e' = e(1 - e_1) \cdots (1 - e_n) \qquad (4.27)$$

for $e, e_i \in E$. (The 1 in the expressions $(1 - e_i)$ of (4.27) is just a convenient formal identity, the right-side of (4.27), when multiplied out, belonging to $\mathbf{C}(S)$.) We include in this definition the case where there are no e_i's, so that $E \subset E'$. As earlier, we can suppose above that every $e_i \leq e$. Note that the zero 0 of $\ell^1(E)$ belongs to E' since $e(1-e) \in E'$ for any e. Let S' be the subset of $\ell^1(S)$ whose elements are of the form $e's$ for some $e' \in E', s \in S$. Clearly, $s = (ss^*)s \in S'$ so that $S \subset S'$. If e' is as in (4.27), then by passing the ss^* through the product (recall that the idempotents commute in S) we have

$$e's = ess^*(1 - e_1 ss^*) \ldots (1 - e_n ss^*)s,$$

and so can suppose that the e, e_i satisfy $e_i \le e \le ss^*$.

Since E is a semilattice, it easily follows that E' is also a semilattice under the multiplication inherited from $\mathbf{C}(S)$. We note also that if $e' = e(1 - e_1) \cdots (1 - e_n) \in E'$ and $s \in S$, then $se's^* = ses^*(1 - se_1 s^*) \cdots (1 - se_n s^*) \in E'$. Further, $(e's)(f't) = (e'sf's^*)st \in S'$, and $(e's)^* = (s^* e's)s^* \in S'$. It follows that S' is a *-subsemigroup of $\ell^1(S)$. Also

$$(e's)^*(e's) \in E'. \tag{4.28}$$

The semigroup S' is actually an inverse semigroup. Perhaps the quickest way to see this is to observe that, using Wordingham's theorem (Theorem 2.1.1), the faithful image of S' under the left regular representation of $\ell^1(S)$ on $\ell^2(S)$ is, using (4.28), a *-semigroup of partial isometries on a Hilbert space and hence (Proposition 2.1.4) is an inverse semigroup. The next proposition extends the action of S on X to one of S' on X.

Proposition 4.3.3   Let $s' = e's \in S'$ with $e' = e(1 - e_1) \cdots (1 - e_n)$ and $e_i < e \le ss^*$. Then

$$D_{s'} = D_e \cap D_{e_1}^c \cap \cdots \cap D_{e_n}^c \tag{4.29}$$

is well-defined, and the map $x \rightarrow x.s = \beta_{s'}(x)$ $(x \in D_{s'})$ defines a right action of S' on X such that (X, S') is a localization. This action restricts to the action of S on X given in Proposition 4.3.2.

Proof.   The proof is modelled on the construction of the left regular representation for an inverse semigroup (**2.1**). For $s \in S$, define $T : S \rightarrow B(\ell^2(X))$ by:

$$T(s)\left(\sum_{x \in X} a_x x\right) = \sum_{x \in D_s} a_x x.s.$$

(Of course, $T(s)(x) = 0$ whenever $x \in D_s^c$.) Direct checking shows that $T(s)$ is a partial isometry from $\ell^2(D_s)$ onto $\ell^2(R_s)$ and that $T(s)^* = T(s^*)$.

The map $s \rightarrow T(s)$ is an antihomomorphism (using Proposition 4.3.2), and it extends to a *-algebra antihomomorphism, also denoted by T, from $\mathbf{C}(S)$ into $B(\ell^2(X))$. If s' and $D_{s'}$ are as in the statement of the proposition, it is easily checked that $T(s')$ is a partial isometry from $\ell^2(D_{s'})$ onto $\ell^2(D_{(s')^*})$. Restricting $T(s')$ to $X \subset \ell^2(X)$ gives $T(s')(x) = x.s'$ for $x \in D_{s'}$ and is 0 otherwise. (In particular, $D_{s'}$ is independent of the decomposition

$s' = e's$.) It now follows that the maps $x \to x.s$ ($x \in D_{s'}$) define a right action of S' on X and this obviously extends the given right action of S on X. That the pair (X, S') is a localization follows since the sets of (4.29) (with $r = n$) are the same as the basis sets of (4.22). □

Note that by considering elements $s'(s')^*$ with $s' \in S'$, we see that $E' = E(S')$.

The existence and indeed construction of the universal groupoid $G_\mathbf{u}$ associated with S now follows from Theorem 3.3.2. Specifically, $G_\mathbf{u}$ is just the r-discrete groupoid $G(X, S')$ for the localization (X, S').

So $G_\mathbf{u} = \Xi / \sim$ where Ξ is the set of pairs (x, s') with $s' \in S'$ and $x \in D_{s'}$, and we define $(x, s') \sim (y, t')$ to mean that $x = y$ and there exists $e' \in E'$ such that $x \leq e'$ and $e's' = e't'$. As in **3.3**, $\psi : \Xi \to G_\mathbf{u}$ is the quotient map. The next lemma enables us to formulate the above construction of $G_\mathbf{u}$ in terms of S rather than S'.

Proposition 4.3.4   Let $\Sigma = \{(x, s) : x \in D_s, s \in S\}$ and define an equivalence relation R on Σ by requiring $(x, s)R(y, t)$ if and only if $x = y$ and there exists $e \in E$ such that $x \leq e$ and $es = et$. Then the inclusion map from Σ into Ξ induces a bijection Θ from Σ/R onto $G_\mathbf{u}$.

Proof.   That R is an equivalence relation on Σ follows from Lemma 3.3.1. Obviously, if $(x, s)R(x, t)$ then, since $E \subset E'$, we have $(x, s) \sim (x, t)$. So the inclusion map from Σ into Ξ induces a map $\Theta : \Sigma/R \to \Xi/\sim = G_\mathbf{u}$. Now Θ maps onto $G_\mathbf{u}$. For let $(x, s') \in \Xi$ where $s' = e's$ with $e' \in E', s \in S$. Then $x \leq e'$ and $e's' = e's$ so that $(x, s') \sim (x, s)$. So Θ is onto.

It remains to show that Θ is one-to-one. Suppose that $(x, s) \sim (x, t)$ in Σ. We have to show that $(x, s)R(x, t)$. Since $(x, s) \sim (x, t)$, there exists $e' \in E'$ such that $x \leq e'$ and $e's = e't$. We can write $e' = e(1-e_1) \cdots (1-e_n)$ where for each i, $e_i < e \leq (ss^*)(tt^*)$. Expanding e' on both sides of the equality $e's = e't$ gives

$$es - \sum \alpha_p g_p s = et - \sum \alpha_p g_p t \qquad (4.30)$$

where for each p, g_p is a product of some e_i's and α_p is ± 1. Since for each p, $g_p < e$, we have $(g_p s)(g_p s)^* = g_p < e = (es)(es)^*$ so that $g_p s \neq es$. Similarly, $g_p t \neq et$. So for the equality in (4.30) to hold in $\mathbf{C}(S)$, we must have $es = et$. It follows that $(x, s)R(x, t)$ and the proof is complete. □

It follows from the above proposition that $(x, s) \sim (x, t)$ in Σ if and only if $(x, s)R(x, t)$. We now reformulate the construction of $G_\mathbf{u}$ in terms of Σ. Firstly, the set Σ has product and involution exactly as in (3.100) and (3.101). We take $G_\mathbf{u} = \Sigma/R$. As in **3.3**, we let $(x, s) \to \overline{(x, s)}$ be the quotient map from Σ into $G_\mathbf{u}$. The map ψ'_u from S' into the power set of $G_\mathbf{u}$, corresponding to the ψ_X of Theorem 3.3.2, is given by: $\psi'_u(s') = \{\overline{(x, s')} : x \in D_{s'}\}$. We sometimes write $D(U, s')$ in place of $\psi'_u(s')$. Let ψ_u be the restriction of ψ'_u to S.

Theorem 4.3.1   *The pair $(G_\mathbf{u}, \psi_u)$ is an S-groupoid, with its set $G_\mathbf{u}{}^2$ of composable pairs being those of the form $(\overline{(x, s)}, \overline{(x.s, t)})$ $(x \in D_s$, $x.s \in D_t$, $s, t \in S)$. The product and involution on $G_\mathbf{u}$ are determined by*

$$\overline{(x, s)}, \overline{(x.s, t)}) \to \overline{(x, st)}$$

and $\overline{(x, s)} \to \overline{(x.s, s^*)}$, respectively. The unit space of $G_\mathbf{u}$ is canonically identified with X. A basis for the topology of $G_\mathbf{u}$ is given by the family of sets of the form $D(U, s)$, where $s \in S$, U is an open subset of D_s, and

$$D(U, s) = \{\overline{(x, s)} : x \in U\}.$$

Proof.   Applying Theorem 3.3.2 to the localization (X, S') and using Proposition 4.3.4, we obtain that $G_\mathbf{u}$ $(= G(X, S'))$ is an r-discrete groupoid under the given product and inversion operations and under the topology determined by the sets $D(U, s) = \{\overline{(x, s)} : x \in D_s\}$ and that the unit space of $G_\mathbf{u}$ can be identified with X. Note that the family of $D(U, s)$'s coincide with the corresponding family of $D(U, s')$'s. We need to establish (i) and (ii) of Definition 4.3.1 for $G_\mathbf{u}$, and to show that $G_\mathbf{u}$ is ample. We shall write in this proof G, ψ, ψ' in place of $G_\mathbf{u}, \psi_u, \psi'_u$.

Firstly, for (i), the map ψ' is a homomorphism from S' into G^{op} with $\psi'(S')$ a basis for the topology of G. So the restriction $\psi : S \to G^{op}$ is a homomorphism, and since $\cup_{s \in S}\psi(s) = \cup_{s' \in S'}\psi'(s')$ and $\psi'(S')$ is a basis for G, it follows that $\cup_{s \in S}\psi(s) = G$. So (i) holds.

Turning to (ii), if $e' = e(1 - e_1) \cdots (1 - e_n) \in E'$ (as in our previous notation) then $\psi'(e') = D_{e'}$, and so by (4.29), with $G^0 = X$, we have

$$\psi'(e') = D_e \cap D^c_{e_i} \cap \cdots \cap D^c_{e_n} = \psi(e) \cap \psi(e_1)^c \cap \cdots \cap \psi(e_n)^c.$$

Now $\psi'(E')$, the family of $D_{e'}$'s, is a basis for X, and, by the above, $\psi'(e') = U_{e,e_1,\cdots,e_n}$ (given in (4.18)). So the family of sets $U_{e,e_1,\cdots,e_n}$ is a basis for the topology of X and (ii) holds.

Now we saw earlier that $\psi'(e') = D_{e'}$ is a compact open subset of X. Since, from the proof of Theorem 3.3.2, r, restricted to $\psi'(e's)$, is a homeomorphism onto the compact open set $D_{e'}$, it follows that $\psi(e's) \in G^a$. In particular $\psi : S \to G^a$. Since the family of sets $\psi(e')$ is a basis for the topology of X, it also follows that the family of sets $\psi(e's)$ forms a basis for the topology of G. In particular, G is ample. This concludes the proof of the theorem.   □

We note that if $x \in D_s$, then

$$\overline{(x.s, s^*)}\,\overline{(x, ss^*)}\,\overline{(x, s)} = \overline{(x.s, s^*s)} = x.s,$$

so that

$$x.s = x.\psi_u(s) = \psi_u(s)^{-1}x\psi_u(s). \qquad (4.31)$$

So the natural action of S on X corresponds to the natural action of the inverse subsemigroup $\psi_u(S)$ of $G_{\mathbf{u}}{}^a$ on the unit space X.

Our next objective is to show that $G_{\mathbf{u}}$ determines all S-groupoids. It is in this sense that it is *universal*.

Let H be a groupoid. A subset Y of H^0 is called *invariant* if for any $z \in H$, $r(z) \in Y$ if and only if $d(z) \in Y$. The simplest invariant subsets of H^0 are given by the *orbits*. The orbit $[u]$ of $u \in H^0$ is the set of elements $v \in H^0$ for which there exists $x \in H$ such that $r(x) = u, d(x) = v$. To see that $[u]$ is invariant, suppose that $y \in H$ and that $r(y) = v_1 \in [u]$. Then there exists $x \in H$ such that $r(x) = u, d(x) = v_1$. But then $r(xy) = u, d(xy) = d(y)$ whence $d(y) \in [u]$. So $[u]$ is invariant. It is easy to check that the relation $u \sim v$ in G^0 if and only if $[u] = [v]$ is an equivalence relation, the *orbit equivalence relation* for G.[7]

Further if, for some inverse semigroup S, (H, ψ) is an S-groupoid and T is a closed invariant subset of H^0, then $(H_T, \psi_{|T})$ is also an S-groupoid, where H_T is the reduction of H by T ((1.12)) and $\psi_{|T}(s) = \psi(s) \cap H_T$. The invariance of T is needed in proving that $\psi_{|T}$ is a homomorphism. Indeed, one only needs to show that if $A, B \in G^a$, then $(A \cap H_T)(B \cap H_T) = (AB \cap H_T)$. Clearly, the left-hand side of this equality is contained in the right-hand side. For the reverse inclusion, if $a \in A, b \in B$ with $ab \in H_T$, then $r(a) = r(ab) \in T$, and the invariance of T gives $a \in H_T$. A similar argument using the d map gives that $b \in H_T$, and the required inclusion follows.

Now let $(G, \psi), (H, \chi)$ be S-groupoids. Then a continuous groupoid homomorphism[8] $\Phi : G \to H$ is called an *S-homomorphism* if Φ is S-equivariant, i.e. $\Phi(\psi(s)) = \chi(s)$ for all $s \in S$.

The next proposition shows that every S-groupoid is obtained from $(G_{\mathbf{u}}, \psi_u)$ by the processes of reduction by a closed invariant subset and taking a suitable open, homomorphic image.

Proposition 4.3.5   *Let (G, ψ) be an S-groupoid. Then the unit space of G can be identified with a closed invariant subset Y of $X = G_{\mathbf{u}}{}^0$. Further for each $e \in E$ and $y \in Y$, $y \leq \psi(e)$ if and only if $y \leq \psi_u(e)$, and there is a continuous, open, surjective, S-homomorphism $\Phi : G_{\mathbf{u}|Y} \to G$ such that the restriction of Φ to Y is the identity map. (Here, $G_{\mathbf{u}|Y}$ is the reduction of $G_{\mathbf{u}}$ by Y.)*

---

[7]The reader can easily check that if G is an equivalence relation on a set X, then the orbit equivalence relation for G coincides with the original equivalence relation on X (when X is identified with G^0).

[8]So Φ is continuous, $(\Phi(x), \Phi(y)) \in H^2$ whenever $(x, y) \in G^2$ and in that case, $\Phi(xy) = \Phi(x)\Phi(y)$ and $\Phi(x)^{-1} = \Phi(x^{-1})$.

Proof. Let $w \in G^0$. Then the function $e \to \chi_{\psi(e)}(w)$ is a semicharacter $\alpha(w)$ on E which is nonzero by (ii) of Definition 4.3.1. So $\alpha(w) \in X$. The map $\alpha : G^0 \to X$ is one-to-one since G^0 is Hausdorff and the sets $\psi(e)$ determine the topology of G^0 (again by (ii) of Definition 4.3.1). Since each $\psi(e)$ is compact and open, we have $\chi_{\psi(e)} \in C_c(G)$, and it follows that if $w_\delta \to w$ in G^0, then $\alpha(w_\delta) \to \alpha(w)$ pointwise on E. So α is continuous from G^0 onto a subset Y of G_u^0. Note that by the definition of α, if $w \in G^0$, then $w \in \psi(e)$ if and only if $y = \alpha(w) \in \psi_u(e)$ (i.e. $y \le e$) for any $e \in E$. Suppose that for some net $\{w_\delta\}$ of G^0, we have $\alpha(w_\delta) \to v$ in G_u^0. Now $v \in \psi_u(e)$ for some $e \in E$. So eventually, $\alpha(w_\delta) \in \psi_u(e)$. So eventually, $w_\delta \in \psi(e)$, and since $\psi(e)$ is compact, we can suppose that $w_\delta \to w$ for some $w \in \psi(e)$. Then $v = \alpha(w)$ and it follows that α is a homeomorphism from G^0 onto Y and that Y is a closed subset of X.

We now show that Y is invariant in $G_u{}^0$. Let $z \in G_{u|Y}$ be such that $r(z) = y \in Y$. Then for some $s \in S$, we have $z = \overline{(y, s)}$ and so $d(z) = y.s$. In the process, we will also show that $y.s = \alpha(w.\psi(s))$ where $\alpha(w) = y$ and we are using the canonical right action of $\psi(S) \subset G^a$ on $Y = G^0$. (So $w.\psi(s) = \psi(s)^{-1}w\psi(s)$.) To this end, let $w \in G^0$ be such that $\alpha(w) = y$. Then $w \in \psi(ss^*)$. Using (3.97) and (4.24), we have $\alpha(\psi(s)^{-1}w\psi(s))(e) = \chi_{\psi(e)}(\psi(s)^{-1}w\psi(s)) = 1 \iff \psi(s^*)w\psi(s) \le \psi(e) \iff w \le \psi(ses^*) \iff \alpha(w)(ses^*) = 1 \iff (\alpha(w).s)(e) = 1$. So $y.s = \alpha(w.\psi(s)) \in Y$, and Y is invariant in $G_u{}^0$.

We will identify G^0 with Y, together with their right actions, for the remainder of the proof.

Let $y \in Y$, $s \in S$ and $z = \overline{(y, s)} \in G_u$. Define $\Phi : G_{u|Y} \to G$ by: $\Phi(z) = y\psi(s)$. We claim that Φ is well defined. Indeed, suppose that $(y, s) \sim (y, t)$. By Proposition 4.3.4, there exists $e \in E$ such that $y \le e$ and $es = et$. Then $y\psi(s) = y\psi(e)\psi(s) = y\psi(t)$ so that Φ is well defined. Since $\cup_{s \in S}\psi(s) = G$, the map Φ takes G_u onto G. Since $y.s = y.\psi(s)$, we have $(y\psi(s))((y.s)\psi(t)) = y\psi(s)\psi(s)^{-1}y\psi(s)\psi(t) = y\psi(st)$, and it follows that Φ is a homomorphism. Since $\Phi(U\psi_u(s)) = U\psi(s)$ for any open subset U of Y, we have that Φ is an open map. Since $\Phi^{-1}(U\psi(s)) = U\psi_u(s)$, the map Φ is continuous. Obviously, Φ is the identity map on Y. $\square$

The above proposition makes precise the sense in which G_u is the *universal S-groupoid*. It is left as a simple exercise to the reader to show directly from its construction that G_u is a *faithful S-groupoid*, i.e. that ψ_u is one-to-one. (In fact ψ_u extends to an isomorphism from $\mathbf{C}(S)$ onto a subalgebra of $C_c(G_u)$ as we shall see in the next section.)

We now discuss the universal groupoid for some of the examples examined in **4.2**. The numbering of the examples in both sections correspond.

Example 1. Monogenic inverse semigroups.

We will use the model GB for the free monogenic inverse semigroup $S = I_1$ described in **4.2**, Example 1. The idempotent set E for S is given by the elements of the form $(p, 0, q)$ $(p, q \geq 0, p + q > 0)$. Using (4.2), if $(p, 0, q), (p', 0, q') \in E$, then

$$(p, 0, q)(p', 0, q') = (p \vee p', 0, q \vee q'). \tag{4.32}$$

So the ordering on E corresponds to reversed ordering on the components, i.e. $(p, 0, q) \leq (p', 0, q')$ if and only if $p \geq p', q \geq q'$. Further, if $s = (k, l, m)$, then $e = (p, 0, q) \leq ss^*$ if and only if $p \geq k, q \geq l + m$. In this case $es = (p, l, q - l)$. Next, we determine the filter completion X of E.

Let A be a filter in E and p_A (resp. q_A) be the sup over all $(p, 0, q) \in A$ of p (resp. q). We allow ∞ as a value for p_A, q_A in the unbounded cases. Then since A is a subsemigroup of E and $f \in A$ whenever $e \in A$ and $f \geq e$, we get, using (4.32), that $A = \{(p, 0, q) : p \leq p_A, q \leq q_A\}$. The filter completion X is then to be identified as a set with $\mathbf{N}^\infty \times \mathbf{N}^\infty$ where $\mathbf{N}^\infty = \mathbf{N} \cup \{\infty\}$. Under this identification, if $e = (p, 0, q)$, then $\bar{e} = (p, q)$. Using (4.22), a basis for the topology on X is given by the singleton subsets $\{\bar{e}\}$ with $e \in E$ together with the sets of the form $\{k\} \times [m, \infty], [k, \infty] \times \{m\}$ and $[k, \infty] \times [m, \infty]$ (using (extended) integer intervals). For convenience, we will write e in place of $\bar{e}$ below.

Next we calculate the right action $x \to x.s$ of S on X. Let $s = (k, l, m)$ and suppose that $x \in D_s$. Suppose firstly that $x = (p, 0, q) \in E$. Then $p \geq k, q \geq (l + m)$ and using (4.2), (4.3) and (4.32), we have

$$x.s = (k + l, -l, l + m)(p, 0, q)(k, l, m) = (p + l, 0, q - l).$$

Identifying x with $(p, q) \in \mathbf{N}^\infty \times \mathbf{N}^\infty$, the action is given by:

$$(p, q).s = (p + l, q - l).$$

Since the action is continuous and every $x \in D_s$ is a limit of $(p, 0, q)$'s (Proposition 4.3.1) the preceding formula holds for $x \in X$ in general. In particular,

$$D_s = \{(p, q) \in \mathbf{N}^\infty \times \mathbf{N}^\infty : p \geq k, q \geq l + m\}.$$

Now suppose that $x = (p, q) \in X$, $t = (k', l', m')$ where $x \leq s, t$ and $(x, s) \sim (x, t)$. Then there exists $e \in E$ with $x \leq e$ and $e \leq ss^*tt^*$, $es = et$. Let $e = (p', 0, q')$. Then a simple calculation shows that $k \vee k' \leq p' \leq p, (l + m) \vee (l' + m') \leq q' \leq q$ and

$$es = (p', l, q' - l) = et = (p', l', q' - l').$$

In other words, we have $l = l'$. By sending $\overline{(x, s)} \to ((p, q), l)$, we can identify the universal groupoid $G_{\mathbf{u}}$ with the groupoid of pairs

$$\{((p, q), l) : p, q \in \mathbf{N}^\infty, l \in \mathbf{Z}\}$$

where $p + l \geq 0, q - l \geq 0$ and $p + q > 0$. Given such a pair $((p, q), l)$, we define $(p, q).l = (p + l, q - l)$ $(l \leq q)$, and then translating the product $\overline{(x, s)(x.s, t)} = \overline{(x, st)}$ and inverse $\overline{(x, s)}^{-1} = \overline{(x.s, s^*)}$ into $((p, q), l)$ terms, we get (with $s = (p, l, q - l), x = (p, q)$) that

$$(x, l)(x.l, l') = (x, l + l'), \quad (x, l)^{-1} = (x.l, -l).$$

Example 2. Toeplitz inverse semigroups.

We will use the terminology of **4.2**, Example 2. Recall that the unit space Ω of the ample groupoid $\mathcal{G}_{G,P}$ is the closure in the pointwise topology on G of the set $\{\tau(p) : p \in P\}$.

Define $\psi : \mathcal{S}_{G,P} \to \mathcal{G}_{G,P}^a$ by:

$$\psi(\alpha) = \{(x_\alpha, \omega) : \omega \in D_\alpha\}.$$

It is easy to check that ψ is a homomorphism. It is obvious that

$$\cup_{\alpha \in \mathcal{S}_{G,P}} \psi(\alpha) = \mathcal{G}_{G,P}.$$

To prove that $(\mathcal{G}_{G,P}, \psi)$ is an $\mathcal{S}_{G,P}$-groupoid, we need to establish (ii) of Definition 4.3.1. This follows from the observation of Nica [188] that the D_α's separate the points of Ω. So $(\mathcal{G}_{G,P}, \psi)$ is an $\mathcal{S}_{G,P}$-groupoid.

Consequently, by Proposition 4.3.5, Ω is identifiable with a closed invariant subset of the unit space X of the universal groupoid $G_\mathbf{u}$ of $\mathcal{S}_{G,P}$. To identify this subset, for each $p \in P$, let ϕ_p be the semicharacter on the set E of idempotents of $\mathcal{S}_{G,P}$ (identified with subsets of P) given by: $\phi_p(A) = \chi_A(p)$. Explicitly, $\phi_p(\beta_x \beta_{x^{-1}}) = 1$ if and only if $x^{-1}p \in P$ if and only if $p \in xP$. Recalling that $\tau(p)(x) = 1$ if and only if $p \in xP$, we see that

$$\phi_p(\beta_x \beta_{x^{-1}}) = \tau(p)(x). \tag{4.33}$$

Next, by (4.6), ϕ_p is determined by its values on the elements $\beta_x \beta_{x^{-1}}$ $(x \in G)$. This is true for any semicharacter of E, and pointwise convergence of a net of $\tau(p)$'s corresponds to the convergence in X of the ϕ_p's. The map $\phi_p \to \tau(p)$, extended by continuity, gives a homeomorphism, denoted by Γ, from a closed invariant subset Y of X onto Ω. One then checks the equivariant condition for $y \in Y$: $\Gamma(y.\alpha) = \beta_{x_\alpha^{-1}}(\Gamma(y))$. The reader is invited to illustrate Proposition 4.3.5 by showing that the map

$$\overline{(y, \alpha)} \to (x_\alpha, \beta_{x_\alpha^{-1}}(\Gamma(y)))$$

is a continuous, open surjective $\mathcal{S}_{G,P}$-homomorphism from $G_{\mathbf{u}|Y}$ onto G.

Example 3. The universal groupoids of the Cuntz inverse semigroups.

We note that an argument similar to that below, but more complicated, can be applied to the inverse semigroup $S_{\mathcal{G}}$ and the vertex groupoid $G_{\mathcal{G}}$ associated with an admissible directed graph (**4.2**, Example 3).

In the present example, S is the *Cuntz semigroup* S_n described in **4.2**, Example 3. We will approach the determination of the S_n-groupoids using the explicit construction of the universal groupoid given in Theorem 4.3.1. This gives a procedure that will work in principle for any inverse semigroup.

We first determine the filter completion of $E = E(S_n)$ and then the universal S_n-groupoid $G_{\mathbf{u}}(n)$. All other S_n-groupoids are obtained from $G_{\mathbf{u}}(n)$ using Proposition 4.3.5. The Cuntz and Cuntz-Toeplitz groupoids are realized quite simply in the formulation of Renault in the cases $n < \infty$. There is a problem with the Cuntz groupoid O_∞ in [230].[9] The approach developed here readily gives the universal groupoid of O_∞.

Let $B_n = \{i \in \mathbf{N} : 1 \le i \le n\}$ when $n < \infty$. Set $B_\infty = \mathbf{P}$. The idempotents of S_n are $z_0, 1$ and the elements $s_\alpha t_\alpha$. Switching (for convenience) from α to y, we will freely identify each finite sequence y where each $y_i \in B_n$ with the idempotent $s_y t_y$. The set of all such y is denoted by Y.

In order to determine the S_n-groupoids, we first calculate the filter completion X of the semilattice E of idempotents in S_n. It is convenient for this purpose to discuss X in terms of filters rather than in terms of semicharacters. From (4.8) and (4.9), the order structure on E is determined by: $y \le y'$ if we can write $y = y'y''$ for some string y'', and for all $e \in E$, $z_0 \le e$ and $e \le 1$. Multiplication for strings is given by $y'.y = y$ if $y \le y'$. If neither $y \le y'$ nor $y' \le y$, then $y'y = z_0$.

For $e \in E$, recall that $\bar{e} = \{f \in E : f \ge e\}$. Clearly, $\bar{1} = \{1\}$. Suppose that A is a filter different from $\bar{1}$. If $z_0 \in A$, then $A = E = \overline{z_0}$. Suppose that z_0 does not belong to A. Then if $y, y' \in A$, we have, since $y.y' \in A$ and can't be z_0, that either $y \le y'$ or $y' \le y$. If A is finite, then there exists a unique $y \in A$ of maximal length, and $A = \bar{y} = \{z \in E : z \ge y\} \cup \{1\}$. Otherwise, there exists an infinite string $z = z_1 z_2 \cdots$ such that $A = \{z_1 \cdots z_r : r \ge 1\} \cup \{1\}$. We identify A with z. Let Z be the set of such elements z.

Regard a filter or set of filters as semicharacters of E by overlining. (For $e \in E$, $\bar{e}$ will stand either for the filter or the semicharacter.) We can then write X as the set of semicharacters:

$$X = \{\bar{1}\} \cup \overline{Y} \cup \overline{Z} \cup \{\overline{z_0}\}.$$

Summarizing the above in terms of semicharacters:

(i) $\bar{1}(e) = 1 \Leftrightarrow e = 1.$

---

[9]The unit space of O_∞ as given in [230, p.140] is not compact as $(\infty, 1, 1, \ldots) = \lim(n, 1, 1, \ldots)$ does not belong to it.

(ii) $\overline{z_0}(e) = 1$ for all e.

(iii) $\overline{y}(e) = 1 \Leftrightarrow e = y_1 \cdots y_r$ for some $r \leq \ell(y)$ or $e = 1$.

(iv) $\overline{z}(e) = 1 \Leftrightarrow e = z_1 \cdots z_r$ for any $r \geq 1$ or $e = 1$.

We now discuss the topology on X. This is determined by the sets $D_e = \{x \in X : x(e) = 1\}$ and their complements. Since $D_{z_0} = \{\overline{z_0}\}$, it follows that the singleton $\{\overline{z_0}\}$ is open in X.

Now suppose that $n < \infty$. Then $\{\overline{1}\}$ is also open in X since it is equal to $D_1 \cap D^c_{s_1 t_1} \cap \cdots \cap D^c_{s_n t_n}$. Further, each $\{\overline{y}\}$, where $y \in Y$, is open in X since it is equal to $D_y \cap D^c_{y1} \cap \cdots \cap D^c_{yn}$.

Let $\{\overline{u(\delta)}\}$ be a net in the open subset $\overline{Y} \cup \overline{Z}$ of X and $z \in Z$. Then $\overline{u(\delta)} \to \overline{z} \in \overline{Z}$ if and only if $\overline{u(\delta)}(y) \to \overline{z}(y)$ for all $y \in Y$. This is equivalent to requiring that for any r, $u(\delta)_i = z_i$ $(1 \leq i \leq r)$ eventually. If $\overline{u(\delta)} \to \overline{y} \in \overline{Y}$, then $u(\delta) = y$ eventually (as it should do since $\{\overline{y}\}$ is open). The upshot is that the relative X topology on $\overline{Y} \cup \overline{Z}$ is that of componentwise convergence. In particular, $\overline{Z}$ is a compact subset of X.

The case $n = 1$ is particularly simple. Each $y \in Y$ is just a finite string of 1's and Z is the singleton $\{111 \cdots\}$. Writing the string with n 1's as n, we have that X in this case is just the one point compactification $\{1, 2, \ldots\}^\infty$ with two open and closed singletons adjoined. (We will look again at S_1 in Example 4 below.)

The case $n = \infty$ exhibits a somewhat different behaviour. Indeed, let $\{\overline{u(\delta)}\}$ be a net in $\overline{Y} \cup \overline{Z}$ converging to some $\overline{u}$ in $X \sim \{\overline{z_0}\}$. If $u \in Z$, then, as above, $u(\delta) \to u$ componentwise. The converse is also true. Suppose then that $u = u_1 \cdots u_r \in Y$ and that $u(\delta) \neq u$ eventually. Then eventually $u(\delta) = u_1 \cdots u_r u(\delta)_{r+1} \cdots$. Now for any $m \geq 1$, $\overline{u}(um) = 0$ so that $\{u(\delta)_{r+1}\}$ does not have a bounded subnet. It follows that $u(\delta)_{r+1} \to \infty$. Conversely, if eventually $\ell(u(\delta)) \geq (r+1)$, then $u(\delta)_{r+1} \to \infty$ and $u(\delta)_i \to u_i$ for $1 \leq i \leq r$. So $\overline{u(\delta)} \to \overline{u}$. Slightly modifying the preceding argument shows that $\overline{u(\delta)} \to \overline{1}$ if and only if $u(\delta)_1 \to \infty$.

Thus topologically X behaves very differently in the case $n = \infty$ from the finite n case. In the former case, the closures of both $\overline{Y}$ and $\overline{Z}$ are $\overline{1} \cup \overline{Y} \cup \overline{Z}$. In the latter case, the closure of $\overline{Y}$ is $\overline{Y} \cup \overline{Z}$ while $\overline{Z}$ is compact.

We now take n to be general and determine the right action of S_n on X. It is obvious that $D_1 = X$ and that $D_{z_0} = \{\overline{z_0}\}$. Let $s = s_y t_{y'}$ where $y, y' \in Y$. Then $ss^* = s_y t_y$ and $D_s = \{\overline{y}\} \cup \{\overline{yu} : u \in Y \cup Z\} \cup \{\overline{z_0}\}$. If $e \leq ss^*$, then $e = s_{y'y''} t_{y'y''}$ for some y'', and using (4.9), (4.8) and (4.10), we have

$$ses^* = s_{yy''} t_{yy''}.$$

Using (4.24), we have $\overline{yu}.s(e) = 1 \iff \overline{yu}(yy'') = 1 \iff \overline{y'u}(y'y'') = 1$. It follows that

$$\overline{yu}.(s_y t_{y'}) = \overline{y'u} \tag{4.34}$$

(and of course $\overline{z_0}.s = \overline{z_0}$). In other words, s acts on $\overline{yu}$ by putting y' in place of the y. If u is empty and y' is empty, then $\overline{y}.s_y = \overline{1}$. (The element 1 can be regarded as the empty word.)

The orbit under this S_n-action of $x \in X$ is the set $\{x.s : x \in D_s, s \in S_n\}$. The closure of the orbits for this action for $n < \infty$ are $\{\overline{1}\}$, $\{\overline{z_0}\}$, $\overline{Z}$, and $\overline{Y} \cup \overline{Z}$. For $n = \infty$, the closed orbits are $\{\overline{1}\}$, $\{\overline{z_0}\}$ and $\{\overline{1}\} \cup \overline{Y} \cup \overline{Z}$ (for elements in $\overline{Y} \cup \overline{Z}$). These determine the closed invariant subspaces of X. For the rest of this example, we will assume that $n < \infty$.

We now proceed to construct the universal S_n-groupoid $G_{\mathbf{u}}(n)$. We have to calculate equivalence classes for pairs $(\overline{u}, s)$ where $\overline{u} \in D_s$, and $(\overline{u}, s) \sim (\overline{u}, t)$ if and only if there exists $e \in E$ such that $\overline{u} \leq e$ and $es = et$. If $u = 1$, then $s = 1 = t$ and we can identify the class of $(\overline{1}, s)$ with $\overline{1}$. If $u = z_0$, then the equivalence class of $(\overline{u}, s)$ is just $\overline{z_0}$ (since $z_0 \leq z_0$ and $z_0.s = z_0.t$ for all $s, t \in S_n$).

We will therefore consider the reduction of $G_{\mathbf{u}}(n)$ to $Y \cup Z$. Suppose then that $u = \{u_i\} \in Y \cup Z$. We can also suppose that $s, t \in S \sim \{z_0, 1\}$ by multiplying both on the left by $s_{u_1} t_{u_1}$. Since $\overline{u} \leq ss^*tt^*$ we can assume without loss of generality that $s = s_{u_1 \cdots u_m} t_{y'}$, $t = s_{u_1 \cdots u_r} t_{y''}$ where $r \geq m$. Let e be such that $\overline{u} \leq e \leq (ss^*)(tt^*)$ and $es = et$. Then we can write $e = s_{u_1 \cdots u_{r'}} t_{u_1 \cdots u_{r'}}$ for some $r' \geq r$. Using (4.8)

$$s_{u_1 \cdots u_{r'}} t_{y' u_{m+1} \cdots u_{r'}} = es = et = s_{u_1 \cdots u_{r'}} t_{y'' u_{r+1} \cdots u_{r'}}.$$

This is equivalent to $y'' = y' u_{m+1} \cdots u_r$.

We can now link this up with Renault's model for the Cuntz groupoid G_n described in **4.2**, Example 3. Indeed, associate $(\overline{u}, s)$ with the triple

$$(u_1 \cdots u_m u_{m+1} \cdots, m - \ell(y'), y' u_{m+1} u_{m+2} \cdots).$$

The triple associated with $(\overline{u}, t)$ coincides with the $(\overline{u}, s)$ triple. Conversely, given two such equal triples associated with pairs $(\overline{u}, s)$, $(\overline{u}, t)$, we can reverse the process to obtain $(\overline{u}, s) \sim (\overline{u}, t)$. This sets up a bijection between equivalence classes $\overline{(u, s)}$ where $u \in Y \cup Z$ and the set of triples (z, k, z') where $k \in \mathbf{Z}$ and for some $y, \tilde{y} \in Y$ and some v, we have $z = yv, z' = \tilde{y}v$. Restricting to the case $z \in Z$ will give the Cuntz groupoid.

We now need to translate the algebraic operations and the topology of $G_{\mathbf{u}}(n)$ (reduced to $Y \cup Z$) over to the family T of triples (together with $\overline{1}, \overline{z_0}$). The product $(\overline{u}, s)(\overline{u}.s, t) = (\overline{u}, st)$ translates over to:

$$(yw, \ell(y) - \ell(y'), y'w,)(y'y''w',$$
$$\ell(y'y'') - \ell(y'''), y''w') = (yy''w', \ell(yy'') - \ell(y'''), y'''w')$$

where $s = s_y t_{y'}, t = s_{y'y''} t_{y'''}$ and $w = y''w'$. This is equivalent to

$$(z, k, z')(z', k', z'') = (z, k + k', z'').$$

for appropriate z, k, z', z''. Similarly one can show that

$$(z, k, z')^{-1} = (z', -k, z).$$

To determine the topology on $G_{\mathbf{u}}(n)$ reduced to $Y \cup Z$, we know that a neighborhood basis of $\overline{(u, s)}$ ($s = s_y t_{y'}$ as above) is given (Theorem 4.3.1) by sets of the form $D(U, s)$ where U runs over a neighborhood basis of $\overline{u}$ in D_s. It is easy to check that we can take the U's to be of the form $\{\overline{yv} : v \in V\}$ where V is an open subset of $Y \cup Z$. Translating into the triples language gives that a neighborhood basis for (z, k, z') is given by sets of the form $\{(yv, k, y'v) : v \in V\}$. When $n < \infty$ and the relative topology is given to $G_u(n)_{|\overline{Z}}$, then we get the same topology for the Cuntz groupoid as that given in (4.13). (Neighborhood bases for both $\overline{1}$ and $\overline{z_0}$ are just singletons (since we are assuming $n < \infty$).)

The S-groupoid homomorphism $\psi_u : S_n \to G_{\mathbf{u}}(n)^a$ is then given by: $\psi_u(1) = X$ and $\psi_u(z_0) = \{\overline{z_0}\}$, while $\psi_u(s_y t_{y'})$ is the set of all triples $(yw, \ell(y) - \ell(y'), y'w)$ where $w \in Y \cup Z$ or is empty, together with $\overline{z_0}$.

When $n < \infty$, the Cuntz groupoid is thus the reduction of $G_u(n)$ to $\overline{Z}$. The reduction (for the same n) of $G_u(n)$ to $\{\overline{1}\} \cup \overline{Y} \cup \overline{Z}$ is the *Cuntz-Toeplitz* groupoid E_n ([187, 139]).

Example 4. Localization inverse semigroups, sheaf groupoids, Clifford semigroups and maximal group homomorphic images.

Let (Z, S) be a localization where S is given as partial homeomorphisms on Z, and such that the domain of each $s \in S$ is compact and open. Let G be the sheaf groupoid $G(Z, S)$ associated with (Z, S) as described following Theorem 3.3.2. Then (G, ψ) is easily checked to be an S-groupoid where $\psi(s) = D(D_s, s)$. Of course we know from Proposition 4.3.5 that the unit space Z of G is identifiable with a closed invariant subset of the unit space X of $G_{\mathbf{u}}$ and G is the image of an open, unit preserving, surjective S-homomorphism defined on $G_{\mathbf{u}|Z}$. The identification of Z with a subset of X is the obvious one: each $z \in Z$ can be identified with a semicharacter on E by setting $z(e) = 1$ if and only if $z \in \psi(e)$. The open surjective homomorphism sends $\overline{(x, s)} \in G_{\mathbf{u}}$ to $x\psi(s)$.

The vertex groupoids $G_{\mathcal{G}}$ of **4.2**, Example 4 and the r-discrete groupoid $G(T_0)$ of a tiling (**4.2**, Example 5) are localization groupoids and so the considerations of the previous paragraph apply to them.

Related to the situation of the preceding paragraph, for a general inverse semigroup S and S-groupoid (G, ψ) there is the natural localization (Y, S'), where S' acts in the canonical way on $Y = G^0$. The situation is different when we consider the inverse semigroup Σ' of maps $y \to y.s'$ on Y. Then (Y, Σ') is a localization and gives rise to the sheaf groupoid $H = G(Y, \Sigma')$

which is a Σ'-groupoid. In the natural way, H can also be regarded as an S-groupoid. When $G = G_u, Y = X$ then the map $T : G \to H$, where $T(x\psi(s))$ is the equivalence class of (x, s), is an open, unit preserving, surjective S-homomorphism. This gives another illustration of Proposition 4.3.5.

Simple examples of inverse semigroups S for which G_u is not the sheaf groupoid of the localization (X, Σ') of the preceding paragraph are provided by Clifford semigroups. Let S be a Clifford semigroup. Then as in **4.2**, Example 4, we can write $S = \cup_{e \in E} H_e$ where H_e is the maximal subgroup of S whose identity is e. Since the idempotents of S are central, if $f \leq e$ in E, then the map $s \to fs$ is a homomorphism $Q_{f,e}$ from H_e into H_f. Let $x \in X$, the space of semicharacters of E. For each basis neighborhood $D_{e,e_1,\dots,e_r}$ of x ((4.22)), set $H_{D_{e,e_1,\dots,e_r}} = H_e$. (That this is well-defined follows from the discussion below.) The family of such basis neighborhoods is a downwards directed set, and if $D_{f,f_1,\dots,f_s} \subset D_{e,e_1,\dots,e_r}$, then $e \geq f$. (This is because $\overline{f} \in D_{f,f_1,\dots,f_s} \subset D_{e,e_1,\dots,e_r} \subset D_e$. Hence by (4.21), we have $\overline{f}(e) = 1$ so that from the definition of $\overline{f}$ as a filter, we obtain $e \geq f$.) Using the above groups $\{H_e\}$ and homomorphisms $Q_{f,e}$, we obtain a direct system of groups whose limit we denote by H_x.

For every e with $x \leq e$, there is a canonical homomorphism from H_e into H_x. In particular, this applies with $e = ss^*$ where $s \in S$ and $x \leq s$. Let s_x be the image of s in H_x under this homomorphism. Note that $H_{\overline{e}} = H_e$. For when $x = \overline{e}$, the basis neighborhoods for x are eventually of the form $D_{e,e_1,\dots,e_r}$ (since e is the smallest $f \in E$ for which $f \geq e$), and so eventually, the homomorphisms of the direct limit are the identity maps on H_e.

Now let $s, t \in S$, $x \in X$ and $e \in E$ be such that $e \leq (ss^*)(tt^*)$. Then $es = et$ if and only if $Q_{e,ss^*}(s) = Q_{e,tt^*}(t)$. So $(x, s) \sim (x, t)$ if and only if $s_x = t_x$. It follows that we can identify the universal groupoid G_u of S with *the union of the groups H_x.* The map ψ_u is given by: $\psi_u(s) = \{s_x : x \leq s\}$. Using the fact that E is contained in the center of S together with (4.24), each of the maps $x \to x.s$ $(x \in D_s)$ is the identity map. Next, $G_u{}^2$ is the set of pairs of the form $((x, s_x), (x, t_x))$ (where $s_x, t_x \in H_x$). The product of such a pair is $(x, s_x t_x)$ while $(x, s_x)^{-1} = (x, s_x^{-1})$. So as a groupoid, G_u is a bundle of groups. For $s \in S$ and a basis set $D_{e,e_1,\dots,e_r}$ for X where $e \leq ss^*$, the sets $\{s_x : x \in D_{e,e_1,\dots,e_r}\}$ give a basis for the topology of G_u.

So if any of the H_e's are non-trivial, then G_u is not the sheaf groupoid for the canonical localization for the inverse semigroup of maps $x \to x.s'$ $(s' \in S')$ – the latter groupoid would be just a groupoid of units.

Lastly, we observe that *every* inverse semigroup S admits a natural S-groupoid which is a group H. This group is the maximal group homomorphic image $G(S)$ of S described in Proposition 2.1.2. The S-groupoid map $\psi : S \to H$ sends s to its σ_S-equivalence class in H. The single element of X, associated with the identity of the group H, is the semicharacter on E which is identically 1.

Of particular interest is the case where $S = S_1 \sim \{0\}$, where S_1 is the Cuntz inverse semigroup with one generator. (As an aside, if $n > 1$, then $S_n \sim \{0\}$ is *not* a subsemigroup of S_n since if $i \neq j$, then $s_i^* s_j$ *has to be* 0!) Apart from the identity 1, every element of S is uniquely expressed in the form $s^m t^n$ where $s_1 = s, t_1 = t$. The analogues of (4.8) and (4.9) are then: $(s^m t^{n+p})(s^n t^q) = s^m t^{p+q}$ and $(s^m t^n)(s^{n+p} t^q) = s^{m+p} t^q$. The inverse semigroup S is the so-called *bicyclic inverse semigroup* ([202, p.113]). A model for this is the inverse semigroup generated by the unilateral shift and its adjoint, where s corresponds to the shift and t to its adjoint. A simple calculation (a more general version of which is effectively in Example 3 above) gives that $s^m t^n$ is σ_S-equivalent to $s^p t^q$ if and only if $m - n = p - q$. So we can identify H in this case with $\mathbf{Z}$ and then $\psi : S \to H$ is given by: $\psi(s^m t^n) = n - m$. When S_1 is identified with the inverse semigroup generated by the unilateral shift and its adjoint, then ψ is just the index map on the subset S of the Fredholm operators.

Let S be an inverse semigroup. We conclude this section by discussing the question: under what conditions is the universal S-groupoid G_u Hausdorff? There are inverse semigroups for which this is not the case. An example is the Clifford semigroup discussed in Appendix C. (In fact it is shown in [197, Example 2] that this Clifford semigroup admits *no* Hausdorff S-groupoid (G, ψ) for which ψ is an isomorphism.) We will show below that if S is E-unitary (**2.1**) then G_u is Hausdorff. The following proposition is strikingly parallel to the characterization of the Hausdorff condition for the case of the holonomy groupoid given earlier in Proposition 2.3.2.

Let S be an inverse semigroup with $E = E(S)$ and with universal S-groupoid (G, ψ). For $s, t \in S$, let

$$D_{s,t} = \{x \in D_s \cap D_t : x\psi(s) = x\psi(t)\}.$$

Note that $x\psi(s) = \overline{(x, s)}$. So $x \in D_{s,t} \iff x \leq (ss^*)(tt^*)$ and $\overline{(x, s)} = \overline{(x, t)}$. Now if $x \in D_{s,t}$, then there exists $e \in E$ such that $x \leq e \leq (ss^*)(tt^*)$ and $es = et$. Also if $y \leq e \leq (ss^*)(tt^*)$, then we have $y\psi(s) = y\psi(es) = y\psi(et) = y\psi(t)$ giving $D_e \subset D_{s,t}$. It follows that if

$$I_{s,t} = \{e \in E : es = et, e \leq ss^* tt^*\} \tag{4.35}$$

then

$$D_{s,t} = \cup\{D_e : e \in I_{s,t}\}. \tag{4.36}$$

Proposition 4.3.6   *Let (G, ψ) be the universal S-groupoid (G_u, ψ_u). Then G is Hausdorff if and only if, for all $s, t \in S$, the set $D_{s,t}$ is closed in G^0.*

Proof.   We will use the notation of the proof of Theorem 4.3.1. Suppose that G is Hausdorff. Let $s, t \in S$ and $x_\delta \to x$ in $X = G^0$ with $x_\delta \in D_{s,t}$ for all

δ. Since $x_\delta \in D_s \cap D_t$, which is compact and therefore closed in X, we have $x \in D_s \cap D_t$. Now suppose that it is false that $(x, s) \sim (x, t)$. Then $x\psi(s) \neq x\psi(t)$, and since G is Hausdorff, there exists an open neighborhood U of x in X with $D(U, s) \cap D(U, t) = \emptyset$. Since $x_\delta \to x$, we can suppose that every $x_\delta \in U$. Since $x_\delta \in D_{s,t}$, we have $x_\delta\psi(s) = x_\delta\psi(t) \in D(U, s) \cap D(U, t)$ giving a contradiction. So $(x, s) \sim (x, t)$ and $x \in D_{s,t}$. So $D_{s,t}$ is closed in G^0.

Conversely, suppose that every $D_{s,t}$ is closed in X and let $u = (y, s)$ and $v = (z, t)$ in Σ be such that $\overline{u} \neq \overline{v}$. If $y \neq z$, then there exist disjoint open neighborhoods $U_y \subset D_s, U_z \subset D_t$ of y and z respectively in X, and we can separate $\overline{u}, \overline{v}$ by the neighborhoods $D(U_y, s)$ and $D(U_z, t)$. Suppose then that $y = z$. Suppose that $\overline{u}, \overline{v}$ cannot be separated by open subsets of G. Then for every open neighborhood W of y in $D_s \cap D_t$, we have $D(W, s) \cap D(W, t) \neq \emptyset$. So there exists $y_W \in W$ with $(y_W, s) \sim (y_W, t)$. So $y_W \in D_{s,t}$. The W's form a net in the obvious way, and $y_W \to y$. Since $D_{s,t}$ is closed in X, we have $y \in D_{s,t}$. But then $\overline{u} = \overline{v}$ giving a contradiction. So G is Hausdorff. $\qquad \square$

Corollary 4.3.1 *Suppose that every non-empty $I_{s,t}$ contains a finite subset $F(s,t)$ such that for each $e \in I_{s,t}$, we have $e \leq f$ for some $f \in F(s,t)$. Then $G_{\mathbf{u}}$ is Hausdorff.*

Proof. Let $s, t \in S$. We show that $D_{s,t}$ is closed in X. We can suppose that $D_{s,t}$ (and therefore $I_{s,t}$) is non-empty. Let $x_\delta \to x$ in X with $x_\delta \in D_{s,t}$ for all δ. Then for each δ, there exists by (4.36) an element $e_\delta \in E$ such that $x_\delta \leq e_\delta \leq ss^*tt^*$ and $e_\delta s = e_\delta t$. Using the finiteness of $F(s,t)$, without loss of generality, we can suppose that there exists $f \in F(s,t) \subset I_{s,t}$ such that $e_\delta \leq f$ for all δ. Then $x_\delta \in D_f$, and since D_f is compact and open in X, we have $x \in D_f \subset D_{s,t}$. So $D_{s,t}$ is closed in X and by Proposition 4.3.6, $G_{\mathbf{u}}$ is Hausdorff. $\qquad \square$

Corollary 4.3.2 *If S is E-unitary then $G_{\mathbf{u}}$ is Hausdorff.*

Proof. Recall (**2.1**) that S is E-unitary if E is a σ_S-equivalence class, where $s\sigma_S t$ in S whenever there exists $e \in E$ such that $es = et$. Let S be E-unitary and $s, t \in S$ be such that $I_{s,t} \neq \emptyset$. Let $e \in I_{s,t}$. Then $e \leq ss^*tt^*$ and $es = et$. So $e(st^*) = e(tt^*) = e \in E$ and $e\sigma_S st^*$. Since S is E-unitary, $st^* \in E$. We also have $e \leq st^*$. Similarly, if $f = s^*es \in E$, then $f(s^*t) = f$ so that $f\sigma_S s^*t$, and $s^*t \in E$. Next, because $st^* \in E$, we have $st^* = ss^*st^*tt^* \leq ss^*tt^*$, and since $s^*t \in E$, we have $t^*s = (s^*t)^* = s^*t$, so that $(st^*)t = (ss^*)(st^*)t = st^*s(s^*t) = st^*s^*t^*s = (st^*)s$. So $st^* \in I_{s,t}$. Since $e \leq st^*$ for all $e \in I_{s,t}$, applying Corollary 4.3.1 gives that $G_{\mathbf{u}}$ is Hausdorff. $\qquad \square$

4.4   Inverse semigroup universal and reduced C*-algebras as groupoid C*-algebras

Let S be an inverse semigroup with universal groupoid $(G_{\mathbf{u}}, \psi_u)$ as in Theorem 4.3.1. The full and reduced C^*-algebras $C^*(S), C^*_{red}(S)$ of S were defined in **2.1**, while the full and reduced C^*-algebras $C^*(G), C^*_{red}(G)$ of a locally compact groupoid G were defined in **3.1**. The objective of this section is to show that $C^*(S) \cong C^*(G_{\mathbf{u}})$ and that $C^*_{red}(S) \cong C^*_{red}(G_{\mathbf{u}})$. These isomorphisms enable us to use the representation theory of groupoids to investigate the representations of S.

For notational convenience, in this section, (G, ψ) **will stand for the universal S-groupoid** $(G_{\mathbf{u}}, \psi_u)$.

We will sometimes regard $\psi : S \to C_c(G)$ by replacing the set $\psi(s) \in G^a$ by its characteristic function. Recall (Proposition 2.2.6) that if $A, B \in G^a$, then $\chi_A, \chi_B \in C_c(G)$, $\chi_A * \chi_B = \chi_{AB}$ and $(\chi_A)^* = \chi_{A^{-1}}$. So ψ, regarded as taking values in $C_c(G)$, is a *-homomorphism. (Of course, this is also true for *every* S-groupoid.)

The key theorem that we will use to establish $C^*(S) \cong C^*(G)$ is Theorem 3.2.1. Before we can use this theorem, we need to relate S more closely to the inverse semigroup G^a. For this reason, we use the inverse semigroups S' and S'' below.

Recall (**4.3**) that S' is the inverse semigroup in $\mathbf{C}(S)$ whose elements are of the form $e's$ where $e, e_i \in E$, $e_i \le e \le ss^*$, $e' = e(1 - e_1) \cdots (1 - e_n)$ and $s \in S$. Its semilattice of idempotents is just the set of such elements e'.

The inverse semigroup S'' is also a *-subsemigroup of $\mathbf{C}(S)$ containing S'. It is the set of elements of the form $\sum_{i=1}^n s_i'$ where $\{s_1', \ldots, s_n'\}$ is a sequence in S' which is *orthogonal* in the sense that if $i \ne j$, then

$$s_i'(s_j')^* = 0 = (s_i')^* s_j'. \tag{4.37}$$

This can be expressed in terms of the idempotents $p_i = (s_i')^* s_i', q_i = s_i'(s_i')^*$ and the corresponding idempotents p_j, q_j:

$$p_i p_j = 0 = q_i q_j. \tag{4.38}$$

(This is the natural version for S' of the orthogonality property that we saw in (3.59) for G^{op}.) Of course, S, S' are *-subsemigroups of S''. The proof that S'' is a *-subsemigroup of $\mathbf{C}(S)$ and is an inverse semigroup follows exactly the corresponding proof (using the left regular representation) for S' in **4.3** together with the fact that the sum of an orthogonal finite sequence of partial isometries is itself a partial isometry.

The next proposition is useful for replacing sums of idempotents in E' by orthogonal sums.

Proposition 4.4.1   *Let $e'_1, \ldots, e'_n \in E'$. Then there exists an orthogonal sequence $\{g'_1, \ldots, g'_m\}$ in E' such that for each k, there exists an i_k such that $g'_k \leq e'_{i_k}$, and for each i, $\sum_{g'_k \leq e'_i} g'_k = e'_i$.*

Proof. We can take $e'_i \neq 0$ for all i. For each i, write

$$e'_i = e_{i,1}(1 - e_{i,2}) \cdots (1 - e_{i,p_i})$$

where $e_{i,j} \in E$ and $e_{i,j} < e_{i,1}$ whenever $j > 1$. Let $B = \{(i,j) : 1 \leq i \leq n, 1 \leq j \leq p_i\}$. For each $\alpha : B \rightarrow \{1, c\}$, $\alpha(B) \neq \{c\}$, define

$$e'_\alpha = \prod_{(i,j) \in B} e_{i,j}^{\alpha(i,j)}$$

where $e_{i,j}^c = 1 - e_{i,j}$. The e'_α form an orthogonal sequence in E' since if $\alpha \neq \beta$, then there exists a (k,l) such that $e_{k,l}$ occurs in one of e'_α, e'_β while $(1 - e_{kl})$ occurs in the other. Let $e'_{i,\alpha} = e'_i e'_\alpha$.

We now claim that the elements $\{e'_{i,\alpha}\}$ are orthogonal in E'. Indeed, suppose that $e'_{i,\alpha} e'_{j,\beta} \neq 0$. Then $\alpha = \beta$. Further, both expressions for e'_i and e'_j must occur in e'_α. So $e'_{i,\alpha} = e'_i e'_j e'_\alpha = e'_j e'_\alpha e'_{i,\alpha} = e'_{j,\beta}$. This proves the orthogonality claim. Note that for fixed i, each $e'_{i,\alpha} \leq e'_i$ and that if $e'_{j,\alpha} \leq e'_i$, then $e'_{j,\alpha} \leq e'_{i,\alpha}$. So we can take the set of elements g'_k to be the set of all $e'_{i,\alpha}$ once we have shown that for fixed i, $\sum_\alpha e'_{i,\alpha} = e'_i$. This follows since $\sum_\alpha e'_{i,\alpha} = e'_i \sum_\alpha e'_\alpha = e'_i [\prod_{k,l}(e_{k,l} + (1 - e_{k,l}))] = e'_i$. $\square$

The sequence $\{g'_k\}$ of the preceding proposition will be called a *refinement* of $\{e'_i\}$. Note that for any k and any i, either $g'_k \leq e'_i$ or $g'_k e'_i = 0$.

The notion of orthogonality for S' is extended to S'' in the obvious way: the elements $a, b \in S''$ are said to be *orthogonal* ($a \perp b$) if $ab^* = 0 = a^*b$.

Lemma 4.4.1   *Let $\{s'_i\}$ and $\{t'_j\}$ be orthogonal sequences in S' and*

$$a = \sum s'_i \qquad b = \sum t'_j.$$

Then $a \perp b$ in S'' if and only if $\{s'_1, \ldots, s'_m, t'_1, \ldots, t'_n\}$ is an orthogonal sequence in S'.

Proof. Suppose that $a \perp b$ in S''. Then

$$\left(\sum s'_i\right)\left(\sum t'^*_j\right) = 0 = \left(\sum s'^*_i\right)\left(\sum t'_j\right).$$

So for $1 \leq p \leq m$ and $1 \leq q \leq n$, we have

$$s'_p s'^*_p \left(\sum s'_i\right)\left(\sum t'^*_j\right) t'_q t'^*_q = 0.$$

Expanding the left-hand side of this equality gives $s'_p t'^*_q = 0$. A similar argument involving $(\sum s'^*_i)(\sum t'_j)$ gives that $s'^*_p t'_q = 0$ and it follows that $\{s'_1, \ldots, s'_m, t'_1, \ldots, t'_n\}$ is an orthogonal sequence in S'. The converse is trivial. $\square$

It follows from Lemma 4.4.1 that if a and b are orthogonal in S'', then the sum $a + b$ of a, b in $\mathbf{C}(S)$ also belongs S''. (We saw in **3.2** that the same holds for orthogonal partial isometries, and for orthogonal sets in G^{op} with $\cup$ in place of $+$.) The proof of the next proposition is trivial.

Proposition 4.4.2   If $\{s'_1, \ldots s'_m\}$ and $\{t'_1, \ldots, t'_n\}$ are orthogonal sequences in S' then the sequences $\{s'_i t'^*_j\}$ and $\{s'^*_i t'_j\}$ are orthogonal in S'.

Recall that in this section, (G, ψ) is the universal groupoid of S. The next result shows that ψ effectively identifies S'' with G^a.

Proposition 4.4.3   The map $\psi : S \to G^a$ extends to an inverse semigroup isomorphism, also denoted by ψ, from S'' onto G^a, and this isomorphism is additive in the sense that if a, b are orthogonal in S'', then $\psi(a), \psi(b)$ are orthogonal in G^a and $\psi(a + b) = \psi(a) \cup \psi(b)$. In addition, the map $\psi^{-1} : G^a \to S''$ is also additive.

Proof.   Regard G^a as a *-subsemigroup of $C_c(G)$ by taking characteristic functions (Proposition 2.2.6). Since the vector space $\mathbf{C}(S)$ is free on S, the homomorphism $\psi : S \to G^a$ extends to a *-homomorphism from $\mathbf{C}(S)$ into $C_c(G)$. This *-homomorphism restricts to a *-homomorphism, also denoted by ψ, from S'' into $C_c(G)$. We now show that ψ is additive from S'' into G^a.

From the proof of Theorem 4.3.1, we have $\psi(S') = \{\psi(e's) : e' \in E', s \in S, e' \leq ss^*\} \subset G^a$. Write $s'' \in S''$ as an orthogonal sum of $s'_i \in S'$. Then applying ψ to (4.37) and using Proposition 3.2.4, the disjointness of the $\psi(s'_i)$ and the linearity of the ψ on $\mathbf{C}(S)$, we obtain $\psi(s'') = \cup \psi(s'_i) \in G^a$. By construction, ψ is additive on S''. For the remainder of the argument we will regard ψ as restricted to S''. We now prove that ψ is one-to-one and maps S'' onto G^a.

First, ψ is one-to-one on E'. For if $e', f' \in E'$ and $\psi(e') = \psi(f')$, then the functions $\chi_{\psi(e')} = \chi_{\psi(f')}$ are the Gelfand transforms of e', f' in $\mathbf{C}(E)$ on X. By Corollary 2.1.1, we have $e' = f'$ and $\psi_{|E'}$ is one-to-one.

Now suppose that $s'' = \sum_{i=1}^m e'_i s_i$, $t'' = \sum_{j=1}^n f'_j t_j$ (orthogonal sums) belong to S'', where $e'_i, f'_j \in E', s_i \in S, t_j \in S, e'_i \leq s_i s^*_i, f'_j \leq t_j t^*_j$ and with all $e'_i s_i, f'_j t_j \neq 0$. Then using (4.38), the sequences $\{e'_i\}, \{f'_j\}$ are orthogonal. Suppose next that $\psi(s'') = \psi(t'')$. We will show that $s'' = t''$. From Proposition 4.4.1, there exists a refinement $\{g'_k\}$ of the set $\{e'_1, \ldots, e'_m, f'_1, \ldots, f'_n\}$. Then using the additivity of ψ together with

Proposition 4.4.2, we have

$$\cup_{k,i}\psi(g_k'e_i's_i) = \psi(s'') = \psi(t'') = \cup_{k,j}\psi(g_k'f_j't_j). \tag{4.39}$$

Remove the empty $\psi(g_k'e_i's_i), \psi(g_k'f_j't_j)$. Since $\psi(g_k'e_i's_i) = \psi(g_k')\psi(e_i's_i)$, we have $\psi(g_l'f_j't_j) = \psi(g_l')\psi(f_j't_j)$ and $\psi(g_k') \cap \psi(g_l') = \emptyset$ if $k \neq l$. It follows from (4.39) that for fixed i, k,

$$\psi(g_k'e_i's_i) \subset \cup_j\psi(g_k'f_j't_j).$$

Now by Proposition 4.4.1, for any j, either $g_k' \leq f_j'$ or $g_k'f_j' = 0$, and the f_j' are orthogonal. So there exists a unique j such that $\psi(g_k'e_i's_i) = \psi(g_k'f_j't_j)$. Then applying the range map r, we have $g_k'e_i' = g_k'f_j' = e'$ for some e', since, as shown above, ψ is one-to-one on E'. It follows that $\psi(e's_i) = \psi(e't_j)$. The equality $s'' = t''$ will now follow if we can show that $e's = e't$ with $s = s_i, t = t_j$. (For then every $g_k'e_i's_i$ is a unique $g_k'f_j't_j$ and conversely, so that from (4.39), we have $s'' = t''$.) As usual, write $e' = e(1-e_1)\cdots(1-e_n)$ where for all l, $e_l < e \leq (ss^*)(tt^*)$. The semicharacter $\bar{e}$ then belongs to $\psi(e') = D_{e,e_1,\dots,e_n}$, and since $\psi(e's) = \psi(e't)$, we have $(\bar{e}, s) \sim (\bar{e}, t)$ in G. So there exists $g \in E$ such that $\bar{e} \leq g$ and $gs = gt$. Since $\bar{e}$, as a filter, is the set $\{h \in E : h \geq e\}$, we have $eg = e$ and so $es = e(gs) = e(gt) = et$. Hence $e's = e't$ and the map ψ is one-to-one.

Second, we show that ψ is onto. Let $A \in G^a$. Since A is compact and open and the sets $\psi(e's)$ form a basis for the topology of G, it follows that there exist $e_i's_i$ ($1 \leq i \leq n$) with every $e_i' \leq s_is_i^*$ such that $A = \cup\psi(e_i's_i)$. By replacing the $\{e_i'\}$ by a refinement we can suppose that $e_i'e_j' = 0$ if $i \neq j$. Note that $r(\psi(e_i's_i)) = \psi(e_i')$, the latter subsets being also disjoint. So the sets $\psi(e_i's_i)$ are disjoint and since A is a G-set, the sets $d(\psi(e_i's_i))$ are also disjoint. It follows that the sets $\{\psi(e_i's_i)\}$ are orthogonal in G^a. Since ψ is one-to-one on S'', we have that the $e_i's_i$ are orthogonal. Then $A = \psi(s'')$ where $s'' = \sum e_i's_i \in S''$. So ψ is onto. The proof of the additivity of ψ^{-1} is left to the reader. $\square$

It will follow from Theorem 4.4.1 below that ψ extends to an isomorphism on $\mathbf{C}(S)$.

Additive representations for S'' are defined as for subsemigroups of G^{op} in **3.2**. Precisely, an *additive* representation of S'' is a representation π of S'' on a Hilbert space $\mathcal{H}$ such that $\pi(0) = 0$ and if $s'' \perp t''$ in S'' then $\pi(s'' + t'') = \pi(s'') + \pi(t'')$.

Proposition 4.4.4 *The restriction map takes the additive representations of S'' onto the representations of S.*

Proof. It is trivial that an (additive) representation of S'' restricts to give a representation of S, non-degeneracy following since $S'' \subset \mathbf{C}(S)$.

Conversely, let π be a representation of S on the Hilbert space $\mathcal{H}$. Then π extends to a representation of $\mathbf{C}(S)$ on $\mathcal{H}$ which restricts to give an additive representation of S'' on $\mathcal{H}$. The latter restricts to S to give π again. $\square$

Recall, from the proof of Proposition 4.4.3 and from Proposition 4.4.4 that the homomorphism (indeed isomorphism) $\psi : S \to G^a$ can also be regarded as a homomorphism from $\mathbf{C}(S)$ into $C_c(G)$ and as an additive isomorphism from S'' onto G^a. The following theorem shows that ψ implements an isomorphism between $C^*(S)$ and $C^*(G)$.

Theorem 4.4.1   *The map* $\psi : \mathbf{C}(S) \to C_c(G)$ *extends to an isomorphism from* $C^*(S)$ *onto* $C^*(G)$.

Proof.   Let $w \in \mathbf{C}(S)$. Then the C^*-norms respectively of w and $\psi(w)$ in $C^*(S), C^*(G)$ are given by:

$$\|w\| = \sup_{\pi} \|\pi(w)\|, \quad \|\psi(w)\| = \sup_{\pi'} \|\pi'(w)\|$$

where π ranges over the representations of $\mathbf{C}(S)$ and π' over the representations of $C_c(G)$. By Proposition 4.4.4, Proposition 2.2.7 and Corollary 3.2.1, such a π is given by extending linearly to $\mathbf{C}(S)$ the restriction to S of an additive representation of S'', and such a π' is given by an additive representation of G^a extended by linearity to the I-norm dense subalgebra W of $C_c(G)$ spanned by the functions χ_A with $A \in G^a$. (See Proposition 2.2.7.) From Proposition 4.4.3, ψ extends to a homomorphism from $\mathbf{C}(S)$ into $C_c(G)$ with range W. Again, by Proposition 4.4.3, the map $\pi \to \pi \circ \psi^{-1}$ is a bijection from the set of additive representations of S'' to the set of additive representations of G^a. By the above, this bijection extends to the level of the set of representations of $\mathbf{C}(S)$ onto the set of representations of $C_c(G)$ restricted to W where ψ is regarded as a homomorphism from $\mathbf{C}(S)$ onto W. It follows that ψ is a C^*-norm isometric isomorphism from $\mathbf{C}(S)$ onto the I-norm dense $*$-subalgebra W of $C^*(G)$, and hence that the map ψ extends by continuity to an isomorphism from $C^*(S)$ onto $C^*(G)$. $\square$

We now turn to showing that $C^*_{red}(S) \cong C^*_{red}(G)$. To this end, we have to relate the left regular representation π_2 of S to that of $G = G_{\mathbf{u}}$. This will be done by first decomposing π_2 into a direct sum of representations over E. This is directly suggested by $Ind\, v$ in the groupoid theory ((3.41)) with S being regarded as the groupoid G_S of Proposition 1.0.1. As far as the author is aware, this groupoid-inspired decomposition for the left regular representation of an inverse semigroup is new.[10] As we will see, the

---

[10]The representations π_2^e for the case of a Clifford semigroup are, however, used in the proof of [281, Theorem 3].

decomposition of π_2 will translate into the elementary $Ind\,v$ decomposition of $Ind\,\mu$ for a natural discrete measure μ on $X = G^0$. The measure μ has dense support, and Proposition 3.1.2 will then give the required isomorphism. We start, then, by discussing the decomposition of the left regular representation π_2 of S on $\ell^2(S)$ into a direct sum of more fundamental representations π_2^e with e ranging over E.

Let $e \in E$ and $S_e = \{t \in S : t^*t = e\}$. For $s \in S$ and $\sum_{t \in S_e} a_t t \in \ell^2(S_e)$, define

$$\pi_2^e(s)\left(\sum_{t \in S_e} a_t t\right) = \sum_{tt^* \leq s^*s} a_t st. \tag{4.40}$$

It is easy to check (as for π_2 in **2.1**) that in (4.40), $st = st'$ if and only if $t = t'$. Further, if $tt^* \leq s^*s$, then $(st)^*(st) = t^*s^*st = t^*(s^*s)t = t^*tt^*t = t^*t = e$. So $st \in S_e$ if $t \in S_e$, and it follows that $\pi_2^e(s) \in B(\ell^2(S_e))$ and has norm 1.

(We need to be careful here. While π_2^e is defined in an essentially similar way to $Ind\,v$ earlier (which is effectively convolution by δ_v) we cannot simply treat π_2^e as "$Ind\,e$" since the elements of S cannot be identified with elements of $C_c(G_S)$ where G_S is regarded as a groupoid with the discrete topology. This is briefly discussed in the paragraph preceding Proposition 4.4.6.)

Proposition 4.4.5 *The map π_2^e is a representation of S on $\ell^2(S_e)$ and the left regular representation π_2 is a direct sum of the representations π_2^e ($e \in E$).*

Proof. It is easy to check that $\pi_2^e : S \rightarrow B(\ell^2(S_e))$ is a representation of S. Since S is the disjoint union of the S_e's, it follows that $\ell^2(S) = \oplus_{e \in E} \ell^2(S_e)$. For any $s, t \in S$, we have $(\oplus_{e \in E} \pi_2^e)(s)(t) = \pi_2^{t^*t}(s)(t)$, the $\pi_2^e(s)(t)$'s being all zero when $e \neq tt^*$. Now $\pi_2^{t^*t}(s)(t)$ equals st if $s^*s \geq tt^*$ and is zero otherwise, and hence coincides with $\pi_2(s)(t)$. This completes the proof. $\square$

Our next objective is to show that G_S can be realized as a dense subgroupoid of the universal S-groupoid G_u. As above (e.g. in the proof of Proposition 4.4.3) the context will determine whether we take $\psi(s)$ to be an element of G^a or as the characteristic function of this element $\in C_c(G)$.

Recall that $\overline{E} = \{\overline{e} : e \in E\}$ where $\overline{e}$ is the filter $\{f \in E : f \geq e\}$, and that $\overline{E}$ is a dense subset of X. Recall also (**4.3**) that a typical element of $G = G_u$ is an equivalence class $\overline{(x, s)}$ ($x \leq ss^*$) where $x \in X$ and $(x, s) \sim (x, t)$ if and only if there exists $e \in E$ such that $x \leq e \leq (ss^*)(tt^*)$ and $es = et$. Further, $\overline{E}$ is an invariant subset of X. Indeed, if $e \in E$ and $z \in G$ is such that $r(z) \in \overline{e}$, then we can write $z = \overline{(\overline{e}, s)}$ where $e \leq ss^*$, and then, by (4.25), $d(z) = \overline{e}.s = \overline{s^*es} \in \overline{E}$.

We define a map $\alpha_S : G_S \to G$ by:

$$\alpha_S(s) = \overline{(ss^*, s)}.$$

The next result identifies G_S as a groupoid with the reduction H of G to $\overline{E}$. Note that while G_S is, of course, discrete, the reduction H (in the relative topology inherited from G) is not in general discrete (even for the case where S is a semilattice). This can be illustrated using the semilattice $E(S)$ where S is the Clifford semigroup discussed and used in Appendix C.

Proposition 4.4.6   *The map α_S is a groupoid isomorphism from G_S onto the reduction H of G to $\overline{E}$, and H is a dense subgroupoid of G.*

Proof.   Write $\alpha = \alpha_S$. Using the invariance of $\overline{E}$,

$$H = \{w \in G : d(w) \in \overline{E}\} = \{w \in G : r(w) \in \overline{E}\}.$$

Let $s \in S$. Then $r(\alpha(s)) = \overline{ss^*} \in \overline{E}$ and $\alpha(G_S) \subset H$.

Next let (s, t) be a composable pair in G_S. So $s^*s = tt^*$. Then $\alpha(st) = \overline{(stt^*s^*, st)} = \overline{(ss^*, st)}$. Now $\overline{ss^*}.s = \overline{s^*s} = \overline{tt^*}$ so that $(\alpha(s), \alpha(t))$ is composable in H. Then $\alpha(s)\alpha(t) = \overline{(ss^*, st)} = \alpha(st)$. Also,

$$\alpha(s)^{-1} = \overline{(s^*s, s^*)} = \alpha(s^*).$$

So α is a groupoid homomorphism.

Next, let $w = \overline{(\overline{e}, s)} \in H$. Then $e \leq ss^*$. Now $\overline{(\overline{e}, s)} = \overline{(\overline{e}, es)} = \overline{((es)(es)^*, es)}$. So $\alpha(es) = w$ and α maps G_S onto H.

Next, the pair $(\overline{(ss^*, s)}, \overline{(tt^*, t)})$ is composable in H if and only if $\overline{ss^*}.s = \overline{tt^*}$, i.e. if and only if $s^*s = tt^*$, i.e. if and only if the pair (s, t) is composable in G_S.

Finally, α is one-to-one. Indeed, suppose that $\alpha(s) = \alpha(t)$. Then $\overline{(ss^*, s)} = \overline{(tt^*, t)}$ so that $\overline{ss^*} = \overline{tt^*}$ and there exists an $e \in E$ such that $ss^* \leq e \leq ss^*tt^*$ and $es = et$. Since $\overline{ss^*} = \overline{tt^*}$, we have $ss^* = tt^*$, and since $ss^* \leq e \leq ss^*$, $tt^* \leq e \leq tt^*$, we also have $ss^* = e = tt^*$ giving $s = es = et = t$. So α is an isomorphism.

Lastly, let $\overline{(x_0, s)} \in G$. Since $\overline{E}$ is dense in X, there exists a net $\{e_\delta\}$ in E such that $\overline{e_\delta} \to x_0$ in D_s. Note that $e_\delta \leq ss^*$ since $\overline{e_\delta} \leq ss^*$. Since the map $x \to \overline{(x, s)}$ is a homeomorphism from D_s onto $D(D_s, s)$ (in the notation of Theorem 3.3.2 and Theorem 4.3.1), it follows that $\overline{(e_\delta, s)} \to \overline{(x_0, s)}$. By an earlier part of the proof, $\overline{(e_\delta, s)} \in H$ and it follows that H is dense in G. $\square$

The map α_S thus makes a groupoid identification of G_S with H, the reduction of G to $\overline{E}$. For example, when S is the Cuntz groupoid S_n, then, in the notation of **4.3**, Example 3, $\overline{E} = \{\overline{1}\} \cup \overline{Y} \cup \{\overline{z_0}\}$, and α_S sends $s_y t_{y'}$ to $(\overline{y}, s_y t_{y'})$, or, in terms of triples, to $(y, \ell(y) - \ell(y'), y')$.

We now show that for general S, the C^*-algebras $C^*_{red}(S)$ and $C^*_{red}(G)$ are isomorphic.

Theorem 4.4.2   *The map* $\psi : C(S) \to C_c(G)$ *of the proof of* Theorem 4.4.1 *extends to an isomorphism from* $C^*_{red}(S)$ *onto* $C^*_{red}(G)$.

Proof.   We will assume that the set E of idempotents of S is infinitely countable. (The case where E is finite is easy. In that case, G is isomorphic to the discrete groupoid G_S.) Let $\{e_n : n \geq 1\}$ be an enumeration of E and ρ be the measure on E given by $\rho = \sum_{n=1}^{\infty} 2^{-n} \delta_{e_n}$. Of course, ρ is equivalent to counting measure on E.

Clearly, $\ell^2(S) = \oplus_{n=1}^{\infty} \ell^2(S_{e_n})$ is isomorphic to $\int_E \ell^2(S_e) \, d\rho(e)$. (It is, of course, a direct sum but we use the integral notation to link this up with the groupoid notation.) Further, by Proposition 4.4.5, $C^*_{red}(S)$ is canonically isomorphic with the C^*-algebra generated by the elements $\pi_2(s) = \int_E \pi_2^e(s) \, d\rho(e)$ $(s \in S)$.

We now switch over to G. Let μ be the measure on $X = G^0$ corresponding to ρ: so $\mu = \sum_{n=1}^{\infty} 2^{-n} \delta_{u_n}$ where $u_n = \overline{e_n}$. By Proposition 4.3.1, $\overline{E}$ is dense in X. Hence by Proposition 3.1.2, $C^*_{red}(\mu)$, the C^*-algebra generated by $Ind \, \mu(C_c(G))$, is canonically isomorphic to $C^*_{red}(G)$. Recall that $\alpha_S : G_S \to G$ is the map defined by $\alpha_S(s) = (\overline{ss^*}, s)$ and that by Proposition 4.4.6, α_S is an isomorphism onto the reduction H of G to $\overline{E}$. Since the left Haar system on G is given by counting measures, the map α_S extends to a Hilbert space isomorphism also denoted by α_S from $\int \ell^2(S_e) \, d\rho(e)$ onto $\int \ell^2(G_u) \, d\mu(u)$. Specifically, for appropriate complex numbers $a_{e,t}$,

$$\alpha_S \Big(\sum_{e \in E} \sum_{t \in S_e} a_{e,t} t \Big) = \sum_{e \in E} \sum_{t \in S_e} a_{e,t} \alpha_S(t).$$

This isomorphism then implements an isomorphism from $C^*_{red}(S)$ onto $C^*_{red}(G)$.

To prove this, the measure $\nu = \int_{\overline{E}} \lambda^u \, d\mu(u)$ is equivalent to counting measure on H and so therefore is ν^{-1}. So for the purpose of using the definition of $Ind \, \mu$ in (3.40), we can take $F = \delta_{\alpha_S(t)}$, and identifying $\chi_{\psi(s)} * \delta_{\alpha_S(t)}$ with the product of two G-sets, we have to show that for $s \in S$, the following equality holds:

$$\alpha_S(\pi_2(s)(t)) = \psi(s)\alpha_S(t). \tag{4.41}$$

Suppose first that s^*s is not $\geq tt^*$. Then the LHS (left-hand side) of (4.41) is 0 (see the proof of Proposition 4.4.5). The RHS (right-hand side) of (4.41) is $\{\overline{(x,s)} : x \leq ss^*\}\overline{(tt^*, t)} = \emptyset$, whose characteristic function is 0. So (4.41) holds in this case.

If $s^*s \geq tt^*$, then the LHS of (4.41) is $\alpha_S(st) = \overline{(stt^*s^*, st)}$, while observing that $\overline{tt^*} = \overline{stt^*s^*}.s$, the RHS of (4.41) is $\{\overline{(x, s)} : x \leq ss^*\}\overline{(tt^*, t)} = \overline{(stt^*s^*, st)}$ and again (4.41) holds.   □

4.5   Amenability of the von Neumann algebra of an inverse semigroup

The main objective of this section is to obtain a very simple condition which ensures that the von Neuman algebra $VN(S)$ generated by the left regular representation of an inverse semigroup S is amenable (=injective): it is that the maximal subgroups of S be amenable (in the classical, discrete group sense). This gives a large variety of injective von Neumann algebras and for a given inverse semigroup S, the condition is usually very easy to check. More specifically, we just have to show that for each $e \in E$, the group $S_e^e = \{s \in S : ss^* = e = s^*s\}$ is amenable. The proof uses the theory developed earlier relating the C^*-algebras of S and $G_\mathbf{u}$.

It is simple to see that the S_e^e's are the maximal subgroups of S. Firstly, if H is a subgroup of S then its identity is some $e \in E$. If $s \in H$, then checking (2.1) gives that s^* is the inverse of s in H. So $ss^* = e = s^*s$, and $H \subset S_e^e$. That S_e^e is itself a subgroup follows either by an easy direct argument or by observing that S_e^e is an isotropy group (Chapter 1) for the groupoid G_S. So the S_e^e's are the maximal subgroups of S.

We will need to use Renault's theory of amenable quasi-invariant measures on a groupoid ([230, Chapter 2, §3]). For the convenience of the reader and also in order to make some slight modifications, we will survey here that part of the theory that we will need.[11]

Let G be a locally compact Hausdorff groupoid and $\mu \in P(G^0)$ be quasi-invariant (**3.1**). To obtain amenability properties for the operator algebras associated with G, we require some kind of amenability property associated with μ. In the case of a locally compact group, the measure μ will be the point mass δ_e, and the amenability of this measure will be equivalent to classical amenability for a locally compact group. (Detailed accounts of the theory of amenable locally compact groups are given in the books [196, 204].)

For the groupoid G, there are many quasi-invariant measures on the unit space, and the amenable ones are those measures μ which combine

---

[11]There is a very recent extensive investigation into amenable groupoids by C. Anantharaman-Delaroche and J. Renault ([5]). Among many other things, this contains definitive results on amenability for r-discrete groupoids. The present section was written without knowledge of that investigation, and it seems likely that the results of [5] will be helpful for resolving the questions raised in this section.

with the left Haar system to give measures ν ((3.1)) on G for which there is "Reiter-type" condition,[12] involving $L^\infty(G, \nu)$ and $L^\infty(G^0, \mu)$, and so ensures, for example, amenability of the von Neumann algebra generated by any representation of G for which the associated quasi-invariant measure is μ. (The definition of amenable measure makes sense even if μ is not assumed to be quasi-invariant, but there are indications that the definition forces quasi-invariance on μ.)

The kind of amenability involved is thus a variant of the *Reiter condition* for locally compact groups. A novel feature (from the point of view of classical amenability for locally compact groups) is that the Reiter-type nets involved are required to converge in the appropriate *weak** topologies. As far as the author can ascertain, this weak* kind of Reiter condition has not been considered for locally compact groups. We will show in Appendix B that such a condition is (reassuringly) equivalent to classical amenability for a locally compact group (with $\mu = \delta_e$).

For the definition of an amenable measure, two functions associated with $g \in C_c(G)$ are required. The first is the function $g^0 \in C_c(G^0)$ used in (ii) of Definition 2.2.2. Next, for $x \in G$, define $F_g : G \to \mathbf{C}$ by:

$$F_g(x) = \int_{G^{r(x)}} \mid g(x^{-1}t) - g(t) \mid d\lambda^{r(x)}(t) \qquad (4.42)$$

$$= \|x * g - g\|_1^{r(x)} \qquad (4.43)$$

where $x * g(t) = g(x^{-1}t)$ is defined for $t \in G^{r(x)}$ (as in (2.18)) and $\|.\|_1^u$ is the $L^1(\lambda^u)$ norm. (So in (4.43), the second function g inside the norm is restricted to $G^{r(x)}$.) The u will be omitted from $\|.\|_1^u$ when its meaning is clear. We will also use the norm $\|.\|_{I,r}$ on $C_c(G)$. This was defined in (2.23). Note that using (2.14),

$$F_g(x) = \int_{G^{d(x)}} \mid g(xt) - g(t) \mid d\lambda^{d(x)}(t). \qquad (4.44)$$

In order to define Renault's notion of amenability for a measure μ, as given below, we need to know that $F_g \in L^\infty(\nu)$. It seems plausible that every F_g is Borel measurable for every locally compact groupoid, but the writer has not been able to prove this. In the Hausdorff case, the function F_g is actually continuous. This need not be the case for a locally compact groupoid in general. However, in the ample case, F_g is Borel measurable. These results are proved in Appendix C. They are not really needed for the purposes of this work since the measure μ whose amenability we will be using is discrete (as are the measures of the left Haar system). However a discussion of it in that Appendix was felt justified in view of its obvious

---

[12]See Appendix B.

importance for groupoid amenability in general. The Borel measurability of F_g will be assumed below where necessary.

Definition 4.5.1. The quasi-invariant measure μ on G^0 is said to be *amenable* if there exists a sequence $\{g_n\} \geq 0$ in $C_c(G)$ such that:

(i) $\sup_n \|g_n\|_{I,r} < \infty$;

(ii) $F_{g_n} \to 0$ weak* in $L^\infty(G, \nu)$;

(iii) $g_n^0 \to 1$ weak* in $L^\infty(G^0, \mu)$.

The definition of amenable measure for the groupoid G given above is essentially due to Renault ([230, p.89]). Condition (i) is implicit in his work. He also uses a net $\{g_\delta\}$ rather than a sequence $\{g_n\}$. However we can get by with the sequence using the fact that both $L^1(G, \nu)$ and $L^1(G^0, \mu)$ are separable.

Implicit in (ii) and (iii) are the facts that for all n, $g_n^0 \in L^\infty(G^0, \mu)$ and $F_{g_n} \in L^\infty(G, \nu)$. The first follows since $g_n^0 \in C_c(G^0)$ by (ii) of Definition 2.2.2. The second follows using (i) as described below. It also follows that the sequences $\{g_n^0\}, \{F_{g_n}\}$ are bounded in $L^\infty(G^0, \mu)$ and $L^\infty(G, \nu)$ respectively.

To see these facts, for all $u \in G^0$ and any m, we have

$$g_m^0(u) \leq \int_{G^u} g_m(t) \, d\lambda^u(t) \leq \sup_n \|g_n\|_{I,r}.$$

So $\{g_n^0\}$ is bounded in $L^\infty(G^0, \mu)$. Also for any $x \in G$, using (2.16),

$$F_{g_n}(x) \leq \int g_n(x^{-1}t) \, d\lambda^{r(x)}(t) + \int g_n(t) \, d\lambda^{r(x)}(t) \leq 2\|g_n\|_{I,r}.$$

In particular, every F_{g_n} is bounded (and by assumption, Borel measurable) and so belongs to $L^\infty(G, \nu)$, and the sequence $\{F_{g_n}\}$ is bounded in $L^\infty(G, \nu)$.

One of the many useful properties that amenability entails for a locally compact group is that any C^*-algebra or von Neumann algebra generated by a representation of the group is itself amenable. In particular, a discrete group H is amenable if and only if the von Neumann algebra $VN(H)$ generated by the left regular representation of G is amenable. We will show that (at least) one direction of this result holds for the inverse semigroup S: if every maximal subgroup of S is amenable, then $VN(S)$ is amenable. This will be proved by using the universal groupoid and the theory of amenable measures. We start with a brief discussion of amenable von Neumann algebras.

Amenability for a von Neumann algebra can be formulated in a number of ways. See, for example, [56, V.7] and [196, (2.35)] for a discussion and for references. The formulation that we shall use here is that of injectivity. A von Neumann algebra $M \subset B(\mathcal{H})$ is called *injective* if there exists a linear, norm 1 projection P from $B(\mathcal{H})$ onto M. Such a projection is automatically positive. Further, the injective property is independent of the Hilbert space $\mathcal{H}$ on which M is realized. The class of injective von Neumann algebras is large. In particular, every abelian von Neumann algebra is injective and the representations of an amenable locally compact group generate injective von Neumann algebras (though the converse is not true in general). An important result which we will need is that M *is injective if and only if its commutant M' is injective.*

Recall (**3.1**) that a representation of a locally compact groupoid G is given by a triple $(\mu, \{H_u\}, L)$ whose integrated form is a representation π_L of $C_c(G)$ on the Hilbert space of sections $L^2(G^0, \{H_u\}, \mu)$. We will require a theorem of Renault ([230, p.90]) which establishes that when the quasi-invariant measure μ is amenable, then *the von Neumann algebra generated by $\pi_L(C_c(G))$ is injective.* In our discussion of this, we will prove this first in the special case where G is a locally compact group and sketch the proof for the general case. The reason why we will look at the special case first is that this will clarify the main lines of the argument and also highlight the roles played by (ii) and (iii) of Definition 4.5.1 in the study of amenable quasi-invariant measures.[13] With the group case as a guide, the reader should also find it instructive to fill out the details of the calculations of the proof below for the general case.

Proposition 4.5.1   *Let H be a locally compact group such that δ_e is amenable and π be a unitary representation of H on a Hilbert space $\mathcal{H}$. Then the von Neumann algebra M generated by $\pi(H)$ (or equivalently, by $\pi(C_c(H))$) is injective.*

Proof.   Let $\{g_n\}$ be as in Definition 4.5.1 (with H in place of G). Note that $H^e = H$. Then (i) is redundant because (iii) gives that $\|g_n\|_{I,r} = g_n^0(e) \to 1$, and we can suppose that

$$\int g_n \, d\lambda = 1 \tag{4.45}$$

for all n, where λ is a left Haar measure on H. Then (i) and (iii) hold and this leaves (ii) which says that $F_{g_n} \to 0$ weak* in $L^\infty(H)$. Since M is injective if and only if M' is injective, it is sufficient to show that M' is injective. Obviously,

$$M' = \{B \in B(\mathcal{H}) : B\pi(x) = \pi(x)B \text{ for all } x \in H\}. \tag{4.46}$$

---

[13]Of course, we could deduce the special case using known results from amenable locally compact groups and the result of Appendix B.

For each n, let $P_n : B(\mathcal{H}) \to B(\mathcal{H})$ be given by:

$$\langle P_n(B)\xi, \eta \rangle = \int g_n(y) \langle \pi(y) B \pi(y)^{-1} \xi, \eta \rangle \, d\lambda(y) \qquad (4.47)$$

where $\xi, \eta \in \mathcal{H}$. Next, the integrated form of the representation π for $f \in C_c(G)$ is given by:

$$\langle \pi(f)\xi, \eta \rangle = \int f(x) \langle \pi(x)\xi, \eta \rangle \, d\lambda(x). \qquad (4.48)$$

It follows from (4.47) and (4.45) that $P_n \in B(B(\mathcal{H}))$, $\|P_n\| = 1$ and $P_n(I) = I$. It is obvious from (4.46) and (4.47) that $P_n(B) = B$ for all $B \in M'$. If we can show that for all $f \in C_c(H), B \in B(\mathcal{H})$ and $\xi, \eta \in \mathcal{H}$, we have

$$\lim_n \langle (\pi(f)P_n(B) - P_n(B)\pi(f))\xi, \eta \rangle = 0 \qquad (4.49)$$

then we can obtain a linear, norm 1 projection P from $B(\mathcal{H})$ onto M' and hence the obtain injectivity of M'. For the bounded net $\{P_n\}$ has a weak operator convergent subnet (effectively by Tychonoff's theorem), and we can take P to be the limit of this subnet.

So we have to establish (4.49). Firstly, using (4.48) and (4.47), we have that

$$
\begin{aligned}
& \langle \pi(f)P_n(B)\xi, \eta \rangle \\
&= \int f(x) \langle \pi(x)P_n(B)\xi, \eta \rangle \, d\lambda(x) \\
&= \int f(x) \, d\lambda(x) \int g_n(y) \langle \pi(y) B \pi(y)^{-1} \xi, \pi(x)^{-1} \eta \rangle \, d\lambda(y).
\end{aligned}
$$

$$(4.50)$$

Next, using (4.48), (4.47), Fubini's theorem and the left invariance of λ, we have

$$
\begin{aligned}
& \langle P_n(B)\pi(f)\xi, \eta \rangle \\
&= \int g_n(y) \langle \pi(y) B \pi(y)^{-1} \pi(f)\xi, \eta \rangle \, d\lambda(y) \\
&= \int g_n(y) \, d\lambda(y) \int f(x) \langle \pi(y) B \pi(y)^{-1} \pi(x)\xi, \eta \rangle \, d\lambda(x) \\
&= \int f(x) \, d\lambda(x) \int g_n(y) \langle \pi(y) B \pi(x^{-1}y)^{-1} \xi, \eta \rangle \, d\lambda(y) \\
&= \int f(x) \, d\lambda(x) \int g_n(xy) \langle \pi(y) B \pi(y)^{-1} \xi, \pi(x)^{-1} \eta \rangle \, d\lambda(y).
\end{aligned}
$$

$$(4.51)$$

Combining (4.50) with (4.51) and using the left invariance of λ and (ii) of Definition 4.5.1, we obtain that $|\langle(\pi(f)P_n(B) - P_n(B)\pi(f))\xi, \eta\rangle|$ equals

$$\left| \int f(x) \, d\lambda(x) \int [g_n(y) - g_n(xy)]\langle\pi(y)B\pi(y)^{-1}\xi, \pi(x)^{-1}\eta\rangle \, d\lambda(y) \right|.$$

Using (ii) of Definition 4.5.1, the preceding expression is

$$\begin{aligned} &\leq \quad \|B\|\|\xi\|\|\eta\| \int |f(x)| \, F_{g_n}(x) \, d\lambda(x) \\ &= \quad \|B\|\|\xi\|\|\eta\|\langle F_{g_n}, |f|\rangle \\ &\rightarrow \quad 0. \end{aligned}$$

This completes the proof. $\square$

Theorem 4.5.1   ([230, p.90]) *Let G be a locally compact groupoid and $(\mu, \{H_u\}, L)$ be a representation of G where μ is amenable. Then the von Neumann algebra M generated by $\pi_L(C_c(G)) \subset B(L^2(G^0, \{H_u\}, L))$ is injective (= amenable).*

Proof.   Let $\{g_n\}$ be as in Definition 4.5.1. The proof is a modified version of that of Proposition 4.5.1. Let $\mathcal{H} = L^2(G^0, \{H_u\}, \mu)$. We want to obtain a linear norm 1 projection P from $B(\mathcal{H})$ onto M'. Unfortunately, an arbitrary $B \in B(\mathcal{H})$ will not relate in any natural way to the $\{H_u\}$'s. (This was not a problem in the group case since then there was only one H_u.) The operators in $B(\mathcal{H})$ that *do* relate to the $\{H_u\}$'s are the decomposable ones, $B = \{B_u\}$. (So for $\xi \in \mathcal{H}$, we have $B(\xi)(u) = B_u(\xi(u))$.) Let D be the algebra of decomposable operators on $\mathcal{H}$. Now D is the commutant of an *abelian* von Neumann algebra, that of the algebra Δ of diagonalizable operators ([72, p.188]). So D is injective as well. Further, $M' \subset D$ since M contains the weak operator closure of $\pi_L(B_c(G^0))$ and this is just Δ. So the injectivity of M' will follow if we can show that there is a linear norm 1 projection P from D onto M', for then we get a linear norm 1 projection from $B(\mathcal{H})$ onto M' by following up a linear norm 1 projection from $B(\mathcal{H})$ onto D by the projection P.

The analogue of (4.46) with $B \in D$ is:

$$M' = \{B \in D : B_{r(x)}L(x) = L(x)B_{d(x)} \ \nu - \text{ a.e. }\}. \tag{4.52}$$

To prove this, use the fact that if $B \in D$, then $B \in M'$ if and only if for all $f \in C_c(G)$ and $\xi, \eta \in \mathcal{H}$, we have $\langle(B\pi_L(f) - \pi_L(f)B)\xi, \eta\rangle = 0$. The latter expression is calculated using (3.17).

In place of (4.47), we define $\langle P_n(B)\xi, \eta\rangle$ to be

$$\int g_n(y)\langle L(y)B_{d(y)}L(y)^{-1}\xi(r(y)), \eta(r(y))\rangle \, d\nu(y). \tag{4.53}$$

We obtain the desired projection $P : D \to M'$ as in the group case by taking a weak operator convergent subnet of $\{P_n\}$. Using (4.52), (4.53) and (iii) of Definition 4.5.1, one obtains for $B \in M'$, that

$$\langle P_n(B)\xi, \eta \rangle = \langle g_n^0, k \rangle \to \langle B\xi, \eta \rangle$$

where $k \in L^1(G^0, \mu)$ is given by: $k(u) = \langle B_u\xi(u), \eta(u) \rangle$. This gives $P(B) = B$ for all $B \in M'$.

In place of (4.48), we take (3.17). It is useful to supplement these two expressions determining $P_n(B)$ and $\pi_L(f)$ by pointwise formulae obtained respectively from them by using $\nu = \int \lambda^u \, d\mu(u)$ and $\nu_0 = \int D^{-1/2}\lambda^u \, d\mu(u)$. These are respectively for μ - a.e. u,

$$P_n(B)\xi(u) = \int g_n(y)L(y)B_{d(y)}L(y)^{-1}\xi(u) \, d\lambda^u(y) \qquad (4.54)$$

and ((3.20))

$$\pi_L(f)\xi(u) \quad = \quad \int f(x)L(x)\xi(d(x))D^{-1/2}(x) \, d\lambda^u(x). \qquad (4.55)$$

In place of (4.50) we then obtain (using (3.17) and (4.54)) that

$$\langle \pi_L(f)P_n(B)\xi, \eta \rangle$$

equals:

$$\int f(x) \, d\nu_0(x) \int g_n(y)\langle L(y)B_{d(y)}L(y)^{-1}\xi(d(x)), L(x)^{-1}\eta(r(x)) \rangle \, d\lambda^{d(x)}(y).$$
$$(4.56)$$

Using (4.53), (4.55) and (2.14), we obtain in place of (4.51) that

$$\langle P_n(B)\pi_L(f)\xi, \eta \rangle$$

equals:

$$\int f(x) \, d\nu_0(x) \int g_n(xy)\langle L(y)B_{d(y)}L(y)^{-1}\xi(d(x)), L(x)^{-1}\eta(r(x)) \rangle \, d\lambda^{d(x)}(y).$$
$$(4.57)$$

Estimating the difference between the left-hand sides of (4.56) and (4.57) and using (iii) of Definition 4.5.1 then gives for $B \in D$ that

$$| \langle (\pi_L(f)P_n(B) - P_n(B)\pi_L(f))\xi, \eta \rangle | \quad \leq \quad \|B\|\langle g, F_{g_n} \rangle$$
$$\to \quad 0$$

where

$$g(x) =| f(x) | \, \|\xi(d(x))\|\|\eta(r(x))\|D^{-1/2}(x).$$

(Note that $g \in L^1(G, \nu)$ by the argument following (3.21).) This gives $P(B) \in M'$ as required.   □

The following easy proposition gives a simple interpretation of amenability in the case where μ is discrete. Of course, this is a very special situation but it occurs remarkably often – indeed as we shall see in the case of the universal groupoid of an inverse semigroup S, the amenability of $VN(S)$ follows from the amenability of the natural discrete measure μ used in the proof of Theorem 4.4.2.

Proposition 4.5.2   *Let μ be a discrete quasi-invariant measure on G^0. Let $T = \{v \in G^0 : \mu(\{v\}) > 0\}$. Then μ is amenable if and only if there exists a sequence $\{g_n\} \geq 0$ in $C_c(G)$ such that $\sup_n \|g_n\|_{I,r} < \infty$, $(F_{g_n})_{|G^u} \to 0$ weak* in $L^\infty(G^u, \lambda^u)$ for all $u \in T$ and $g_{n|T}^0 \to 1$ pointwise.*

Proof.   Recalling that $\nu = \int \lambda^u \, d\mu(u)$, the space $L^1(G, \nu)$ can be identified with the weighted vector-valued ℓ^1- space of functions f on T where $f(t) \in L^1(G^t, \lambda^t)$ for all $t \in T$ and

$$\sum_{t \in T} \mu(\{t\}) \|f(t)\|_1 < \infty.$$

Then $L^\infty(\nu)$ is identified with the vector-valued ℓ^∞- space of functions F on T for which $F(t) \in L^\infty(G^t, \lambda^t)$ and

$$\sup_{t \in T} \|F(t)\|_\infty < \infty.$$

The duality between these L^1- and L^∞- spaces is given by:

$$\langle F, f \rangle = \sum_{t \in T} \mu(\{t\}) \langle F(t), f(t) \rangle.$$

Weak* convergence of a bounded sequence $\{F_n\}$ to 0 in $L^\infty(G, \nu)$ is then equivalent to $F_{n|G^t} \to 0$ weak* in $L^\infty(G^t, \lambda^t)$ for all $t \in T$. Similar considerations apply to the case of $L^1(G^0, \mu)$ which is the scalar-valued version of the above.

If μ is amenable and $\{g_n\}$ satisfies the properties (i), (ii) and (iii) of Definition 4.5.1, then the boundedness of the sequences $\{g_n^0\}$, $\{F_{g_n}\}$ in the appropriate L^∞- spaces together with the preceding paragraph give the first implication of the proposition. The equivalences of that paragraph give the other implication.   □

Amenability for a locally compact group is often defined in terms of the existence of an invariant mean on the group (e.g. [196, p.5]). In the groupoid case with which we will be concerned, the notion of an invariant mean will be replaced by an appropriately invariant *family* $\{m_v\}$ of means on the G^v's.

Recall from **3.1** that for each x in a locally compact groupoid G, the map $f \to x*f$ is a linear isometry from $L^1(G^{d(x)}, \lambda^{d(x)})$ onto $L^1(G^{r(x)}, \lambda^{r(x)})$, where $(x * f)(t) = f(x^{-1}t)$. The Banach space dual of this map gives a right isometric action $\phi \to \phi x$ from $L^\infty(G^{r(x)}, \lambda^{r(x)})$ onto $L^\infty(G^{d(x)}, \lambda^{d(x)})$. To calculate this action, for a function ϕ in $L^\infty(G^{r(x)}, \lambda^{r(x)})$, we have, using (2.14),

$$\langle x * f, \phi \rangle = \int f(x^{-1}t)\phi(t) \, d\lambda^{r(x)}(t) = \int f(t)\phi(xt) \, d\lambda^{d(x)}(t)$$

so that

$$\langle x * f, \phi \rangle = \langle f, \phi x \rangle \qquad (4.58)$$

where $\phi x(t) = \phi(xt)$. So this dual action from $L^\infty(G^{r(x)}, \lambda^{r(x)})$ into $L^\infty(G^{d(x)}, \lambda^{d(x)})$ is given by: $\phi \to \phi x$.

The dual of the map $\phi \to \phi x$ in turn gives a left isometric map $m \to xm$ from the second dual space $L^1(G^{d(x)}, \lambda^{d(x)})^{**}$ onto $L^1(G^{r(x)}, \lambda^{r(x)})^{**}$, the latter extending the initial left action on $L^1(G^{d(x)}, \lambda^{d(x)})$. Of course by "action" in such contexts we mean that inverses and products of composable pairs satisfy the obvious analogues for group actions. For example, if $r(x) = d(y)$ and $m \in L^1(G^{d(x)}, \lambda^{d(x)})^{**}$, then $(yx)m = y(xm)$. When G is a group, then the above actions of G are the standard ones.

A *mean* on a set X is a state m on the C^*-algebra $\ell^\infty(X)$. Alternatively a mean is a finitely additive positive measure μ on the family of all subsets of X with $\mu(X) = 1$. We shall use this in the proof of Proposition 4.5.4. The two formulations of *mean* are shown to be equivalent by defining $m(\chi_A) = \mu(A)$ for all $A \subset X$. When H is a discrete group, a *left invariant mean* on H is a mean $m \in \ell^\infty(H)^* = \ell^1(H)^{**}$ such that $xm = m$ for all $x \in H$. Regarding m as a finitely additive measure μ, the left invariance is equivalent to the condition that $\mu(x^{-1}A) = \mu(A)$ for all $A \subset H$ (since $\chi_A x = \chi_{x^{-1}A}$). The group H is defined to be *amenable* if there exists a left invariant mean on H.

We now let G be an ample groupoid. Recall that for each $u \in G^0$, λ^u is counting measure so that $L^1(G^u, \lambda^u) = \ell^1(G^u)$. Let T be an invariant subset of G^0. A family $\{m_v : v \in T\}$ is called a *left invariant mean for T* (in G) if each m_v is a mean on G^v and $xm_{d(x)} = m_{r(x)}$ for all $x \in G$ with $r(x)$ (and therefore $d(x)$) in T. This is the natural extension of the notion of a left invariant mean on a group H discussed in the preceding paragraph.

The next proposition gives an invariant mean characterization for a discrete probability measure to be quasi-invariant.

Proposition 4.5.3   Let G be an ample groupoid and μ be a discrete quasi-invariant measure on G^0. Let $T = \{v \in G^0 : \mu(\{v\}) > 0\}$. Then T

is an invariant subset of G^0 and μ is amenable if and only if there exists a left invariant mean for T in G.

Proof. Let $v \in T$ and $x \in G^v$. Let $x \in G^v$ where $v \in T$. Then $\nu(\{x\}) = \mu(\{r(x)\}) > 0$. Since μ is quasi-invariant, we have $\mu(\{d(x)\}) = \nu^{-1}(\{x\}) > 0$. So T is invariant.

Let W be the reduction groupoid G_T. Let μ be amenable and $\{g_n\}$ be as in Proposition 4.5.2. For $u \in T$, let k_n^u be the restriction of $(g_n^0)^{-1}g_n$ to G^u. (This makes sense eventually since $g_n^0(u) \to 1$.) Each k_n^u is a mean on G^u, and by Tychonoff's theorem, there exists a subnet $g_{\alpha(\delta)}$ of g_n such that $k_{\alpha(\delta)}^u \to m_u$ weak* for some mean m_u on G^u and for all $u \in T$. For $u \in G^0$ and $f \in \ell^1(G^u)$, let $f^\wedge \in \ell^1(G^u)^{**}$ be given by: $f^\wedge(\phi) = \phi(f)$ for $\phi \in \ell^\infty(G^u)$. Let $x \in W$. Then using (4.58), for $\phi \in \ell^\infty(G^{r(x)})$ and any n, $\lim_{n \to \infty} | (x * k_n^{d(x)} - k_n^{r(x)})^\wedge(\phi) | = \lim_{n \to \infty} | (x * g_n^{d(x)} - g_n^{r(x)})^\wedge(\phi) | \le \lim_{n \to \infty} F_{g_n}(x) \|\phi\|_\infty = 0$. This argument follows using firstly the fact that $g_n^0(u) \to 1$ for all $u \in T$ and the ℓ^1-boundedness of the functions $x * g_n^{d(x)}, g_n^{r(x)}$, and secondly the fact that weak* convergence of a bounded sequence in any $\ell^\infty(G^u)$ is equivalent to its pointwise convergence. It then follows that $(xm_{d(x)} - m_{r(x)})(\phi) = \lim_\delta (x * k_{\alpha(\delta)}^{d(x)} - k_{\alpha(\delta)}^{r(x)})^\wedge(\phi) = 0$ so that $xm_{d(x)} = m_{r(x)}$. So $\{m_v : v \in T\}$ is a left invariant mean for T.

Conversely, suppose that $\{m_v : v \in T\}$ is a left invariant mean for T. For any set A let $P(A)$ be the convex set of finite means $\sum_{a \in A} \alpha_a a$, where $\alpha_a \ge 0$ and $\sum_a \alpha_a = 1$, in $\ell^1(A)$. Then $P(A)$ is weak* dense in the set of means on A. Hence there exists a directed set Δ and for each $v \in T$, a net f_δ^v ($\delta \in \Delta$) in $P(G^v)$ such that $(f_\delta^v)^\wedge \to m_v$ weak* in $\ell^\infty(G^v)^*$. Now W is countable and for any $x \in W$, using (4.58), $(x * f_\delta^{d(x)} - f_\delta^{r(x)}) \to 0$ weakly in $\ell^1(G^{r(x)})$. Day's condition of "strong convergence to invariance" (see, for example, a classic argument of Namioka presented in [196, p.8]) then gives that there exist nets $\{k_\sigma^v\}$ built out of convex combination of the f_δ^v's "far out" such that

$$\|x * k_\sigma^{d(x)} - k_\sigma^{r(x)}\|_1 \to 0$$

for all $x \in W$.

Let $\{F_n\}$ be an increasing sequence of finite subsets of W with $\cup_{n=1}^\infty F_n = W$ and $F_n^{-1} = F_n$. For each n, choose a σ so that the $h_n^v = f_\sigma^v$ for $v \in r(F_n) = d(F_n)$ satisfy:

$$\|x * h_n^{d(x)} - h_n^{r(x)}\|_1 < 1/n \tag{4.59}$$

for all $x \in F_n$. By construction, each h_n^v ($v \in r(F_n)$) has finite support. Enumerate $r(F_n) = \{v_1, \ldots, v_m\}$ and write $h_i = h_n^{v_i} = \sum_{j=1}^{N_i} \alpha(i,j)x_{i,j}$. (So h_i is a convex combination of $x_{i,j} \in G^{v_i}$.) Since G is ample and the v_i's are distinct, we can find $U_{i,j} \in G^a$ and open subsets V_i of G^0 such that $x_{i,j} \in U_{i,j}, U_{i,j} \cap U_{i,k} = \emptyset$ if $j \neq k$, $r(U_{i,j}) = r(U_{i,k}) = V_i$ for all j, k and $r(U_{i,j}) \cap r(U_{i',k}) = \emptyset$ if $i \neq i'$ for all j, k. The $U_{i,j}$'s are then pairwise disjoint, and we set

$$g_n = \sum_{i,j} \alpha(i,j) \chi_{U_{i,j}} \in C_c(G).$$

If $v \in G^0$, then $g_n^0(v) = 0$ if v does not belong to $\cup_{i=1}^m V_i$. If $v \in V_i$ for some i, then $g_n^0(v) = \sum_{j=1}^{N_i} \alpha_{i,j} = 1$. So $\|g_n\|_{I,r} \leq 1$ and $g_n^0(v_i) = 1$ for all i. Also, $g_n^0 \to 1$ pointwise on T. Further g_n and $h_n^{v_i}$ coincide on each G^{v_i} so that (4.59) gives $F_{g_n}(x) < 1/n$ for all $x \in F_n$. So μ is amenable by Proposition 4.5.2.   □

We keep the same notation as that of the above proposition.

Proposition 4.5.4   *The discrete measure μ is amenable if and only if every isotropy group $G_v^v = \{x \in G : r(x) = v = d(x)\}$ for $v \in T$ is an amenable group.*

Proof.   Suppose that μ is amenable. By Proposition 4.5.3, there exists a left invariant mean $\{m_v\}$ for T. Let $v \in T$ and $H = G_v^v$. Then H acts by left multiplication on the set G^v and this action is free. Then G^v is a disjoint union of orbits Hb, $(b \in B)$. Since G^v is countable, so also is B. We will suppose that B is infinite leaving the case where B is finite to the reader. Enumerate $B = \{x_n : n \geq 1\}$. Regarding m_v as a finitely additive measure, we define a function p on the subsets A of H by:

$$p(A) = m_v(\cup_{n=1}^{\infty} A x_n).$$

Using the freeness of the H-action on G^v, it is easy to check that p is a mean on H. Further, since, for all $x \in H$, $xm_v = m_v$, we have $m_v(x^{-1}Y) = m_v(Y)$ for all $Y \subset G^v$, and so $p(x^{-1}A) = p(A)$. But this is equivalent to p being a left invariant mean on H. So H is amenable. (The above argument is, of course, a slight variant on von Neumann's classical proof that a subgroup of an amenable group is amenable.)

Conversely, suppose that for every $v \in T$, H is amenable. Let $v \in T$. We will use the notation of the preceding paragraph. Let n_v be a left invariant mean on H. Define a mean p_v on G^v by setting

$$p_v(Q) = \sum_{n=1}^{\infty} 2^{-n} n_v((Q \cap H x_n) x_n^{-1}).$$

For example, $p_v(G^v) = 1$ since $n_v((G^v \cap H x_n) x_n^{-1}) = n_v(H) = 1$. Also for $h \in H$, we have $p_v(h^{-1}Q) = p_v(Q)$ by the H-invariance of n_v. So p_v is a left H-invariant mean on G^v.

Define a relation $\sim$ on T by specifying $v \sim w$ if and only if there exists $x \in G$ such that $d(x) = v, r(x) = w$. It is easy to check that $\sim$ is an equivalence relation on T. (In fact, $\sim$ is just the restriction to T of the orbit equivalence relation for G.) The equivalence classes for $\sim$ are just the orbits $\{Z_\gamma\}$ ($\gamma \in \Gamma$) for T. For each γ, choose $v_\gamma \in Z_\gamma$ and set $H_\gamma = G_{v_\gamma}^{v_\gamma}$. Let $p_\gamma = p_{v_\gamma}$ be the left H_γ-invariant mean on G^{v_γ} constructed in the preceding paragraph. Let $v \in Z_\gamma$ and $x \in G$ be such that $d(x) = v_\gamma, r(x) = v$. Let $p_v = xp_\gamma$. Then p_v is a mean on G^v. We claim that p_v is independent of the choice of x. Indeed, suppose that $y \in G$ with $d(y) = v_\gamma, r(y) = v$. Then since $y^{-1}x \in H_\gamma$ and p_γ is left H_γ invariant, we have $y^{-1}xp_\gamma = p_\gamma$ so that $xp_\gamma = yp_\gamma$. It is easily proved that if $v, w \in Z_\gamma$ and $x' \in G$ is such that $d(x') = v, r(x') = w$ then $d(x'x) = v_\gamma, r(x'x) = w$, so that $x'p_v = x'xp_\gamma = p_w$. The set of p_v's for all $v \in Z_\gamma$ and all $\gamma \in \Gamma$ then gives a left invariant mean for T. So μ is amenable by Proposition 4.5.3. □

We now come to the main result of this section. Recall that $VN(S)$ is the von Neumann algebra generated by the left regular representation of S.

Theorem 4.5.2   *Let S be an inverse semigroup, every maximal subgroup of which is amenable. Then $VN(S)$ is amenable.*

Proof.   Let G be the universal groupoid for S. Let $T = \overline{E} \subset X = G^0$ and μ be the measure in the proof of Theorem 4.4.2. Then by definition, $T = \{u \in X : \mu(\{u\}) > 0\}$. By Proposition 4.4.6, G_S is isomorphic to the reduction of G by $\overline{E}$. Now by hypothesis, the isotropy groups G_v^v, each of which is isomorphic to a maximal subgroup of S, are amenable ($v \in \overline{E}$). So by Proposition 4.5.4, the measure μ is amenable. Theorem 4.5.1 then gives that for every representation of the form $(\mu, \{H_u\}, L)$, the algebra $\pi_L(C_c(G))$ generates an injective (=amenable) von Neumann algebra. In particular, $Ind\,\mu(C_c(G))$ generates an amenable von Neumann algebra A. But from the proof of Theorem 4.4.2, the map α_S implements a spatial isomorphism from $VN(S)$ onto A, and $VN(S)$ is amenable.   □

The author does not know if the amenability of $VN(S)$ implies that every maximal subgroup of S is amenable.

As an illustration of the above theorem, consider, for example, the case where S is the Cuntz semigroup S_n. For $e \in E(S)$, let $G(e)$ be the maximal subgroup S_e^e of S_n. So $G(e) = \{s \in S_n : ss^* = e = s^*s\}$. Clearly $G(e)$ is trivial if $e = 0$ or 1. Suppose then that $e = s_\alpha t_\alpha$. Let $s_\beta t_\gamma \in G(e)$. Then $\beta = \alpha = \gamma$ and again $G(e)$ is trivial. So every $G(e)$ is trivially amenable. Hence $VN(S_n)$ is amenable.

Of course in the particular case of S_n, we know much more than that. Since $C^*_{red}(S_n)$ is amenable, being (**2.1**) a product of the Cuntz-Toeplitz algebra with $\mathbf{C}$, any von Neumann algebra that it generates is an amenable von Neumann algebra. It would be interesting to know if the amenability of every maximal subgroup of S (S an arbitrary inverse semigroup) implies the amenability of $C^*_{red}(S)$ and conversely. A similar question can be raised for the amenability of $C^*(S)$. In the case where E is *finite*, then by a result of Duncan and Namioka (Theorem A.0.3), if the maximal subgroups of S are amenable, then $\ell^1(S)$ is an amenable Banach algebra, and hence its homomorphic image $C^*_{red}(S)$ is also amenable (as is also $C^*(S)$). So in that special case, the amenability of the maximal subgroups of S entails that $C^*_{red}(S)$ is amenable.

It seems plausible that there should be a direct inverse semigroup argument that establishes Theorem 4.5.2. This would probably involve an inverse semigroup version of Theorem 4.5.1.

The results of this section leave open important natural questions on the amenability of inverse semigroups and groupoids. Renault has introduced a concept of amenability for a locally compact groupoid ([230, p.92]). This can reformulated in terms akin to Definition 4.5.1, the weak* topology involved in that definition being replaced by the topology of uniform convergence on compact Hausdorff sets. Among other results, Renault shows that the Cuntz groupoid is amenable in this sense, and that amenability for a locally compact groupoid G entails that $C^*(G)$ is amenable. It seems likely that the amenability of the universal groupoid $G_{\mathbf{u}}$ for an inverse semigroup S corresponds to some Følner type condition ([196, Chapter 4]) on S which ensures the amenability of $C^*(S)$.

Appendix A

Amenability for Inverse Semigroups

The results of this Appendix discuss amenability for an inverse semigroup and its group algebra. They are included because of their relevance to **4.5**. In **4.5**, we discussed the problem of when the operator algebras associated with an inverse semigroup S are amenable. Two initially plausible suggestions for solving this problem will be looked at below. Unfortunately, the proposed solution of the first suggestion is too weak, while that of the second suggestion, is too strong. A partial solution to the problem was given in Theorem 4.5.2, which says that $VN(S)$ is amenable if all of the maximal subgroups of S are amenable.

The first suggestion is inspired by the well known result ([137, 42]), discussed in [196, (1.31),(2.35)], that a discrete group H is amenable if and only if the von Neumann algebra $VN(H)$ is amenable (if and only if $C^*(H)$ is amenable). Amenability for H can be defined in many ways, but the most popular and elegant is that there should be an invariant mean on H (**4.5**). Now (as discussed below) invariant means make sense for semigroups just as for groups (see, for example, [196, p.16]). So we might speculate that for an inverse semigroup S, the amenability of S could be equivalent to, or at least imply, the amenability of $VN(S)$. The following result of Duncan and Namioka shows that amenability of S is much too weak for that implication.

Turning to the details, every semigroup T has natural right and left actions on $\ell^\infty(T)$ where if $\phi \in \ell^\infty(T)$ and $t \in T$, the functions ϕt and $t\phi$ are given by:

$$\phi t(x) = \phi(tx) \qquad t\phi(x) = \phi(xt). \tag{A.1}$$

A mean m on T (i.e. a state on $\ell^\infty(T)$) is called *left invariant* if $m(\phi t) =$

$m(\phi)$ for all $t \in T$ and all $\phi \in \ell^\infty(T)$, or equivalently, if m is left invariant under the left action of T on $\ell^\infty(T)^*$ which is dual to the above right action. The semigroup T is *left amenable* if there exists a left invariant mean on T. The notions of *right amenability* and (two-sided) *amenability* for T are defined in the obvious ways. As for groups ([196, p.49]) (using the involution in place of inversion and the Arens product) the three kinds of amenability are all equivalent for an inverse semigroup S.

Amenability for an inverse semigroup S turns out to be disappointing since it depends only on the *maximal group homomorphic image* $G(S)$ of S (Proposition 2.1.2). The following result is due to Duncan and Namioka ([77]).

Proposition A.0.5 *The inverse semigroup S is amenable if and only if the group $G(S)$ is amenable.*

Proof. Let $G = G(S)$ and $\psi : S \rightarrow G$ be the canonical homomorphism of Proposition 2.1.2 from S onto G. If m is a left invariant mean on S, then the functional $\phi \rightarrow m(\phi \circ \psi)$ is a left invariant mean on G so that G is amenable. Conversely, suppose that G is amenable. Let $s_1, \ldots, s_n \in S$, $g_i = \psi(s_i) \in G$ and $C = \{g_i : 1 \le i \le n\}$. For $x \in G$ and $f \in \ell^1(G)$, the function $x * f \in \ell^\infty(G)$ is (cf. **4.5** for the groupoid version) given by: $x * f(t) = f(x^{-1}t)$. Let m be a left invariant mean on $\ell^\infty(G)$ and let $\{f_\delta\}$ be a net of finite means (i.e. each is a convex combination of point masses on G) in $\ell^1(G)$ with $f_\delta^\wedge \rightarrow m$ weak* in $\ell^\infty(G)^*$. Using Day's condition as in the proof of Proposition 4.5.3, we can suppose that $\|x * f_\delta - f_\delta\|_1 \rightarrow 0$ for all $x \in G$. Let $\epsilon > 0$. Then there exists a finite mean h on G such that $\|g_i * h - h\|_1 < \epsilon$ for $1 \le i \le n$. (Take h to be one of the f_δ's.) Let K be the (finite) support of h and $L \subset S$ be such that $\psi(L) = K$ and $\psi_{|L}$ is one-to-one. Let $K' = \{g_i k : 1 \le i \le n, k \in K\} \cup K$ and $L' = \{s_i l : 1 \le i \le n, l \in L\} \cup L$. By replacing L by Le for small enough $e \in E$, we can suppose that $\psi_{|L'}$ is a bijection onto K'. (Note that multiplying an element of S on the right by some $e \in E$ does not change the σ_S-equivalence class.) For each $k \in K$ let $l_k \in L$ be such that $\psi(l_k) = k$. Let $f = \sum_{k \in K} h(k) l_k$. Then f is a finite mean on S, and $\|s_i * f - f\|_1 = \|g_i * h - h\|_1 < \epsilon$ for $1 \le i \le n$. (The reason is that any simplification in the expansion of one of $s_i * f - f, g_i * h - h$ exactly matches a corresponding simplification in the other since $\psi_{|L'}$ is one-to-one.) Such finite means f form a net indexed by the natural ordering on the pairs $(\{s_i : 1 \le i \le n\}, \epsilon)$ (subsets increasing, ϵ decreasing), and any weak* cluster point of the net is a left invariant mean on S. So S is amenable. $\square$

Clearly, then, amenability for S is a very weak condition since it only reflects the group structure of S associated with "small" identity elements

$e \in E$. Very simple examples show that S can be amenable with $C^*(S)$ and $VN(S)$ not amenable. (Take, for example, the Clifford semigroup $S = H \cup \{z\}$, where z is the zero and H is a non-abelian free group.)

We turn then to the second suggestion alluded to earlier. This gives a stronger condition which will ensure that $C^*(S)$, and hence von Neumann algebras such as $VN(S)$ generated by representations of S (cf. [196, p.79])), are amenable. The condition is that of the amenability of the Banach algebra $\ell^1(S)$, a quotient of which is $C^*(S)$. The concept of an amenable Banach algebra was introduced by B. E. Johnson ([137]). Other accounts are given in [196, 204, 205]. We briefly summarize the relevant details.

Let A be a Banach algebra. Then the projective tensor product $A \hat{\otimes} A$ is a Banach A-module with actions determined by:

$$a(b \otimes c) = ab \otimes c \quad (b \otimes c)a = b \otimes ca. \tag{A.2}$$

We now discuss amenability for A. The definition given below is not the original one but Johnson ([138]) showed that it is equivalent to amenability. It is the Banach algebra analogue of the *Reiter net* in the group context (Appendix B), which is equivalent to amenability for a locally compact group.

A *bounded* net $\{m^\alpha\}$ in $A \hat{\otimes} A$ is called an *approximate diagonal* if for all $a \in A$, we have

$$\|am^\alpha - m^\alpha a\| \to 0, \quad \|\pi(m^\alpha)a - a\| \to 0. \tag{A.3}$$

Here π is the (bilinear) product map extended canonically to $A \hat{\otimes} A$, so $\pi(a \otimes b) = ab$. The algebra A is said to be *amenable* if it admits an approximate diagonal.

Johnson showed ([137]) that if G is a locally compact group, then $L^1(G)$ is amenable if and only if G is amenable. The next theorem characterizes those inverse semigroups S for which $\ell^1(S)$ is amenable. It follows, in particular, that the amenability of $\ell^1(S)$ has very strong consequences indeed. It implies that $E(S)$ *is finite*. The theorem is a consequence of a more general result due to Duncan and Paterson ([80]) after earlier work by Duncan and Namioka ([77]) and N. Groenbaek ([104]). Let S be an inverse semigroup and set $E = E(S)$.

Lemma A.0.1 *For each $e \in E$, let*

$$Z(e) = \{(s, t) \in S \times S : s \in eS, ste = e\}.$$

Then the sets $\{Z(e) : e \in E\}$ are pairwise disjoint.

Proof. Let $e, f \in E$ and suppose that $(s, t) \in Z(e) \cap Z(f)$. So $s \in eS \cap fS$, $ste = e$ and $stf = f$. Then $e(st)e = e, (st)e(st) = est = st$ and the same equations hold with f in place of e. Considering (2.1), we see that $(st)^* = e = f$. So $e = f$ and the lemma follows.   □

Proposition A.0.6   *If $\ell^1(S)$ is amenable, then E is finite.*

Proof.   Suppose that $\ell^1(S)$ is amenable. Let $\{m^\alpha\}$ be an approximate diagonal for $\ell^1(S)$. We can identify $\ell^1(S) \hat{\otimes} \ell^1(S)$ with $\ell^1(S \times S)$, where $s \otimes t$ is identified with the pair (s, t). Write $m^\alpha = \sum_{(s,t) \in S \times S} \beta^\alpha_{s,t}(s, t)$ where each $\beta^\alpha_{s,t} \in \mathbf{C}$. Let $v \in S$. Substituting for m^α in (A.3) with $a = v$ gives that

$$\lim_\alpha \| \sum_{(s,t) \in S \times S} \beta^\alpha_{s,t}[(s, tv) - (vs, t)] \|_1 = 0, \quad \lim_\alpha \sum_{(s,t) \in S \times S} \beta^\alpha_{s,t} stv = v.$$

$$(A.4)$$

Restricting the summation in the first equality of (A.4) gives

$$\lim_\alpha \sum_{s \in S \sim vS} \sum_{x \in S} | \sum_{tv=x} \beta^\alpha_{s,t} | \leq \lim_\alpha \|vm^\alpha - m^\alpha v\|_1 = 0 \qquad (A.5)$$

since, if $s \in S \sim vS$ and $x \in S$, then the coefficient of (s, x) in $vm^\alpha - m^\alpha v$ is $\sum_{tv=x} \beta^\alpha_{s,t}$. Restricting summation over x in (A.5) to those $x \in S$ for which $sx = v$, we obtain

$$\lim_\alpha \sum \{| \sum_{tv=x} \beta^\alpha_{s,t} | : s \in S \sim vS, sx = v\} = 0. \qquad (A.6)$$

Substituting $x = tv$ in (A.6) and removing absolute values gives

$$\lim_\alpha \sum_{s \in S \sim vS, stv=v} \beta^\alpha_{s,t} = 0. \qquad (A.7)$$

Now the second limit of (A.4) gives

$$\lim_\alpha \sum_{s \in S, stv=v} \beta^\alpha_{s,t} = 1. \qquad (A.8)$$

From (A.7) and (A.8) we obtain

$$\lim_\alpha \sum_{s \in vS, stv=v} \beta^\alpha_{s,t} = 1. \qquad (A.9)$$

Now let $v = e \in E$. Then (A.9) becomes

$$\lim_\alpha \sum_{(s,t) \in Z(e)} \beta^\alpha_{s,t} = 1. \qquad (A.10)$$

So using Lemma A.0.1,

$$\|m^\alpha\|_1 \geq \left\| \sum_{e\in E} \sum_{(s,t)\in Z(e)} \beta_{s,t}^\alpha(s,t) \right\|_1 \geq \sum_{e\in E} \left| \sum_{(s,t)\in Z_e} \beta_{s,t}^\alpha \right|.$$

It follows using (A.10) that if E is infinite then $\|m^\alpha\|_1 \to \infty$. Since the net $\{m^\alpha\}$ is norm bounded, the set E is finite. $\square$

Theorem A.0.3   *Let S be an inverse semigroup. Then $\ell^1(S)$ is amenable if and only if $E(S)$ is finite and every maximal subgroup of S is amenable.*

Proof.   The proof is due to Duncan and Namioka ([77]) and uses the theory of principal series and Brandt semigroups ([50]).

By Proposition A.0.6, we need only consider the case where $E(S)$ is finite. Let $E(S)$ have n elements $e_1, \ldots, e_n$. Let I be an ideal in S. Then I is an inverse subsemigroup of S since, if $a \in I$, then $a^* = a^*aa^* \in I$. Recall that S/I is the Rees factor semigroup (**2.1**). Two ideals I, J are different if and only if their idempotent sets $E(I), E(J)$ are different. (Indeed, if $s \in I \sim J$, then the idempotent s^*s does not belong to J, since otherwise, $s = s(s^*s) \in J$.) Since $E(S)$ is finite and the map $I \to E(I)$ is one-to-one on the ideals of S, simple algebra gives a *principal series* ([50, p.73]) for S, i.e. a chain $S = S_1 \supset S_2 \supset \cdots \supset S_m \supset S_{m+1} = \emptyset$ of distinct ideals S_i of S such that there is no ideal of S strictly between S_i and S_{i+1}. (For example, take S_m to be any minimal ideal I (i.e. an ideal I with $E(I)$ minimal), then take S_{m-1} to be the inverse image in S of a minimal ideal $J \neq \{0\}$ in the Rees factor semigroup S/S_m.) Each Rees factor semigroup S_i/S_{i+1} is *0-simple* ([50, p.67]) in the sense that it contains no ideal strictly between 0 and S_i/S_{i+1} and, since it contains a nonzero idempotent, it has nonzero square.

Now if I is an ideal in S, then canonically, $\ell^1(I)$ is an ideal in the convolution Banach algebra $\ell^1(S)$, and $\ell^1(S)/\ell^1(I)$ is, also canonically, isometrically isomorphic to the Banach algebra $\ell^1(S/I)/\mathbf{C}z$ where $z = I$ is the zero of S/I. (The point about quotienting out by the ideal $\mathbf{C}z$ is to make the zero z of S/I coincide with the vector space zero.) A result of B. E. Johnson states that a Banach algebra A is amenable if and only if a closed ideal L and the quotient A/L are amenable ([137, p.61]), and as $\mathbf{C}$ is trivially amenable, it follows that $\ell^1(S)$ is amenable if and only if both $\ell^1(I)$ and $\ell^1(S/I)$ are amenable. Applying this result recursively to the ideals S_{i+1} of the S_i, and noting that every non-zero maximal subgroup of S_i/S_{i+1} is a maximal subgroup of S, and that every maximal subgroup of S is in some S_i/S_{i+1} (the maximal subgroup is in an S_i once its identity belongs to $E(S_i)$) we can suppose that S is one of the S_i/S_{i+1}, i.e. is itself 0-simple. (We need also to consider the case of a simple inverse semigroup to cope with $S_m = S_m/\emptyset$ but this is an easier version of the 0-simple case.) Since there are only finitely many idempotents in S, there exists a nonzero $e \in E(S)$ which is *primitive* in the sense that if $f \in E(S)$ and $f \leq e$, then $f = 0$ or $f = e$. In semigroup terminology, S is thus a *Brandt semigroup*, a completely 0-simple inverse semigroup ([202, p.92]).

A well-known theorem (e.g. [202, p.93]) then expresses S as a special kind of "Rees matrix semigroup". Explicitly, let $H \neq \{0\}$ be a maximal subgroup of S. Then there is a positive integer k such that S is isomorphic to the following semigroup T associated with the pair (H, k). Then T consists of those $k \times k$ matrices each of which has exactly one nonzero entry from H, together with the zero $k \times k$ matrix. One just multiplies the matrices and takes adjoints in the usual way. Following a remark by the referee of [77], the semigroup algebra $\mathbf{C}(S)$ is thus isomorphic to $(\mathbf{C}(H) \otimes M_k(\mathbf{C})) \oplus \mathbf{C}$ so that its projective tensor product is $(\ell^1(H) \hat{\otimes} M_k(\mathbf{C})) \oplus \mathbf{C}$. Since $M_k(\mathbf{C})$ and $\mathbf{C}$ are amenable, a result of Johnson ([137, p.63]) gives that $\ell^1(S)$ is amenable if and only if $\ell^1(H)$ is amenable. But by another theorem of Johnson's alluded to earlier, $\ell^1(H)$ is amenable if and only if H is amenable. So $\ell^1(S)$ is amenable if and only if every maximal subgroup of S is amenable.   □

As we saw in Theorem 4.5.2, the amenability of the von Neumann algebra $VN(S)$ generated by the left regular representation of S is also related to the amenability of the maximal subgroups of S. In fact, in that theorem, we proved that the amenability of the maximal subgroups of S implies the amenability of $VN(S)$. The extremely strong finiteness condition on E, required for the amenability of $\ell^1(S)$ above, is thus not needed for the amenability of $VN(S)$. But under this finiteness condition, we know that $C^*(S)$ is amenable. The author does not know if $C^*(S)$ is amenable in the situation of Theorem 4.5.2.

Appendix B

Groupoid Amenability and Locally Compact Groups

Recall Renault's notion of an amenable measure for a locally compact groupoid G given in Definition 4.5.1. It is clearly desirable that in the case where H is a locally compact group, the amenability of the point mass at the identity e should be equivalent to the amenability of H. We will show this in Proposition B.0.8 below. For the discussion below, [196, Chapter 4] is particularly relevant to our present purposes.

Let H be a locally compact group with left Haar measure λ and

$$P_1(H) = \{f \in L^1(H) : f \geq 0, \int_H f \, d\lambda = 1\}. \tag{B.1}$$

The following condition of (B.2) is often called *Reiter's condition*. A proof is given in [196, p.127]. As in **4.5**, for $g \in L^1(H)$ and $x \in H$, $x * g \in L^1(H)$ is given by: $x * g(t) = g(x^{-1}t)$.

Proposition B.0.7   *The locally compact group H is amenable if and only if given $C \in C(H)$ and $\epsilon > 0$, there exists $g \in P_1(H)$ such that for all $x \in C$,*

$$\|x * g - g\|_1 < \epsilon. \tag{B.2}$$

For the locally compact group H we have, for every $g \in L^1(H)$, $F_g(x) = \|x * g - g\|_1$. Then F_g is a continuous function on H since the map $x \rightarrow x * g$ is norm continuous ([120, 19.27]).[1]

If H is σ-compact, then it is obvious from the proof below that the net $\{g_\delta\}$ can be taken to be a sequence. The conditions (i) and (ii) in the definition of *amenable measure* (**4.5**) follow immediately from this proof

---

[1]This is generalized to locally compact Hausdorff groupoids in Appendix C.

together with the fact that the g_δ's are in $P_1(H)$. So Proposition B.0.8 below gives that H is amenable if and only if the point mass δ_e is amenable.

Proposition B.0.8   *The locally compact group H is amenable if and only if there exists a net $\{g_\delta\}$ in $P_1(H)$ such that $F_{g_\delta} \to 0$ weak* in $L^\infty(H)$.*

Proof.   Suppose that H is amenable and let $\{g_\delta\}$ be a net in $P_1(H)$ such that $\|x * g_\delta - g_\delta\|_1 \to 0$ uniformly on compacta. (Such a net exists by Proposition B.0.7.) Let $F_\delta = F_{g_\delta}$. Let $f \in C_c(H)$ with support C and $\epsilon > 0$. Then

$$| \langle F_\delta, f \rangle | = | \int_C \|x * g_\delta - g_\delta\|_1 f(x) \, d\lambda(x) | < \epsilon \|f\|_1$$

eventually. So $\langle F_\delta, f \rangle \to 0$ for $f \in C_c(H)$. Since $C_c(H)$ is norm dense in $L^1(H)$ and the F_δ's are uniformly bounded, it follows that $F_\delta \to 0$ weak* in $L^\infty(H)$.

Conversely, suppose that there exists a net $\{g_\delta\}$ in $P_1(H)$ such that $F_\delta \to 0$ weak* in $L^\infty(H)$ where $F_\delta = F_{g_\delta}$. Let $C \in \mathcal{C}(H)$. By making C bigger if necessary, we can suppose that C is not null. Let $\epsilon > 0$. By hypothesis, $\int F_\delta(x) \chi_C(x) \, d\lambda(x) \to 0$. For each δ, let $U_\delta = \{x \in C : F_\delta(x) \geq \epsilon\}$. Then U_δ is compact since F_δ is continuous, and

$$\int F_\delta(x) \chi_C(x) \, d\lambda(x) \geq \epsilon \lambda(C \cap U_\delta) \geq 0$$

so that $\lambda(C \cap U_\delta) \to 0$. So given $\epsilon, \eta > 0$, there exists a compact set $N \subset C$ and $g \in P_1(H)$ such that $\lambda(N) < \eta$ and $\|x * g - g\|_1 < \epsilon$ for all $x \in C \sim N$. (We can take the g to be some g_δ and $N = C \cap U_\delta$ as above.)

We now make a simple adaptation of the proof of [196, Theorem 4.10]. Let C be as above and $D = C \cup C^2 \in \mathcal{C}(H)$. By the above applied to D with ϵ, η replaced by $\epsilon/2, \lambda(C)/2$, there exists $g \in P_1(H)$ and a compact subset N of D such that $\lambda(N) < \lambda(C)/2$ and

$$\|x * g - g\|_1 < \epsilon/2$$

for all $x \in D \sim N$. If $x, y \in D \sim N$, then

$$
\begin{aligned}
\|xy^{-1} * g - g\|_1 &= \|y^{-1} * g - x^{-1} * g\|_1 \\
&\leq \|y^{-1} * g - g\|_1 + \|x^{-1} * g - g\|_1 \\
&= \|y * g - g\|_1 + \|x * g - g\|_1 \\
&< \epsilon. \tag{B.3}
\end{aligned}
$$

One then shows ([196, p.132]) that $C \subset (D \sim N)(D \sim N)^{-1}$, and it follows from (B.3) that

$$\|x * g - g\|_1 < \epsilon$$

for all $x \in C$. So H is amenable. $\square$

Appendix C

The Measurability of F_g

In this appendix we will show that if G is a locally compact Hausdorff groupoid, then the function F_g ((4.42)) is continuous for all $g \in C_c(G)$. An example will then be given to show that this fails for a general locally compact groupoid. Finally we will show that if the groupoid is ample, then the function F_g is Borel measurable.

The following lemma will be used to show that F_g is continuous in the locally compact Hausdorff case. The proof is an adaptation of the approach used by Renault ([230, pp.48-49]) to prove that $C_c(G)$ is closed under convolution. (To obtain that result, which was referred to in the proof of Theorem 2.2.1, just take $f(x,y) = h(xy)g(y^{-1})$ for $h, g \in C_c(G)$.) Like Renault's proof, it uses a device of Connes in [53, 2.2].

Lemma C.0.2   *Let G be a locally compact Hausdorff groupoid. Let $f : G^2 \to \mathbf{C}$ be a continuous bounded function such that if $C \in \mathcal{C}(G)$ and $W_C = \{(x,y) \in G^2 : x \in C\}$, then $f_{|W_C} \in C_c(W_C)$. Then the map $F : G \to \mathbf{C}$, where*

$$F(x) = \int f(x,y) \, d\lambda^{d(x)}(y) \tag{C.1}$$

is continuous.

Proof.   Note that W_C in the above statement is a closed subset of the closed set $G^2 \subset G \times G$ and hence is locally compact in the relative topology. Note also that the function F in (C.1) is well defined since for given $x \in G$, taking C above to be the singleton $\{x\}$, we have $W_C = \{x\} \times G^{d(x)}$ and by assumption, the function $y \to f(x,y)$ on $G^{d(x)}$ is continuous with compact support.

Let $x_0 \in G$. We shall show that F is continuous at x_0. Let C be a compact neighborhood of x_0 in G and $K_0 = C \times d(C)$. Note that since d is an open map, the set K_0 is a compact neighborhood of $(x_0, d(x_0))$ in $G \times G^0$.

Let V_C be the (compact) support of the restriction of f to W_C. By Tietze's extension theorem, we can extend this restriction to a function $k \in C_c(G \times G)$. Let $h \in C_c(G)$ be positive and such that $h(y) = 1$ if $k(x, y) \neq 0$. So for all $(x, y) \in G \times G$, we have $k(x, y)h(y) = k(x, y)$. Define a complex-valued function H on K_0 by:

$$H(x, u) = \int k(x, y) \, d\lambda^u(y).$$

Then for $(x, u), (x_1, u_1) \in K_0$, we have

$$
\begin{aligned}
& | H(x, u) - H(x_1, u_1) | \\
= \ & | \int k(x, y) \, d\lambda^u(y) - \int k(x_1, y) \, d\lambda^{u_1}(y) | \\
\leq \ & \int | k(x, y) - k(x_1, y) | \, h(y) \, d\lambda^u(y) \\
& + | \int k(x_1, y) \, d\lambda^u(y) - \int k(x_1, y) \, d\lambda^{u_1}(y) | \\
\leq \ & \sup_y | k(x, y) - k(x_1, y) | \int h(y) \, d\lambda^u(y) \\
& + | \int k(x_1, y) \, d\lambda^u(y) - \int k(x_1, y) \, d\lambda^{u_1}(y) | .
\end{aligned}
$$

The uniform continuity of k and (ii) of Definition 2.2.2 then gives the continuity of H on K_0.

It follows that the map $x \rightarrow H(x, d(x))$ is continuous on C. Since

$$H(x, d(x)) = \int k(x, y) \, d\lambda^{d(x)}(y) = \int f(x, y) \, d\lambda^{d(x)}(y) = F(x),$$

we obtain that F is continuous at x_0. □

Proposition C.0.9 *Let G be a locally compact Hausdorff groupoid and $g \in C_c(G)$. Then $F_g \in C_c(G)$.*

Proof. We will apply Lemma C.0.2 to the function $f : G^2 \rightarrow \mathbf{C}$ given by

$$f(x, y) = | g(xy) - g(y) | .$$

Then f is a continuous bounded function on G^2. Let $C \in \mathcal{C}(G)$ and $W_C = \{(x, y) \in G^2 : x \in C\}$. Then f vanishes off the compact subset

$$\{(x, y) \in W_C : x \in C, y \in C^{-1}S \cup S\}$$

of W_C where S is the support of g. Hence the lemma applies and the function $H \in C_c(G)$ where

$$H(x) = \int_{G^{d(x)}} f(x,y) \, d\lambda^{d(x)}(y).$$

But then $F_g = H$.   □

We now give an example of a locally compact groupoid G and a $g \in C_c(G)$ such that F_g is not in $C_c(G)$. The groupoid is the universal groupoid of the inverse semigroup S of Example 2 of [197]. The space $\overline{E(S)}$ for this inverse semigroup also does not have the discrete topology. In [197], S was shown not to admit a faithful Hausdorff S-groupoid. (The author is grateful to J. Duncan for pointing out the usefulness of this inverse semigroup for counterexamples.)

In detail, S is the Clifford semigroup whose elements are the identity 1, the zero 0, idempotents e_n $(n \geq 1)$ and another element z. The product on S is abelian and is determined by the expressions:

$$ze_n = e_n, \quad e_n e_m = 0 \ \ (n \neq m), \quad z^2 = 1.$$

The semilattice $E = E(S)$ is thus $\{1\} \cup \{e_n : n \geq 1\} \cup \{0\}$. The ordering on E is given by: $0 \leq e_n \leq 1$ for all n. Note that if $n \neq m$, then we *neither* have $e_n \leq e_m$ *nor* $e_m \leq e_n$.

The space X of nonzero semicharacters on E is equal to $\overline{E}$ as a set. Indeed, if $x \in X$, then $x = \overline{0}$ if $x(0) = 1$. If $x(0) = 0$ and does not vanish on all of the e_n's, then there is a unique n such that $x(e_n) = 1$. (For if x was 1 on two different e_n's, then it would be 1 on their product, the zero of S.) Since $1 \geq e_n$ we obtain that $x(e) = 1$ for $e \in E$ if and only if $e = e_n$ or $e = 1$, and it follows that $x = \overline{e_n}$. Lastly if x vanishes on 0 and on all of the $e'_n s$ then obviously $x = \overline{1}$. So $X = \overline{E}$.

Let G be the universal groupoid for S and $\overline{z} = \overline{(z,1)} \in G$. As a groupoid, the universal groupoid G of S is just isomorphic to G_S, a bundle of groups (cf. Example 4 of **4.3**) for which G_u is $\{1, \overline{z}\}$ if $u = \overline{1}$ and is $\{\overline{e}\}$ otherwise. The canonical homomorphism $\psi = \psi_u : S \to G^a$ is given by: $\psi(1) = X, \psi(z) = \{\overline{z}\} \cup (X \sim \{\overline{1}\}), \psi(e_n) = \{\overline{e_n}, \overline{0}\}$, and $\psi(0) = \{\overline{0}\}$. By considering the sets $D(U, s)$ of Theorem 4.3.1, we obtain that a basis for the topology of G consists of the singletons $\{\overline{e}\}$ for $e \in E(S) \sim \{1\}$ together with sets of the form $\{\overline{1}\} \cup \{\overline{e_n} : n \geq N\}$ and $\{\overline{z}\} \cup \{\overline{e_n} : n \geq N\}$ for some N. In particular, the relative topology on $\overline{E}$ is not discrete since $\overline{e_n} \to \overline{1}$. Since $\overline{e_n} \to \overline{z}$ also, it follows that it is not possible to separate the points $\overline{z}$ and $\overline{1}$ in G.

Let $g = \chi_{\{\overline{z}\}} - \chi_{\{\overline{1}\}}$. Then $g = \chi_{\psi(z)} - \chi_{\psi(1)} \in C_c(G)$. Now calculating from (4.42), we obtain

$$F_g(\overline{z}) = 2 \mid g(\overline{1}) - g(\overline{z}) \mid = 4$$

while $F_g(\bar{e}) = 0$ for all $e \in E$. So $F_g = 4\chi_{\{\bar{z}\}}$. But $f = \chi_{\{\bar{z}\}}$ is not in $C_c(G)$. For suppose on the contrary that $f = \sum_{i=1}^{n} f_i$ where $f_i \in C_c(U_i)$ with U_i an open Hausdorff set in G. Let I, J and K be respectively the sets of indices i for which $\bar{z} \in U_i$, $\bar{1} \in U_i$ and neither $\bar{1}$ nor $\bar{z}$ is in U_i. Since $\bar{1}$ and $\bar{z}$ cannot be separated in G, each i is in exactly one of the sets I, J, K. If $i \in I \cup J$, then U_i contains a neighborhood of $\bar{z}$ (or $\bar{1}$) and so contains all but finitely many $\bar{e_i}$. Similarly if $i \in K$, then $U_i \subset \{\bar{e_1}, \bar{e_2}, \ldots, \bar{e_k}\} \cup \{\bar{0}\}$ for some k. Then $1 = f(\bar{z}) = \lim_{r \to \infty} \sum_{i \in I} f_i(\bar{e_r})$, $0 = f(\bar{1}) = \lim_{r \to \infty} \sum_{i \in J} f_i(\bar{e_r})$ and a contradiction follows since then $\lim_{r \to \infty} \sum_{i=1}^{n} f_i(\bar{e_r}) = \lim_{r \to \infty} \sum_{i \in I \cup J} f_i(\bar{e_r}) = 1$ yet $\sum_{i=1}^{n} f_i(\bar{e_r}) = f(\bar{e_r}) = 0$ for all r.

However for any ample groupoid (such as that above) F_g is always Borel measurable.

Proposition C.0.10   *Let G be an ample groupoid and $g \in C_c(G)$. Then F_g is Borel measurable on G.*

Proof.   Suppose firstly that g is a linear combination of characteristic functions of sets $C_k \in G^a$. By "disjointifying" the C_k, we can write

$$g = \sum_{i=1}^{n} \lambda_i \chi_{A_i} \tag{C.2}$$

where $\lambda_i \in \mathbf{C}$, the A_i are pairwise disjoint, and each A_i is a measurable G-set. Define sets A_{ij}, B_i and C_i as follows:

$$
\begin{aligned}
A_{ij} &= \{x \in G : \exists y \in G^{r(x)} \text{ such that } x^{-1}y \in A_i, y \in A_j\} \\
B_i &= \{x \in G : \exists y \in G^{r(x)} \text{ such that } x^{-1}y \in A_i, y \in (\cup_j A_j)^c\} \\
C_i &= \{x \in G : \exists y \in G^{r(x)} \text{ such that } x^{-1}y \in (\cup_j A_j)^c, y \in A_i\}.
\end{aligned}
$$

(Here for $A \subset G$, we take $A^c = G \sim A$.) It is easily checked that $A_{ij} = A_j A_i^{-1}$, $B_i = (\cup_j A_j)^c A_i^{-1}$ and $C_i = A_i((\cup_j A_j)^c)^{-1}$. By Proposition 2.2.8, the A_{ij}, B_i and C_i are measurable.

Let $\mathcal{A}$ be the set of A_{ij}'s, B_i's and C_i's and $\epsilon : \mathcal{A} \to \{1, c\}$. Let

$$A_\epsilon = \cap_{A \in \mathcal{A}} A^{\epsilon(A)}. \tag{C.3}$$

(Here, $A^1 = A$.) Then A_ϵ is measurable and F_g is of the form $\sum_\epsilon \alpha_\epsilon A_\epsilon$. Indeed, for $F_g(x)$ to be nonzero, we must have from (C.2) that at least one $x^{-1}y$ or y is in some A_j for some $y \in G^{r(x)}$. Hence x has to be in exactly one of the (disjoint) sets A_ϵ. To illustrate the α_ϵ in (C.3), suppose that $\epsilon(A) = 1$ for all $A \in \mathcal{A}$, and that $x \in A_\epsilon$. Then for each $A_{i,j}$ there is a $y \in G^{r(x)}$ such that $x^{-1}y \in A_i, y \in A_j$. Similar considerations apply to the B_i, C_i. For given $A \in \mathcal{A}$, the y is unique since the A_k's are G-sets. Further the y's change as we change the set $A \in \mathcal{A}$ since the A_k's are disjoint. Recalling that each λ^u is counting measure,

$$F_g(x) = \sum_{i,j} |\lambda_i - \lambda_j| + \sum_i |\lambda_i - 0| + \sum_i |0 - \lambda_i|$$

$$= \alpha_\epsilon.$$

From (C.3), the function F_g is measurable.

Now let $g \in C_c(G)$ be general. So $g = \sum_{i=1}^N g_i$ where for each i, $g_i \in C_c(U_i)$ and U_i is an open Hausdorff subset of G. Let $\eta > 0$. By the proof of Proposition 2.2.7, there exists, for each i, a function $h_i \in C_c(U_i)$ of the form $\sum_{j=1}^{M_i} \beta_j \chi_{C_j}$, where each $C_j \in G^a$ and $C_j \subset U_i$, and such that $\|g_i - h_i\|_I < \eta$. Note that $h = \sum_{i=1}^N h_i$ is a $C_c(G)$-function of the kind considered earlier in the proof.

Then for any $x \in G$, using (2.14),

$$| F_g(x) - F_h(x) |$$

$$\leq \int_{G^{r(x)}} | g(x^{-1}y) - h(x^{-1}y) - g(y) + h(y) | \, d\lambda^{r(x)}(y)$$

$$\leq \sum_{i=1}^N [\int | g_i(x^{-1}y) - h_i(x^{-1}y) | \, d\lambda^{r(x)}(y)$$

$$+ \int | g_i(y) - h_i(y) | \, d\lambda^{r(x)}(y)]$$

$$\leq \sum_{i=1}^N \|g_i - h_i\|_I \leq N\eta.$$

So F_g is the pointwise limit of a sequence of F_h's and so is measurable by the earlier part of the proof. $\square$

Appendix D

Ind μ As an Induced Representation

The groupoid representation $Ind\,\mu$, where μ is a probability measure on the unit space of a locally compact groupoid, was defined in (3.40). In this Appendix, we discuss briefly how $Ind\,\mu$ is actually an induced representation in the sense of Rieffel. We will base the account on [238, §4] and [230, pp.81-82].

A dense *-subalgebra of a C^*-algebra is called a *pre-C^*-algebra*. Let A and B be pre-C^*-algebras. We suppose that B acts as an algebra of right multipliers on A. The action of B on A will be denoted by : $(a, b) \rightarrow a.b$. The action is assumed to be continuous in the sense that the map $(a, b) \rightarrow a.b$ is norm jointly continuous.

A *generalized conditional expectation* from A to B is a linear, self-adjoint, positive map $P : A \rightarrow B$ such that :

(i) $P(a.b) = P(a)b$ for all $a \in A$ and $b \in B$;

(ii) for all $c \in A$, the linear map $a \rightarrow P(c^*ac)$ from A to B is bounded:

(iii) for every $a \in A$ and every $\epsilon > 0$, there exists c in the span A^2 of elements a_1a_2 $(a_i \in A)$ such that

$$\|P((a - c)^*(a - c))\| < \epsilon; \tag{D.1}$$

(iv) $P(A)$ generates a dense subalgebra of B.

In the above, *positive* for P means that $P(a^*a) \geq 0$ for all $a \in A$.

Given such a generalized conditional expectation P, Rieffel shows that we can "induce" a continuous *-representation π of B on a Hilbert (or even pre-Hilbert) space $\mathcal{H}$ up to a continuous *-representation $Ind\,\pi$ of A as follows. We can regard $\mathcal{H}$ as a left Hilbert B-module in the obvious way by defining $b\xi = \pi(b)\xi$ for $b \in B$ and $\xi \in \mathcal{H}$. Recall that A is a right B-module since B acts on it as an algebra of right multipliers. We form the algebraic tensor product $A \otimes_B \mathcal{H}$ (so that $a \otimes b\xi$ is identified with $a.b \otimes \xi$ for $a \in A, b \in B$ and $\xi \in \mathcal{H}$). One then shows that $A \otimes_B \mathcal{H}$ becomes a pre-Hilbert space, the (possibly degenerate) inner product $\langle, \rangle$ being defined by

$$\langle a \otimes \xi, a' \otimes \eta \rangle = \langle P((a')^*a)(\xi), \eta \rangle \qquad \text{(D.2)}$$

with $\langle, \rangle$ on the right-hand side being the inner product on $\mathcal{H}$. A Hilbert space is then obtained from $A \otimes_B \mathcal{H}$ by quotienting it out by the subspace of vectors of zero length and completing. We obtain a representation $Ind\,\pi$ of A on this Hilbert space by defining

$$Ind\,\pi(a)(a' \otimes \xi) = aa' \otimes \xi.$$

Then $Ind\,\pi$ is the *induced representation* of A associated with π.

Rieffel shows ([238, p.228]) that the process of inducing up a unitary representation of a closed subgroup K of a locally compact group H gives an example of the above construction of $Ind\,\pi$. The special case of this where $K = \{e\}$ is, as we shall see, particularly pertinent to the interpretation of $Ind\,\mu$ in the groupoid situation as an induced representation in the above sense of Rieffel, and will also serve as a useful illustration of the notion of a generalized conditional expectation.

In the case where $K = \{e\}$, we have $B = C_c(\{e\}) = \mathbf{C}$ and $A = C_c(H)$ regarded as a pre-C^*-algebra under the norm inherited from $C^*(H)$. Of course B acts as an algebra of right multipliers on A by scalar multiplication! Recall that the involution on $C_c(H)$ is ([73, p.282]) given by $f \rightarrow f^*$, where $f^*(x) = \overline{f(x^{-1})}\Delta(x)^{-1}$ with Δ the modular function on H. We take P to be just the "restriction" map $f \rightarrow f(e)$ where e is the identity of H. The map P is self-adjoint since if $f^* = f$ in $C_c(H)$, then $f(e)$ is real. The map P is also positive since for $f \in C_c(H)$, we have

$$P(f^* * f) = \int \overline{f(t^{-1})}f(t^{-1}e)\Delta(t)^{-1}\,d\lambda(t) = \int |\,f(t)\,|^2\,d\lambda(t) \geq 0.$$

Conditions (i) and (iv) for P above are trivially true. To check (ii), let $g \in C_c(H)$. Then for $f \in C_c(H)$, we have, using Fubini's theorem, that

$$\begin{aligned} P(g^* * f * g) &= g^* * f * g(e) \\ &= \int g^*(t)\,d\lambda(t) \int f(u)g(u^{-1}t^{-1})\,d\lambda(u) \\ &= \langle f, h \rangle, \end{aligned}$$

where $h \in C_c(H) \subset L^\infty(H)$ is given by:

$$h(u) = \int g^*(t)g(u^{-1}t^{-1})\,d\lambda(t) = \int \overline{g(t)}g(u^{-1}t)\,d\lambda(t).$$

Direct checking then shows that $h(u) = \langle \pi_2(u)g, g \rangle$ where π_2 is the left regular representation of H, so that

$$\langle f, h \rangle = \int f(u)\langle \pi_2(u)g, g \rangle \, d\lambda(u) = \langle \pi_2(f)g, g \rangle$$

and

$$| P(g^* * f * g) | = | \langle \pi_2(f)g, g \rangle | \le \|g\|_2^2 \|f\|$$

and (ii) follows. For (iii), let $\{e_\delta\}$ be a bounded approximate identity for $L^1(H)$ in $P_1(H) \cap C_c(H)$ ((B.1)) whose supports contract to $\{e\}$. Then from the preceding paragraph, we have

$$| P((f - e_\delta * f)^*(f - e_\delta * f)) | = \|f - e_\delta * f\|_2^2 \to 0$$

(the latter convergence to 0 following as in the proof for the L^1-case).

Now identify $C_c(H) \otimes_{\mathbf{C}} \mathbf{C}$ with $C_c(H)$. Then it is obvious that if π is the trivial representation of $\mathbf{C}$ on itself, then the induced representation $Ind\,\pi$ is determined by: $Ind\,\pi(f)(g) = f * g$, and the inner product on $C_c(H)$ is given by (D.2):

$$\langle f, g \rangle = P(g^* * f) = g^* * f(e) = \int g^*(t)f(t^{-1}e) \, d\lambda(t) = \langle f, g \rangle$$

where $\langle, \rangle$ is the restricted $L^2(H)$-inner product. It follows that $Ind\,\pi$ is just the left regular representation of G as it should be.

The above is a special case of the induced representations $Ind\,\mu$ for a locally compact Hausdorff groupoid G with unit space G^0. For the general case, the group identity e is replaced by the unit space G^0. Here, A is the convolution algebra $C_c(G)$. Then A is a pre-C^*-algebra under the $C^*(G)$-norm. Next, $B = C_c(G^0)$ which is a pre-C^*-algebra when regarded as a dense *-subalgebra of $C^*(G^0) = C_0(G^0)$. Then B acts as an algebra of right multipliers on A by defining $f.\phi(x) = f(x)\phi(d(x))$ for $x \in G$, $f \in A$ and $\phi \in B$. Indeed, if $f, g \in A$ and $\phi \in B$, then for $x \in G$,

$$\begin{aligned} f * (g.\phi)(x) &= \left(\int f(xy)g(y^{-1})\phi(d(y^{-1})) \, d\lambda^{d(x)}(y) \right) \\ &= ((f * g).\phi)(x) \end{aligned}$$

since if $r(y) = d(x)$, then $d(y^{-1}) = r(y) = d(x)$.

To check that this multiplier action is continuous, let π_L be a representation of $C_c(G)$ which gives the $C^*(G)$-norm on $C_c(G)$. Then from (3.17), we have

$$\begin{aligned} &\langle \pi_L(f.\phi)\xi, \eta \rangle \\ &= \int f(x)\phi(d(x))\langle L(x)(\xi(d(x))), \eta(r(x)) \rangle \, d\nu_0(x) \\ &= \langle \pi_L(f)(T_\phi \xi), \eta \rangle \end{aligned}$$

where, of course, T_ϕ is the diagonalizable operator associated with ϕ on the space $L^2(X, \mathcal{K}, \mu)$ of the representation. So

$$\langle \pi_L(f.\phi)\xi, \eta \rangle \le \|f\| \|T_\phi \xi\| \|\eta\| \le \|f\| \|\phi\|_\infty \|\xi\| \|\eta\|,$$

giving that $\|f.\phi\| \le \|f\| \|\phi\|_\infty$. So the multiplier action is continuous.

We take P to be the restriction map $f \to f_{|G^0}$ from $A = C_c(G)$ onto $B = C_c(G^0)$. The proof that the map P is a generalized conditional expectation is an adaptation of the corresponding proof above for the group case $G = H$. In particular, the self-adjointness and positivity of P follow that H-case, and (i) and (iv) are easy. For (ii), we replace the unit e of H by a unit $v \in G^0$ and show (using (2.21) and (3.42)) that $P(g^* * f * g)(v) = \langle Ind\, v(f)g, g \rangle$. (So the $Ind\, v$'s replace the π_2 in the group case.) Then take the sup over v to get, using (3.44) and (ii) of Definition 2.2.2, that $\|P(g^* * f * g)\|_\infty \le M\|f\|_{red} \le M\|f\|$ for some M. Condition (iii) for P to be a generalized conditional expectation seems to need more work. This condition can be proved using the result of Renault, Muhly and Williams ([230, p.56], [184]) that $C_c(G)$ does always have an approximate identity for the inductive limit topology. We conclude this Appendix by showing that $Ind\, \mu$ is an induced representation in the sense of Rieffel.

Suppose that $\mu \in M(G)$ is a probability measure on G^0 and let π be the multiplication representation of $B = C_c(G^0)$ on the pre-Hilbert space $\mathcal{H}$, the image of $C_c(G^0)$ in $L^2(\mu)$, i.e. for $h \in B$ and $\xi \in \mathcal{H}$, we define $\pi(h)\xi = h\xi$ (pointwise product). Recall that $A = C_c(G)$. We claim that $A \otimes_B \mathcal{H} = A$ as left A-modules. Indeed, the natural map $f \otimes \phi \to f.\phi$, where $f \in A, \phi \in B$, clearly takes $f.\phi_1 \otimes \phi$ and $f \otimes \phi_1\phi$ to the same element for any $\phi_1 \in B$. So by the universality of the tensor product, this natural map defines a linear left A-module mapping from $A \otimes_B \mathcal{H}$ onto A. The map is onto since for any $f \in A$, any $\phi \in B$ which is 1 on the set $\{d(x) : f(x) \ne 0\}$ satisfies $f.\phi = f$. The map is also one-to-one since if $\sum_{i=1}^n f_i.\phi_i = 0$ in A, then we can find a $\phi \in B$ which is 1 on the union of the supports of the ϕ_i's, and then in $A \otimes_B \mathcal{H}$, $\sum_{i=1}^n f_i \otimes \phi_i = \sum_{i=1}^n f_i.\phi_i \otimes \phi = 0$. We can thus identify $f \otimes \phi \in A \otimes_B \mathcal{H}$ with $f.\phi \in A$.

We now compute the inner product (defined in (D.2)) on $A \otimes_B \mathcal{H}$. Let $f, g \in A$, $\phi, \psi \in B$ and $F = f.\phi$, $F' = g.\psi$. Let ν, ν^{-1} be the canonical measures on G associated with μ **(3.1)**. Then making the substitution $y \to y^{-1}$, we have

$$
\begin{aligned}
\langle F, G \rangle &= \int P(g^* * f)(u)\phi(u)\overline{\psi(u)}\, d\mu(u) \\
&= \int (g^* * f)(u)\phi(u)\overline{\psi(u)}\, d\mu(u) \\
&= \int \phi(u)\overline{\psi(u)}\, d\mu(u) \int \overline{g(y^{-1})}f(y^{-1})\, d\lambda^u(y)
\end{aligned}
$$

$$= \int \phi(u)\overline{\psi(u)}\, d\mu(u) \int \overline{g(y)}f(y)\, d\lambda_u(y)$$

$$= \int d\mu(u) \int f(y)\phi(d(y))\overline{g(y)\psi(d(y))}\, d\lambda_u(y)$$

$$= \int_G F(y)\overline{F'(y)}\, d\nu^{-1}(y).$$

So the Hilbert space of the induced representation $Ind\,\mu$ is $L^2(\nu^{-1})$ and for $f, F \in A$,

$$Ind\,\mu(f)(F) = f * F \qquad\qquad (D.3)$$

as in (3.40).

Appendix E

Guichardet's Disintegration Theorem

The form of Guichardet's theorem given below follows that of Renault in [230, p.67] and is a straight-forward generalization of the original result in [106]. Guichardet's proof is presented below.

Let $(X, \mu), (X', \mu')$ be finite measure spaces and $\mathcal{M}, \mathcal{M}'$ be the σ-algebras of μ-measurable and μ'-measurable subsets of X, X'. Let ϕ be a bijection from X onto X' such that $\phi(\mathcal{M}) = \mathcal{M}'$. Then $\mu \circ \phi$ is a finite measure on $\mathcal{M}'$ where

$$\mu \circ \phi(E) = \mu(\phi^{-1}(E)) \quad (E \in \mathcal{M}').$$

We suppose also that $\mu \circ \phi$ and μ' are equivalent on $\mathcal{M}'$. Let

$$r = \frac{d(\mu \circ \phi)}{d\mu'}. \tag{E.1}$$

Then $\mu' \circ \phi^{-1} \sim \mu$ and

$$(r \circ \phi)^{-1} = \frac{d(\mu' \circ \phi^{-1})}{d\mu}. \tag{E.2}$$

So for appropriate measurable functions g on X, we have ((2.38)) that

$$\int g(\phi^{-1}(y)) \, d\mu'(y) \;=\; \int g(x) r^{-1}(\phi(x)) \, d\mu(x) \tag{E.3}$$

$$=\; \int g \, d(\mu' \circ \phi^{-1}). \tag{E.4}$$

Next suppose that $(X, \mathcal{K}, \mu)$ and $(X', \mathcal{K}', \mu')$ are Hilbert bundles. Let $\mathcal{K} = \{H_x\}$ and $\mathcal{K}' = \{H'_y\}$. Let

$$U : L^2(X, \mathcal{K}, \mu) \to L^2(X', \mathcal{K}', \mu')$$

be a unitary linear operator such that for every $F \in L^\infty(X, \mu)$, we have

$$UT_F U^{-1} = T_{F \circ \phi^{-1}} \tag{E.5}$$

in $B(L^2(X', \mathcal{K}', \mu'))$. (Here, $T_F, T_{F \circ \phi^{-1}}$ are the multiplication operators on $L^2(X, \mathcal{K}, \mu)$, $L^2(X', \mathcal{K}', \mu')$ associated with F, $F \circ \phi^{-1}$.) Guichardet's theorem is then the following.

Theorem E.0.4   *Under the above assumptions, there exist unitary linear operators* $u(y) : H_{\phi^{-1}(y)} \to H'_y$ *for* μ'-*a.e.* $y \in X'$ *such that for every* $f \in L^2(X, \mathcal{K}, \mu),$

$$Uf(y) = r^{1/2}(y)u(y)(f(\phi^{-1}(y)))   \mu' - a.e.. \tag{E.6}$$

The maps $u(y)$ *satisfying* (E.6) *for all* f *are unique* μ'-*a.e..*

Proof.   In the following proof, we shall have occasion to remove null sets from X or X'. To avoid obfuscating notation, we shall continue to write X or X' after such a removal. A brief discussion of Hilbert bundle theory is given in **3.1**.

Let $\{f_n\}$ be an orthonormal fundamental sequence for $L^2(X, \mathcal{K}, \mu)$. Let $g_n = Uf_n$ for each n. We claim first that, after removing null sets from X and X', $\{g_n\}$ is a fundamental sequence for $L^2(X', \mathcal{K}', \mu')$.

To this end, for each $y \in X'$, let H''_y be the closed subspace of H'_y spanned by $\{g_n(y)\}$. Then $\{g_n\}$ is a fundamental sequence for a Hilbert bundle $(X', \{H''_y\}, \mu')$. Since the span of the sections $T_F f_n$ is dense in $L^2(X, \mathcal{K}, \mu)$, so also is the span of the sections $U(T_F f_n)$ in $L^2(X', \mathcal{K}', \mu')$. But from (E.5), $U(T_F f_n) = T_{F \circ \phi^{-1}} g_n \in L^2(X', \{H''_y\}, \mu')$ so that

$$L^2(X', \{H''_y\}, \mu') = L^2(X', \mathcal{K}', \mu').$$

So if $\{k_n\}$ is a fundamental sequence for $(X', \{H_u\}', \mu')$, then for all n and for a.e. y, $k_n(y) \in H''_y$ for all n so that $H''_y = H'_y$ a.e. as asserted. So by removing null sets from X, X', we can take $\{g_n\}$ to be fundamental for $L^2(X', \mathcal{K}', \mu')$.

Define measurable sections h_n of $L^2(X', \mathcal{K}', \mu')$ by :

$$h_n(y) = r^{-1/2}(y)g_n(y). \tag{E.7}$$

Then for all $y \in X'$, we have that H''_y is the closed subspace of H'_y generated by the h_n's and $\{h_n\}$ is fundamental for $L^2(X', \mathcal{K}', \mu')$.

Let $m, n \geq 1$. Using (E.5) and (E.3), we have for $F \in L^\infty(X, \mu)$,

$$\int |F(x)|^2 \langle f_m(x), f_n(x) \rangle d\mu(x)$$
$$= \langle T_F f_m, T_F f_n \rangle$$

$$\begin{aligned}
&= \langle UT_F f_m, UT_F f_n \rangle \\
&= \langle T_{F\circ\phi^{-1}} U f_m, T_{F\circ\phi^{-1}} U f_n \rangle \\
&= \int \mid F(\phi^{-1}(y)) \mid^2 \langle U f_m(y), U f_n(y) \rangle \, d\mu'(y) \\
&= \int \mid F(x) \mid^2 \langle U f_m(\phi(x)), U f_n(\phi(x)) \rangle r^{-1}(\phi(x)) \, d\mu(x) \\
&= \int \mid F(x) \mid^2 \langle h_m(\phi(x)), h_n(\phi(x)) \rangle \, d\mu(x).
\end{aligned}$$

Since the span of such functions $\mid F \mid^2$ is the whole of $L^\infty(X, \mu)$ and since two L^1-functions that give the same linear functional on $L^\infty(X, \mu)$ are equal almost everywhere, it follows after removing null sets that we can suppose that

$$\langle f_m(x), f_n(x) \rangle = \langle (h_m \circ \phi)(x), (h_n \circ \phi)(x) \rangle \quad (x \in X). \tag{E.8}$$

Since $\{f_n\}$ is an orthonormal fundamental sequence for $L^2(X, \mathcal{K}, \mu)$, it follows from (E.8) that $\{h_n\}$ is an orthonormal fundamental sequence for $L^2(X', \mathcal{K}', \mu')$. Further, again from (E.8), for each $x \in X$, there is a unitary linear operator $v_x : H_x \to H'_{\phi(x)}$ defined by

$$v_x(f_n(x)) = h_n(\phi(x)). \tag{E.9}$$

Let $f \in L^2(X, \mathcal{K}, \mu)$ and define a section g of $L^2(X', \mathcal{K}', \mu')$ by :

$$g(y) = r^{1/2}(y) u(y) (f(\phi^{-1}(y))) \quad (y \in X') \tag{E.10}$$

where $u(y) = v_{\phi^{-1}(y)} : H_{\phi^{-1}(y)} \to H'_y$. For each n and each $y \in X'$, we have, by (E.9),

$$\begin{aligned}
\langle g(y), h_n(y) \rangle &= r^{1/2}(y) \langle u(y)(f(\phi^{-1}(y))), h_n(y) \rangle \\
&= r^{1/2}(y) \langle f(\phi^{-1}(y)), v_{\phi^{-1}(y)}^{-1}(h_n(y)) \rangle \\
&= r^{1/2}(y) \langle f(\phi^{-1}(y)), f_n(\phi^{-1}(y)) \rangle.
\end{aligned}$$

It follows that $y \to \langle g(y), h_n(y) \rangle$ is measurable, and since the h_n's are fundamental for $(X', \mathcal{K}', \mu')$, it follows by definition that g is a measurable section for X'. Further, as $u(y)$ is isometric, we have using (E.3),

$$\begin{aligned}
\int \|g(y)\|^2 \, d\mu'(y) &= \int \|f \circ \phi^{-1}(y)\|^2 r(y) \, d\mu'(y) \\
&= \int \|f \circ \phi^{-1}(y)\|^2 (r \circ \phi) \circ \phi^{-1}(y) \, d\mu'(y) \\
&= \int \|f(x)\|^2 r(\phi(x)) r^{-1}(\phi(x)) \, d\mu(x) \\
&= \int \|f(x)\|^2 \, d\mu(x).
\end{aligned}$$

Hence $g \in L^2(X', \mathcal{K}', \mu')$ and $\|g\|_2 = \|f\|_2$. Clearly there is a well-defined isometric linear map

$$U_1 : L^2(X, \mathcal{K}, \mu) \rightarrow L^2(X', \mathcal{K}', \mu')$$

where $U_1(f) = g$. Since, by (E.9), (E.10) and (E.7), we have each $g_n = U_1(f_n)$ in the range of U_1, it follows that U_1 is unitary from $L^2(X, \mathcal{K}, \mu)$ onto $L^2(X', \mathcal{K}', \mu')$. Comparing (E.6) with (E.10), we only have to show that $U = U_1$ and that the maps $u(y)$ are unique a.e..

For the first, we have $U(f_n) = g_n = U_1(f_n)$. Now let $F \in L^\infty(X, \mu)$. Then by (E.5) and (E.10), for $y \in X'$, we have

$$
\begin{aligned}
U(T_F f_n)(y) &= T_{F \circ \phi^{-1}}(U f_n(y)) \\
&= F \circ \phi^{-1}(y) g_n(y) \\
&= F \circ \phi^{-1}(y) U_1 f_n(y) \\
&= F \circ \phi^{-1}(y) r^{1/2}(y) u(y) (f_n(\phi^{-1}(y))) \\
&= r^{1/2}(y) u(y) ((T_F f_n)(\phi^{-1}(y))) \\
&= U_1(T_F f_n)(y).
\end{aligned}
$$

So U and U_1 coincide on the span of the sections $T_F f_n$, and since the latter span is dense in $L^2(X, \mathcal{K}, \mu)$, we have $U = U_1$.

Lastly, suppose that $u'(y)$ satisfy the same requirements (E.6) as the $u(y)$. Then for each $f \in L^2(X, \mathcal{K}, \mu)$, we have

$$u(y)(f(\phi^{-1}(y))) = u'(y)(f(\phi^{-1}(y)))$$

for μ'-a.e. $y \in X'$, and every function in $L^2(X', \mathcal{K}', \mu')$ is of the form $r^{1/2}(f \circ \phi^{-1})$ (using (E.3)). Using a fundamental sequence for $L^2(X', \mathcal{K}', \mu')$ we have $u(y) = u'(y)$ μ'-a.e.. This concludes the proof. $\square$

In the case where $L^2(X, \mathcal{K}, \mu) = L^2(X', \mathcal{K}', \mu')$ and ϕ is the identity map, then (E.5) says that U commutes with every T_F and Guichardet's result then reduces to the well-known theorem that a bounded linear operator is decomposable if and only if it commutes with all diagonalizable operators.

Appendix F

Some Differential Topology

Section **2.3** uses some concepts and results from differential topology. For the benefit of readers whose background (like that of the writer) is not in that area, sketches of some of these are included in this appendix.

Topics briefly discussed include manifolds, submanifolds, submersions, vector bundles, the tangent and cotangent bundles, the vector bundles $\Lambda^r E$, r-forms and densities. The books by Bott and Tu ([17]) and Lang ([154]) are useful sources for this material. The appendix concludes with a discussion of *foliations*.

We start by recalling the notion of a *manifold*. An *n-dimensional manifold* is a second countable, *Hausdorff* topological space M provided with an n-dimensional smooth structure. The latter is defined by an *atlas*, i.e. a family of pairs (U_α, ϕ_α) $(\alpha \in A)$, where $\{U_\alpha : \alpha \in A\}$ is an open cover of M, $\phi_\alpha : U_\alpha \rightarrow \mathbf{R}^n$ is a homeomorphism onto an open subset $\phi_\alpha(U_\alpha)$ of $\mathbf{R}^n$ and such that the *transition functions*

$$g_{\alpha\beta} = \phi_\alpha \phi_\beta^{-1} : \phi_\beta(U_\alpha \cap U_\beta) \rightarrow \phi_\alpha(U_\alpha \cap U_\beta) \tag{F.1}$$

are C^∞. The atlas can be taken to be maximal. The pairs (U, ϕ) belonging to the atlas, are called *charts*. (Sometimes the chart will be identified just with U.) The scalar-valued functions x_i, where $\phi = (x_1, \ldots, x_n)$ are called *coordinates* for U.

A continuous map from one manifold to the other is said to be *smooth* or C^∞ if it is C^∞ as a map between charts (regarded, using coordinates, as open subsets of Euclidean spaces). The notion of a diffeomorphism is defined using the Euclidean case in a similar way.

Let M be a manifold of dimension n. A subset Z of M is called a *submanifold* of dimension k if for each $z \in Z$, there exists a chart (U, ϕ)

with $z \in U$ such that $\phi(U) = V \times W$ where $0 \in V \subset \mathbf{R}^k$, $0 \in W \subset \mathbf{R}^{n-k}$, V, W are open balls in $\mathbf{R}^k, \mathbf{R}^{n-k}$ respectively, and $\phi(Z \cap U) = V \times \{0\}$ (e.g. [154, p.23], [44, p.16]). Let ϕ_1 be the restriction of ϕ to $Z \cap U$. Then it is readily checked that the collection of pairs $(Z \cap U, \phi_1)$ give an atlas for Z so that it becomes a k-dimensional manifold. Note that the topology of Z is the relative topology inherited from M.

Turning next to the notion of a submersion, let N be an m-dimensional manifold and $f : M \rightarrow N$ be a smooth map, i.e. C^∞ when expressed in terms of local coordinates. For each $p \in M$, the *derivative* of f at p is a linear map $df_p : T_p M \rightarrow T_{f(p)} N$ where $T_p M, T_{f(p)} N$ are the tangent spaces to M at p and N at $f(p)$. The map f is called a *submersion* if, for all $p \in M$, df_p is *surjective*, i.e. $df_p(T_p M) = T_{f(p)} N$. (In terms of local coordinates, $T_p M = \mathbf{R}^n$ and df_p is the matrix of partial derivatives $[\frac{\partial f_i}{\partial x_j}(p)]$.) The fundamental submersion is the projection map $f : \mathbf{R}^n \rightarrow \mathbf{R}^m$ for $n \geq m$, where

$$f(x,y) = x \quad (x \in \mathbf{R}^m, y \in \mathbf{R}^{n-m}) \tag{F.2}$$

in which case $df_{x,y} = \begin{bmatrix} I & 0 \end{bmatrix}$ is trivially surjective. It is a consequence of the inverse mapping theorem that every submersion is locally equivalent to such a projection map f ([154, p.25]). Since every such f is trivially an open map, it follows that every submersion is an open map between manifolds. Now with f as in (F.2) and S a submanifold of $\mathbf{R}^m$, $Z = f^{-1}(S) = S \times \mathbf{R}^{n-m}$, and we have trivially that Z is a submanifold of $\mathbf{R}^n$. Since every submersion is locally equivalent to such an f, we have that *if $g : M \rightarrow N$ is a submersion and S is a submanifold of N, then $g^{-1}(S)$ is a submanifold of M*. Further, as is obvious in the projection case, we have

$$n - \dim g^{-1}(S) = m - \dim S. \tag{F.3}$$

(This is a special case of a result in transversality theory ([2]).)

Next, we need to discuss briefly real vector bundles ([17, p.53f.], [154, Ch.3]). (Of course, similar considerations apply in the complex case.) Such a bundle is a manifold E together with a surjective, C^∞-map $\pi : E \rightarrow M$, called the *bundle map*, such that for each $p \in M$, the fiber $E_p = \pi^{-1}(\{p\})$ is a real m-dimensional vector space (m independent of x, called the *rank* of E) and there is an open cover $\{U_\alpha\}$ of M such that for each α, there is a fiber preserving diffeomorphism $\tau_\alpha : \pi^{-1}(U_\alpha) \cong U_\alpha \times \mathbf{R}^m$ which restricts to a linear isomorphism from $\pi^{-1}(\{p\})$ to $\{p\} \times \mathbf{R}^m$ ($x \in U_\alpha$). A *section* of E is a smooth map $s : M \rightarrow E$ such that $\pi \circ s$ is the identity on M. So $s(p) \in E_p$ for all $p \in M$. If there exist sections $\{\sigma_1, \ldots, \sigma_m\}$ of E such that for each $p \in M$, the vectors $\sigma_i(p)$ in E_p are linearly independent, then in the natural way, $E \cong M \times \mathbf{R}^m$. In this case, the bundle E is called *trivial* and the sections $\{\sigma_i\}$ give a *trivialization* for E. Every vector bundle E is locally trivial (in that there are sections σ_i^α trivializing $E_{|U_\alpha} = \pi^{-1}(U_\alpha)$).

With E, U_α, τ_α as in the preceding paragraph, we can define $\tau_{\alpha\beta}$: $U_\alpha \cap U_\beta \rightarrow GL(\mathbf{R}^m)$ by: $\tau_{\alpha\beta}(x) = (\tau_\alpha \tau_\beta^{-1})_{|\{x\} \times \mathbf{R}^m}$. Then the C^∞-maps $\tau_{\alpha\beta}$ satisfy the cocycle condition:

$$\tau_{\alpha\beta}\tau_{\beta\gamma} = \tau_{\alpha\gamma}. \tag{F.4}$$

The maps $\tau_{\alpha\beta}$ are called the *transition maps* (associated with the pairs (U_α, τ_α)).

Conversely, if we start off with an open cover $\{U_\alpha\}$ of M and associated maps $\tau_{\alpha\beta}$ as above satisfying the cocycle condition above, then these are transition maps for a vector bundle E of rank m. The bundle E is constructed as follows. Form the disjoint union topological space $\sqcup_\alpha(U_\alpha \times \mathbf{R}^m)$ and quotient out by the equivalence relation $\sim$ on this space by requiring that if $(x, v) \in U_\alpha \times \mathbf{R}^m$ and $(y, w) \in U_\beta \times \mathbf{R}^m$ then $(x, v) \sim (y, w)$ if and only if $x = y$ and $\tau_{\alpha\beta}(x)(w) = v$. What $\sim$ is doing is sticking together the two copies of $\{x\} \times \mathbf{R}^m$ to get the fiber E_x. The space E is a vector bundle over M in the natural way.

If E is a vector bundle over M and $f : N \rightarrow M$ is smooth, then $f^{-1}E = \{(n, e) \in N \times E : f(n) = \pi(e)\}$ is a vector bundle on N in the natural way. Note that $(f^{-1}E)_n = E_{f(n)}$. The vector bundle $f^{-1}E$ is called the *pullback bundle of E by f*. Vector bundles over M give other vector bundles over M through the processes of taking direct sums and tensor products.

We now discuss three kinds of vector bundles used in **2.3**. These are the tangent bundles, the bundles $\Lambda^r E$ and density bundles $\Omega^s E$.

We start first with the *tangent bundle* $TM = \cup_{p \in M} T_p M$. We will use the terminology of (F.1). If M is an open subset of $\mathbf{R}^n$, then $TM = M \times \mathbf{R}^n$ in the canonical way, and for each $p \in M$, the directional derivatives $(\partial/\partial x_i)_p$ are identified with the standard basis for $\{p\} \times \mathbf{R}^n$. For general M, let (U_α, ϕ_α), (U_β, ϕ_β) be charts in an atlas of M, and $g_{\alpha\beta}$ be as in (F.1). We then define $\tau_{\alpha\beta} : U_\alpha \cap U_\beta \rightarrow GL(\mathbf{R}^n)$ by:

$$\tau_{\alpha\beta}(x) = D(g_{\alpha\beta})(\phi_\beta(x)) \tag{F.5}$$

where D stands for the differential map. Clearly the maps $\tau_{\alpha\beta}$ satisfy the cocycle condition, and these are taken to be transition maps for the tangent bundle TM. (The "$D(g_{\alpha\beta})$" is natural when we think of the change of variables formula between partial derivatives in different coordinate systems.)

The cotangent bundle $T^*M = \cup_{p \in M} T_p^* M$ (where $T_p^* M$ is the dual space of $T_p M$) is also a vector bundle over M. For this bundle, the transition functions are given by:

$$x \rightarrow (\tau_{\alpha\beta}(x)')^{-1}. \tag{F.6}$$

(Here, $'$ denotes transpose and $\tau_{\alpha\beta}$ is as in (F.5).) Again it is easy to check that these functions satisfy the cocycle condition.

A local trivialization for TM is given by the basis of sections $\{(\partial/\partial x_j)_p\}$ for T_pM. We take $\{(dx_j)_p\}$ to be the basis of T_p^*M dual to this: so $(dx_j)_p(\partial/\partial x_k)_p = \delta_{jk}$.

We now discuss a way of getting another vector bundle from a given one E by means of the functor Λ^r. Let V be a finite-dimensional vector space over $\mathbf{R}$ with dual space V^*. Then the r-th alternating product $\Lambda^r V$ is the vector space of alternating, multilinear maps $w : V^* \times \cdots \times V^* \to \mathbf{R}$ (r copies of V^*). (For w to be alternating means that when a permutation is applied to an r-tuple in V^*, then the value of w changes by a factor equal to the sign of the permutation.) If $\{u_1, \ldots, u_n\}$ is a basis for V, then a basis for $\Lambda^r V$ is given by elements of the form $w = u_{p_1} \wedge \cdots \wedge u_{p_r}$ where $p_1 < p_2 < \cdots < p_r$ and for $g_i \in V^*$,

$$w(g_1, \ldots, g_r) = det([g_i(u_{p_j})]). \tag{F.7}$$

Of course, the above definition of w makes sense when the u_{p_j}'s are replaced by *any* r elements v_j in V and gives an element of $\Lambda^r V$. It is obvious from the definition of w that each $v_1 \wedge \cdots \wedge v_r$ is itself alternating when we permute the v_j's. When $r = \dim V$, the space $\Lambda^r V$ is one-dimensional (since then its basis above has only one element).

The functor Λ^r extends in the obvious way from vector spaces to vector bundles E over M. Indeed, with U_α, τ_α as earlier, there is a natural map $\Lambda^r \tau_\alpha$ which can be used to define the vector bundle structure of $\Lambda^r E$.

This applies, in particular, when $E = T^*M$. We write $\Lambda^r T^*M = D^r M$. An r-*form* on M is a section of $D^r M$. In terms of local coordinates x_i for M, an r-form is a linear combination of elements of the form $f \, dx_{i_1} \wedge \cdots \wedge dx_{i_r}$ where f is a C^∞-function.

Of particular interest is the one-dimensional vector bundle $D^n(M)$. In that case, the section $dx_1 \wedge \cdots \wedge dx_n$ provides a trivialization over U_α and we obtain that the transition functions for the line bundle $\Lambda^n T^*M$ are given by:

$$\tau_{\alpha\beta} = J(g_{\alpha\beta})^{-1}, \tag{F.8}$$

where J stands for the Jacobian.

Further, as discussed above, the product $dx_1 \wedge \cdots \wedge dx_n$ is alternating, and the $\pm$ that arises when permuting the dx_i's reflects how the orientation of the corresponding basis for T_p^*M changes, e.g. if $n = 2$, then $dx_1 \wedge dx_2 = -dx_2 \wedge dx_1$ and the bases $\{dx_1, dx_2\}$, $\{dx_2, dx_1\}$ at p have different orientations. In general, $D^n M$ is not trivial since different orientations of local n-forms can obstruct the construction of a global *non-vanishing* section from these local forms. The orientability of M is equivalent to the triviality of $D^n(M)$ (cf. [100, p.46]).

We now come to the notion of an *s-density*. For motivation, when $s = 1$, we obtain a 1-density from $dx_1 \wedge \cdots \wedge dx_n$ by sending it to $dx_1 \cdots dx_n$, giving us a measure to integrate against. Note that interchanging two dx_i's makes no difference by Fubini's theorem, so that moving from the form to the density is like putting an "absolute value" on the form, removing the $\pm$ arising from the orientation. (The precise formulation of this is that the 1-density bundle is the tensor product of $\Lambda^n T^* M$ with the orientation bundle of M. For details, see the book of Bott and Tu ([17, p.85]).) The "absolute value" idea above, when made precise, leads to the approach to s-densities described below. This account is based on the brief description of the concept by Connes ([56, p.119]), which we have supplemented using Lang's account for the special case of positive densities on TM ([154, p.304f.]).

An s-density on an n-dimensional vector space V over $\mathbf{R}$ is a map $\alpha : \Lambda^n V^* \sim \{0\} \to \mathbf{C}$ such that for all non-zero $c \in \mathbf{R}$ and non-zero $z \in \Lambda^n V^*$, we have

$$\alpha(cz) = |c|^s \alpha(z). \tag{F.9}$$

(Since s could be negative, we have to exclude c (and also z) from being zero.) Let $\Omega^s V$ (or $\Omega^s(V)$) be the set of s-densities on V. Note that under pointwise product,

$$\Omega^s(V) \Omega^t(V) \subset \Omega^{s+t}(V). \tag{F.10}$$

Clearly, $\Omega^s V$ is a vector space over $\mathbf{C}$ under pointwise operations. Further, if $\alpha, \beta \in \Omega^s V$, $z \in \Lambda^n V^*$ is non-zero and such that $\alpha(z) \neq 0$, and $k \in \mathbf{R}$ is such that $\beta(z) = k\alpha(z)$, then by (F.9) and the one-dimensionality of $\Lambda^n V^*$, we have $\beta = k\alpha$. So $\Omega^s V$ is one-dimensional, and using this, it follows that all of the elements of $\Omega^s V$ are of the following form α. Fix non-zero $\omega \in \Lambda^n V$ and $\lambda \in \mathbf{C}$. Let $\{e_i\}$ be a basis for V with dual basis $\{e_i^*\}$. Then for $c \in \mathbf{R} \sim \{0\}$, define α by:

$$\alpha(ce_1^* \wedge \cdots \wedge e_n^*) = \lambda \, | \, \omega(ce_1^*, \ldots, e_n^*) \, |^s \tag{F.11}$$

We write $\alpha = \lambda \, | \, \omega \, |^s$. An s-density α is called *positive* if $\alpha(w) \geq 0$ for all w. For α to be positive, we require $\lambda \geq 0$.

In a manner similar to the Λ^r-functor, we can form the s-density *complex* vector bundle $\Omega^s E$ (or $\Omega^s(E)$) over M for a real vector bundle E over M. We take $(\Omega^s E)_p$ to be the one-dimensional complex vector space $\Omega^s(E_p)$ for $p \in M$. The section $| \, \omega \, |^s$ provides a local trivialization over U_α, where ω trivializes $\Lambda^n E^*$. We obtain that if $\{\tau_{\alpha\beta}^L\}$ is the family of transition functions for $\Lambda^n E$, then the transition functions $\tau_{\alpha\beta}^s$ for $\Omega^s E$ are given by:

$$\tau_{\alpha\beta}^s = | \, \tau_{\alpha\beta}^L \, |^s . \tag{F.12}$$

A 1-density on E is sometimes just called a *density* on E.

The sections of $\Omega^s E$ are called *s-densities (on E)*. An *s*-density α on E is called *positive* if $\alpha(p)$ is a positive *s*-density on the vector space E_p for all $p \in M$. If, in addition, each $\alpha(p) \neq 0$, then α is called *strictly positive*. By considering $\mid \omega \mid^s$ as in the preceding paragraph, there always exist *local* strictly positive *s*-densities on E. Indeed we can use a standard "partition of unity" argument (cf. [17, p.29]) on such local strictly positive *s*-densities to build up a *global* strictly positive *s*-density on E. So in contrast to $D^n(M) = \Lambda^n T^* M$, the (one-dimensional) vector bundle $\Omega^s E$ is *always* trivial. Note that orientation is not a problem as it was for $D^n(M)$ since in the density case, $\mid \omega \mid^s$ is independent of the orientation of the *n*-form ω.

As for the functor Λ^n, of special interest is the case where $E = T^* M$. In that case, the sections are called the *s-densities on M*. Using (F.8) and (F.12), we see that the transition functions for $\Omega^s T^* M$ are given by

$$\tau_{\alpha\beta} = \mid J(g_{\alpha\beta}) \mid^{-s} . \tag{F.13}$$

In local coordinates, we write $dx^s = \mid dx_1 \wedge \cdots \wedge dx_n \mid^s$.

Specializing further to the case where $s = 1$, a 1-density is, in terms of local coordinates, of the form $f \mid dx_1 \wedge \cdots \wedge dx_n \mid = f(x)\,dx$ for some smooth complex-valued function f, and this defines a complex regular Borel measure μ on any compact subset of M. In particular, it defines an integral on $C_c(M)$, where $\int g\,d\mu = \int g(x) f(x)\,dx$ for such a local function g. The measure μ is *smooth* in the sense that in local coordinates, the Radon-Nikodym derivative $d\mu/dx = f$ is smooth. (The preceding construction is closely related to using a nowhere vanishing *n*-form on an (orientable) manifold to integrate over the manifold using a partition of unity ([17, p.29]).) When the 1-density on M is positive, we can take f to be ≥ 0, and μ is a positive regular Borel measure on M. When the density is strictly positive, then μ is locally equivalent to Lebesgue measure, and is a "strictly positive smooth measure" on M (in the sense of [279, p.108]).[1]

If U, V are open subsets of $\mathbf{R}^n$ and $T : U \to V$ is a diffeomorphism then the "change of coordinates" map $T^* : \Omega^1(V) \to \Omega^1(U)$ is given by (cf. [17, p.86]):

$$T^*(g\,dy) = (g \circ T) \mid J(T) \mid dx. \tag{F.14}$$

In the positive case where $g\,dy$ can be identified with a smooth measure μ on M, then $T^* \mu$ is the natural measure $\mu \circ T$ on U (cf. (2.38)). We now turn to the topic of *foliations*.

There is a close connection between foliations and Lie groupoids. Indeed, among the most important Lie groupoids that arise in noncommutative geometry are the *holonomy groupoids* of foliations. To facilitate our discussion of these groupoids in **2.3**, this appendix will sketch some of the

---

[1] Strictly positive densities on M are called *densities* in [100, p.139].

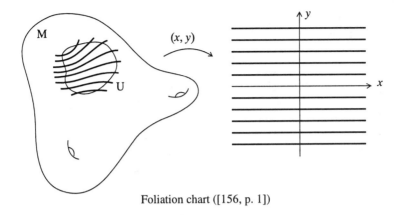

Foliation chart ([156, p. 1])

Figure F.1

rudiments of foliation theory. Accounts of foliation theory are given in the books by Camacho and Neto ([44]), Lawson ([156]), Molino ([177]), Moore and Schochet ([178]), Reinhart ([229]) and Tamura ([259]). The author found the account in [44] particularly helpful. Foliations are beautiful geometrical objects, and the reader is encouraged to consult the preceding texts for helpful illustrative geometrical diagrams to supplement the basic account given here.

Intuitively, a foliation of a (Hausdorff) manifold M of dimension n is a slicing up of M into connected manifolds of fixed dimension k which locally stack up as the sets $\mathbf{R}^k \times \{x\}$ ($x \in \mathbf{R}^{n-k}$) do in $\mathbf{R}^k \times \mathbf{R}^{n-k} = \mathbf{R}^n$. We now give a precise definition (illustrated by Fig. F.1.). Examples of foliations will be disussed later in the appendix.

Definition A6.1 A k-dimensional *foliation* of M is a pair $(M, \mathcal{F})$ where $\mathcal{F}$ is an atlas of charts of M maximal with respect to the following properties:

(i) if $(U, \phi) \in \mathcal{F}$, then $\phi(U) = U_1 \times U_2$ where U_1, U_2 are respectively open balls of $\mathbf{R}^k$, $\mathbf{R}^{n-k}$;

(ii) if $(U, \phi), (V, \psi) \in \mathcal{F}$ are such that $U \cap V \neq \emptyset$, then the diffeomorphism map $\psi \circ \phi^{-1} : \phi(U \cap V) \rightarrow \psi(U \cap V)$ is of the form

$$(x, y) \rightarrow (h_1(x, y), h_2(y)). \tag{F.15}$$

So in (ii) above, each part of the "slice" $U_1 \times \{y\}$ lying in $\phi(U \cap V)$ gets taken into a corresponding part of the slice of $V_1 \times \{h(y)\}$ in $\psi(U \cap V)$. The "horizontal" component gets deformed into the corresponding component in a way that depends on both x and y.

The elements $(U, \phi) \in \mathcal{F}$ are called *foliation charts*. Coordinates for U are given by $x \to (t, u) \in U_1 \times U_2$ where $\phi(x) = (t, u)$. We write $k = \dim \mathcal{F}$. The sets $\phi^{-1}(U_1 \times \{c\})$ $(c \in U_2)$ are called the *plaques* of U. The set U is the disjoint union of its plaques, and ϕ takes a plaque into a set of the form $U_1 \times \{c\}$. So via the plaques, U is made by stacking up copies of the k-dimensional ball U_1. The plaques of foliation charts connect up to give a partition of the manifold M. The associated equivalence relation on M is defined as follows ([44, p.23]). Points p, q of M are equivalent if there exist paths $\alpha_1, \ldots, \alpha_l$ in M with each α_r lying in some plaque of some U such that $p \in \alpha_1, q \in \alpha_l$ and for each $i < l$, $\alpha_i \cap \alpha_{i+1} \neq \emptyset$. (We identify here each α_i with its range in M.) Each equivalence class in M is called a *leaf* of M. Each leaf L is clearly a maximal connected union of plaques.

The leaf L is a k-dimensional manifold with coordinate charts determined by pairs (A, Φ) where $A = \phi^{-1}(U_1 \times \{c\})$ is a plaque of (U, ϕ) contained in L, and $\phi(a) = (\Phi(a), c)$. Further, M has a natural *leaf topology* $\mathcal{T}_{\mathcal{F}}$ ([229, 259]). A basis for $\mathcal{T}_{\mathcal{F}}$ is the family of sets of the form A above as c ranges over U_2 and (U, ϕ) over the foliation charts of M. It is easy to see that $(M, \mathcal{T}_{\mathcal{F}})$ satisfies all of the conditions of a manifold except that of second countability. The leaves are precisely the connected components of $(M, \mathcal{T}_{\mathcal{F}})$, and while $\mathcal{T}_{\mathcal{F}}$ induces on a leaf the structure of a k-dimensional manifold, this is not true for the original topology of M.

Also it is important to stress that a leaf L can intersect a given foliation chart U in *many* plaques which can "cluster up" inside the chart. In fact in many cases, a given leaf can be dense in the whole of M. The Kronecker foliation discussed below and in **2.3** gives a good illustration of this.

We will need the notion of a transverse section (Fig. F.2) in our discussion of holonomy in **2.3**. (The usual notion of *transverse section* ([44, p.48]) is more general than that presented here, but the latter is adequate for our purposes.) Let (U, ϕ) be a foliation chart for $\mathcal{F}$, $p \in U$ and $\phi(p) = (p_1, p_2)$ where each $p_i \in U_i$. The *transverse section for U passing through p* is the submanifold $A = \phi^{-1}(\{p_1\} \times U_2)$ which we can identify with U_2. Clearly, each plaque P is transverse to A in the sense that $T_x M = T_x P + T_x A$ when $x \in P \cap A$. Of course, the submanifold A is of dimension $(n - k)$, and each plaque in U meets A in exactly one point. Also, each leaf intersects a transverse section in countably many points.

In investigating the behavior of the leaves of $\mathcal{F}$ in M, it is reasonable to look at the quotient space M/R where R is the leaf equivalence relation on M: xRy in M if and only if x and y belong to the same leaf L. The space M/R is often called the *leaf space* since trivially every point of it corresponds to a leaf. There are the obvious quotient topological and measurable structures on the leaf space but these are often trivial and therefore

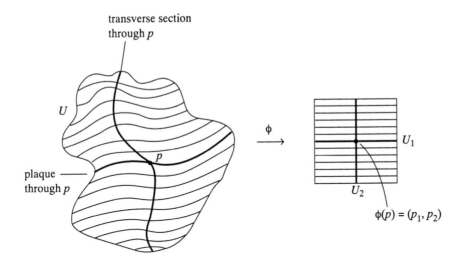

Figure F.2

useless. (An example of this is provided by the Kronecker foliation discussed below, see [56, p.50].) This is a prime situation for noncommutative geometry. Instead of looking at the leaf space M/R, one looks instead at the equivalence relation R itself. This is a groupoid, and if we can make it into a locally compact groupoid, then we can use its C^*-algebra A as a substitute for the space "$C_0(M/R)$" when the latter does not give useful information. (A similar situation holds for the noncommutative space of Penrose tilings (**2.2**).)

So the problem comes down to being able to obtain a locally compact groupoid topology on $R \subset M \times M$. The relative topology on R is usually bad (again as it was in the case of the equivalence relation giving the Penrose tilings) and instead we have to replace R by a related groupoid, the *holonomy groupoid* of $\mathcal{F}$. This is a Lie groupoid, the noncommutative version of a manifold, and is discussed in detail in **2.3**.

Indeed, foliations are, in a sense, intrinsic to Lie groupoids. To see this, we note first a simple source of foliations, that of *submersions*. Let M, N be manifolds, and $f : M \to N$ be a submersion. Then M is foliated by the connected components of the submanifolds $L_n = f^{-1}(\{n\})$ where $n \in f(M)$. Conditions (i) and (ii) of Definition A6.1 follow since a submersion is locally equivalent to a projection map. (See [44, pp.23-25] for the easy details.) By considering the canonical map from a foliation chart onto its transverse section, we see that locally, *all* foliations are given by submersions as above. Since $r : G \to G^0$ is a submersion in any Lie groupoid G, it follows (at least in the Hausdorff case) that G is foliated by the connected components of the G^u's. What we called in **2.3** an *r-fiberwise product* open set is a slightly more general version of a foliation chart.

There is a wide variety of examples of foliated manifolds. These include (e.g. [155, pp.6f.], [44, pp.23f.], [267]) fibrations, manifolds admitting a wide class of Lie group actions, bundles with discrete structure groups, Reeb foliations, Reinhart foliations, Anosov foliations and symplectic manifolds. We will be content with briefly discussing two one dimensional foliations, which are particularly instructive and easily envisaged, and which are used in **2.3**. These foliations are the *Kronecker foliation* and a *Reeb foliation*.

Let **T** be the torus $\mathbf{S}^1 \times \mathbf{S}^1$ regarded as the square $D = [0,1] \times [0,1]$ with opposite sides identified. Let $\theta \in \mathbf{R}$ be irrational. For each point $(x_0, y_0) \in D$ consider the "line" passing through (x_0, y_0) with slope θ. When the line hits a side at a point, we move over to the point on the opposite side with which the former point is identified and continue with the same slope. (See Fig. F.3 below where A', B' are the points on the opposite side corresponding to the points A, B.)

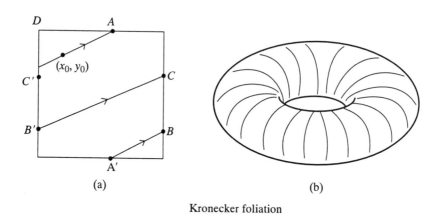

Kronecker foliation

Figure F.3

By the irrationality of θ, the "line" is dense in D. When we identify opposite sides of D to obtain **T**, the "line" goes over to a copy of **R**. These "lines" wind round the torus like strings round the inflated inner tube of a tire. Varying (x_0, y_0) over D gives a one-dimensional foliation of **T**, whose leaves (in the leaf topology) are all diffeomorphic to **R**. This foliation is known as the Kronecker foliation (of the torus). (See, for example, [56, p.50].) That we obtain a foliation is clear by taking foliation charts as pairs (U, ϕ) in D^o where U is the interior of a parallelogram with two opposite sides having slope θ and the other two horizontal. The plaques are the line segments in the parallelogram with slope θ.

As discussed above, one studies the equivalence relation R on **T** which is a Lie groupoid, and its C^*-algebra is an irrational rotation algebra ([56, p.123]).

The Kronecker foliation has trivial *holonomy* (since the leaves are simply connected and holonomy factors through homotopy) and we need a foliation to illustrate both the important holonomy concept and the fact that the holonomy groupoid need not be Hausdorff. For these purposes, we will use one of the Reeb foliations on **T**. (This is used, for example, in [54, §6].)

Two (one-dimensional) Reeb foliations are represented in Fig. F.4 below.

In (a) of Fig. F.4, the torus **T** is, of course, obtained by identifying opposite sides of the rectangle D'. The foliation has a band of circular leaves lying between the extremes of AA and BB. Every other leaf is a copy of **R** bent into the shape of a "round hat" which runs away in both directions, winding round the torus to flatten out against the extreme circular leaves AA and BB. When a non-compact leaf hits the top horizontal edge of D' at a point C, then it continues, of course, at the corresponding point C' on the bottom horizontal edge. As in the Kronecker case, it is easy to check that this indeed does give a foliation of the torus.

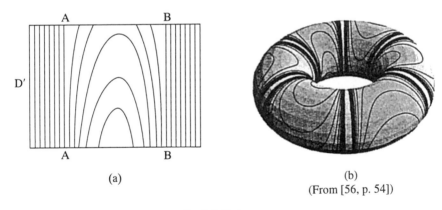

(a)

(b)
(From [56, p. 54])

Reeb foliations

Figure F.4

The case where we have six bands of circular leaves (rather than just the one) is easier to represent and appears in (b) of Fig. F.4.

Bibliography

[1] B. Abadie and R. Exel, *Deformation Quantization via Fell Bundles*, University of Sao Paulo preprint, 1997 (21pp.).

[2] R. Abraham and J. Robbin, *Transversal mappings and flows*, W. A. Benjamin, New York, 1967.

[3] C. Albert and P. Dazord, *Groupoïdes de Lie et groupoïdes symplectiques*, pp. 1-11 of *P. Dazord and A. Weinstein, infra*.

[4] C. Anantharaman-Delaroche, *Purely infinite C^*-algebras arising from dynamical systems*, Bull. Soc. Math. France 125(1997), 199-225.

[5] C. Anantharaman-Delaroche and J. Renault, *Amenable Groupoids*, with an appendix by E. Germain, Laboratoire de Mathématiques, Applications, et Physique Mathématique d'Orléans, 1998, 118pp..

[6] M. A. Aof, *Topological aspects of holonomy groupoids*, With comments by J. Pradines, U.N.C.W. Pure Maths Preprint, 147 pp., 1987.

[7] M. A. Aof and R. Brown, *The holonomy groupoid of a locally topological groupoid*, Topology Appl. 47(1992), 97-113.

[8] V. Arzumanian and J. Renault, *Examples of pseudogroups and their C^*-algebras*, preprint, 1997.

[9] N. W. Ashcroft and N. D. Mermin, *Solid State Physics*, Holt, Rinehart and Winston, New York, 1976.

[10] M. F. Atiyah and I. Singer, *The index of elliptic operators, I*, Ann. of Math. 87(1968), 484-530.

[11] B. A. Barnes, *Representations of the ℓ_1-algebra of an inverse semigroup*, Trans. Amer. Math. Soc. 218(1976), 361-396.

[12] P. Baum, W. Fulton and R. Macpherson, *Riemann-Roch for singular varieties*, Inst. Hautes Études Sci. Publ. Math. 45(1975), 101-145.

[13] F. Bayen, M. Flato, C. Fronsdal, A. Lichnerowicz and D. Stern-heimer, *Quantum mechanics as deformation of classical mechanics.* Lett. Math. Phys. 1(1975/77), 521-530.

[14] J. Belissard, *Schrödinger's operators with an almost periodic potential*, in: R. Schrader and R. Seiler, editors, *Mathematical Problems in Theoretical Physics*, Lecture Notes in Physics, 153(356-359), Springer-Verlag, New York, 1982.

[15] B. Blackadar, *K-Theory for Operator Algebras*, MSRI Publications 5, Springer-Verlag, New York, 1986.

[16] R. Bott and A. Haefliger, *On characteristic classes of Γ-foliations*, Bull. Amer. Math. Soc. 78(1972), 1039-1044.

[17] R. Bott and L. W. Tu, *Differential forms in algebraic topology*, Graduate Texts in Mathematics, Vol. 82, Springer-Verlag, New York 1982.

[18] N. Bourbaki, Lie groups and Lie algebras. Part 1:Chapters 1-3, Hermann, Paris, 1975.

[19] N. Bourbaki, *Intégration.*, Hermann, Paris, 1963, 1965.

[20] W. Brandt, *Über eine Verallgemeinerung des Gruppenbegriffes*, Math. Ann. 96(1927), 360-366.

[21] R. Brown, *Groupoids and van Kampen's theorem*, Proc. London Math. Soc. 17(1967), 385-401.

[22] R. Brown, *Fibrations of groupoids*, J. Alg. 15(1970), 103-132.

[23] R. Brown, *From groups to groupoids: a brief survey*, Bull. London Math. Soc. 19(1987), 113-134.

[24] R. Brown, *Topology: A Geometric Account of General Topology, Homotopy Types and the Fundamental Groupoid*, Ellis Howard, Chichester, England, 1988.

[25] R. Brown, *Symmetry, groupoids and higher-dimensional analogues. Symmetry 2: unifying human understanding Part 1*, Comput. Math. Appl. 17(1989), 49-57.

[26] R. Brown, *Homotopy theory and change of base for groupoids and multiple groupoids*, The European Colloquium of Category Theory (Tours, 1994), Appl. Categ. Structures 4(1996), 175-193.

[27] R. Brown and G. Danesh-Naruie, *The fundamental groupoid as a topological groupoid*, Proc. Edinburgh Math. Soc. 19(1975), 237-244.

[28] R. Brown and J. P. L. Hardy, *Topological groupoid. I. Universal constructions.*, Math. Nachr. 71(1976), 273-286.

[29] R. Brown, G. Danesh-Naruie and J. P. L. Hardy, *Topological groupoid. II. Covering morphisms and G-spaces*, Math. Nachr. 74(1976), 143-156.

[30] R. Brown, P. R. Heath and H. K. Kamps, *Groupoids and Mayer-Vietoris sequence*, J. Pure Appl. Alg. 30(1983), 109-129.

[31] R. Brown, P. R. Heath and H. K. Kamps, *Coverings of groupoids and Mayer-Vietoris type sequences*, Categorical topology: Proc. Conf. Toledo, Ohio, 1983, H. L. Bentley *et al.* (eds), Heldermann, Berlin, (1984), 147-162.

[32] R. Brown and P. J. Higgins, *Tensor products and homotopies for ω-groupoids and crossed complexes*, J. Pure Appl. Algebra 47(1987), 1-33.

[33] R. Brown and K. C. H. Mackenzie, *Determinations of a double Lie groupoid by its core diagram*, J. Pure Appl. Algebra 80(1992), 237-272.

[34] R. Brown and O. Mucuk, *The monodromy groupoid of a Lie groupoid*, Cahiers Topologie Géom. Différentielle Catég. 36(1995), 345-369.

[35] R. Brown and O. Mucuk, *Foliations, locally Lie groupoids and holonomy*, Cahiers Topologie Géom. Différentielle Catég. 37(1996), 61-71.

[36] R. Brown and C. B. Spencer, *Double groupoid and crossed modules*, Cahiers Topologie Géom. Différentielle 17(1976), 343-362.

[37] R. Brown and C. B. Spencer, Erratum to the article: *Double groupoid and crossed modules*, Cahiers Topologie Géom. Différentielle 18(1977), 430.

[38] R. L. Bryant, *An introduction to Lie groups and Symplectic Geometry*, pp. 1-179 of *Geometry and Quantum Field Theory*, ed. D. S. Freed and K. K. Uhlenbeck, IAS/Park City Mathematics Series, Vol. 1, American Mathematical Society, Providence, 1995.

[39] J-L. Brylinski, *Loop Spaces, Characteristic Classes and Geometric Quantization*, Progress in Mathematics, Vol. 107, Birkhäuser, Boston, 1993.

[40] J-L. Brylinski and V. Nistor, *Cyclic cohomology of étale groupoids*, K-theory 8(1994), 341-365.

[41] M. J. Buerger, *Introduction to crystal geometry*, McGraw-Hill, New York, 1971.

[42] J. W. Bunce, *Finite operators and amenable C^*-algebras*, Proc. Amer. Math. Soc. 56(1976), 145-151.

[43] J. W. Bunce and J. A. Deddens, *C^*-algebras generated by weighted shifts*, Indiana Univ. Math. J. 23(1973), 257-271.

[44] C. Camacho and A. L. Neto, *Geometric Theory of Foliations*, Birkhäuser, Boston, 1985.

[45] A. Cannas da Silva and A. Weinstein, *Lectures on Geometric Models for Noncommutative Algebras*, in preparation.

[46] A. H. Chamseddine, *The spectral action principle in noncommutative geometry and superstrings*, Physics Letters B, 400(1997), 87-96.

[47] A. H. Chamseddine and A. Connes, *Universal formula for noncommutative geometry actions: unification of gravity and the standard model*, Phys. Rev. Lett. 77(1996), 4868-4871.

[48] A. H. Chamseddine and A. Connes, *The spectral action principle*, Commun. Math. Phys. 186(1997), 731-750.

[49] A. H. Chamseddine, A. H. Felder and G. Fröhlich, *Gravity in noncommutative geometry*, Commun. Math. Phys. 155(1993), 205-217.

[50] A. H. Clifford and G. B. Preston, *The algebraic theory of semigroups*, Vol. 1, Amer. Math. Soc., Providence, R. I., 1961.

[51] A. H. Clifford and G. B. Preston, *The algebraic theory of semigroups*, Vol. 2, Amer. Math. Soc., Providence, R. I., 1967.

[52] A. Connes, *Classification of injective factors*, Ann. of Math. 104(1976), 73-115.

[53] A. Connes, *Sur la théorie non commutative de l'intégration*, Lecture Notes in Mathematics, 725(1979), 19-143.

[54] A. Connes, *A survey of foliations and operator algebras*, Operator algebras and applications, Part 1, pp. 521-628, Proc. Sympos. Pure Math. 38(1982), American Mathematical Society, Providence, R. I..

[55] A. Connes, *Essay on physics and noncommutative geometry*, The interface of mathematics and particle physics (Oxford, 1988), Inst. Math. Appl. Conf. Ser. New Ser. 24, Oxford University Press, New York, 1990.

[56] A. Connes, *Noncommutative Geometry*, Academic Press, Inc., New York, 1994.

[57] A. Connes, *Non-commutative geometry and physics*, Gravitation et quantifications (Les Houches, 1992), 805-950, North-Holland, Amsterdam, 1995.

[58] A. Connes, *Gravity Coupled with Matter and the Foundation of Noncommuative Geometry*, Commun. Math. Phys. 182(1996), 155-176.

[59] A. Connes, *Noncommutative differential geometry and the structure of space time*, in *Deformation theory and symplectic geometry (Ascona, 1996)*, 1-33, Math. Phys. Stud. 20, Kluwer Acad. Publ., Dordrecht, 1997.

[60] A. Connes and N. Higson, *Déformations, morphismes asymptotiques et K-théorie bivariante*, C. R. Acad. Sci. Paris Sér. I Math. 311(1990), 101-106.

[61] A. Connes and G. Skandalis, *The Longitudinal Index Theorem for Foliations*, Publ. RIMS, Kyoto University 20(1984), 1139-1183.

[62] J. B. Conway, J. Duncan and A. L. T. Paterson, *Monogenic inverse semigroups and their C*-algebras*, Proc. Roy. Soc. Edinburgh 98A(1984), 13-24.

[63] A. Coste, P. Dazord and A. Weinstein, *Groupoïdes symplectiques*, Publications du Département de Mathématiques, Université Claude Bernard-Lyon 12A(1987), 1-62.

[64] D. Crocker, A. Kumjian, I. Raeburn and D. Williams, *An equivariant Brauer Group and Actions of Groups on C*-algebras*, preprint, 1998 (28pp.).

[65] J. Cuntz, *Simple C*-algebras generated by isometries*, Comm. Math. Phys. 57(1977), 173-185.

[66] J. Cuntz and W. Krieger, *A class of C*-algebras and topological Markov chains*, Invent. Math. 56(1980), 251-268.

[67] J. Cuntz, *A class of C*-algebras and topological Markov chains II: reducible chains and the Ext-functor for C*-algebras*, Invent. Math. 63(1981), 25-40.

[68] M. K. Dakin and A. K. Seda, *G-spaces and topological groupoids*, Glasnik Mat. Ser. III 12(32)(1977), 191-198.

[69] K. R. Davidson, *C*-algebras by Example*, Fields Institute Monographs, American Mathematical Society, Providence, R.I., 1996.

[70] P. Dazord, *Groupoïdes symplectiques et troisième théorème de Lie "non linéaire"*, Lecture Notes in Mathematics, Vol. 1416, 39-74, Springer-Verlag, New York, 1990.

[71] P. Dazard and A. Weinstein (editors), *Symplectic Geometry, Groupoids and Integrable Systems*, Séminaire Sud. Rhodanien de Géometrie à Berkeley, MSRI Publications, Vol. 20, Springer-Verlag, New York, 1989.

[72] J. Dixmier, *Von Neumann Algebras*, North-Holland Publishing Company, Amsterdam, 1981.

[73] J. Dixmier, *C*-algebras*, North-Holland Publishing Company, Amsterdam, 1977.

[74] S. Doplicher and J. E. Roberts, *Duals of compact Lie groups realised in the Cuntz algebras and their actions on C*-algebras*, J. Functional Analysis, 74(1987), 96-120.

[75] S. Doplicher and J. E. Roberts, *Endomorphisms of C*-algebras, cross products and duality for compact groups*, Ann. of Math. 130(1989), 75-119.

[76] V. G. Drinfeld, *Quantum groups*. Proc. International Congress of Math., Berkeley, 1986, 798-820, Amer. Math. Soc., Providence, R. I., 1987.

[77] J. Duncan and I. Namioka, *Amenability of inverse semigroups and their semigroup algebras*, Proc. Roy. Soc. Edinburgh Sect. A 80(1978), 309-321.

[78] J. Duncan and A. L. T. Paterson, *C*-algebras of inverse semigroups*, Proc. Edinburgh Math. Soc. 28(1985), 41-58.

[79] J. Duncan and A. L. T. Paterson, *C*-algebras of Clifford semigroups*, Proc. Edinburgh Math. Soc. A111 (1989), 129-145.

[80] J. Duncan and A. L. T. Paterson, *Amenability for discrete convolution semigroup algebras*, Math. Scand. 66(1990), 141-146.

[81] J. Duncan and A. L. T. Paterson, *C*-algebras of Clifford semigroups*, Proc. Roy. Soc. Edinburgh, 111(A)(1989), 129-145.

[82] E. Effros, *Global structure in von Neumann algebras*, Trans. Amer. Math. Soc. 121(1966), 434-454.

[83] E. Effros and F. Hahn, *Locally compact transformation groups and C*-algebras*, Mem. Amer. Math. Soc. 75(1967).

[84] C. Ehresmann, *Oevres complètes et commentées*, Imprimerie, Evrard, Amiens, 1982.

[85] C. Ehresmann, *Catégories topologiques et catégories différentiables*, Colloq. Géom. Diff. Globale (Bruxelles, 1959), 137-150, Centre Belge. Rech. Math. Louvain, 1959.

[86] C. Ehresmann, *Sur les catégories différentiables*, Atti. Conv. Int. Geom. Diff. (Bologna, 1967), 31-40, Zanichelli, Bologna, 1970.

[87] C. Ehresmann, *Structures feuilletées*, Proc. 5th Canadian Math. Congress (1961), University of Toronto Press, 1963, 109-172.

[88] C. Ehresmann and W. B. Shih, *Sur les espaces feuilletées: théorème de stabilité*, C. R. Acad. Sci. Paris 243(1956), 344-346.

[89] D. B. A. Epstein, K. C. Millet and D. Tischler, *Leaves without holonomy*, J. London Math. Soc. 16(1977), 548-552.

[90] R. Exel, *Approximately Finite C*-algebras and Partial Automorphisms*, preprint, (10pp.).

[91] R. Exel, *The Bunce-Deddens Algebras as Crossed Products by Partial Automorphisms*, preprint, (6pp.).

[92] R. Exel, *Circle actions on C*-algebras, Partial Automorphisms and a Generalized Pimsner-Voiculescu Exact Sequence*, J. Functional Analysis, 122(1994), 361-401.

[93] R. Exel, *Twisted Partial Actions, A Classification of Stable C*-Algebraic Bundles (Preliminary Version)*, preprint, (28pp.).

[94] R. Exel, *Partial actions of groups and actions of inverse semigroups*, preprint, 1995.

[95] R. Exel, M. Laca and J. Quigg, *Partial dynamical systems and C*-algebras generated by partial isometries*, preprint, 1998.

[96] T. Fack and G. Skandalis, *Sur les représentations et idéaux de la C*-algebra d'un feuilletage*, J. Operator Theory 8(1982), 95-129.

[97] L. D. Fadeev, N. Y. Reshetikhin and L. A. Takhtajan, *Quantization of Lie groups and Lie algebras.*, Preprint, LOMI, 1987.

[98] J. Feldman and C. C. Moore, *Ergodic equivalence relations, cohomology and von Neumann algebras, I and II,* Trans. Amer. Math. Soc. 234(1977), 289-359.

[99] W. Fulton, *Intersection theory,* Ergebn. der Math. und ihrer Grenzgeb., 2, Springer-Verlag, Berlin, 1984.

[100] S. Gallot, D. Hulin and J. Lafontaine, *Riemannian Geometry,* Springer-Verlag, Berlin, 1990.

[101] M. Gardner, *Extraordinary nonperiodic tiling that enriches the theory of tiles.,* Scientific American (1977), 110-121.

[102] M. Gerstenhaber, *On the deformation of rings and algebras: II,* Annals of Math. 84(1966), 1-19.

[103] G. Gierz *et al, A compendium of continuous lattices,* Springer-Verlag, New York, 1980.

[104] N. Groenbaek, *Amenability of discrete convolution algebras, the commutative case,* Pacific J. Math. 143(1990), 243-249.

[105] B. Grünbaum and G. C. Shephard, *Tilings and Patterns,* W. H. Freeman, New York, 1987.

[106] A. Guichardet, *Une caractérisation des algèbres de von Neumann discrètes,* Bull. Soc. Math. France 89(1961), 77-101.

[107] K. Guruprasad, J. Huebschmann, L. Jeffrey and A. Weinstein, *Group systems, groupoids and moduli spaces of parabolic bundles,* Duke Math. J. 89(1997), 377-412.

[108] R. K. Guy, *The Penrose pieces,* Bull. London Math. Soc. 8(1976), 9-10.

[109] U. Haagerup, *All nuclear C^*-algebras are amenable,* Invent. Math. 74(1983), 305-319.

[110] A. Haefliger, *Structures feuilletées et cohomologie à valeur dans un faisceau de groupoïdes,* Comment. Math. Helv. 32(1958), 249-329.

[111] A. Haefliger, *Variétées Feuiletées,* Ann. Scuola Norm. Sup. Pisa 161(1962), 367-397.

[112] A. Haefliger, *Feuilletages sur les variétés ouvertes,* Topology 9(1970), 183-194.

[113] A. Haefliger, *Groupoïdes d'holonomie et classifiants, Structures Transverses des Feuilletages, Toulouse (1982)*, Astéristique No. 116(1984), 70-97.

[114] P. Hahn, *The regular representation of measure groupoids*, Trans. Amer. Math. Soc. 242(1978), 35-72.

[115] P. Hahn, *Haar measure for measure groupoids*, Trans. Amer. Math. Soc. 242(1978), 1-33.

[116] P. Hahn, *The σ-representations of amenable groupoids*, Rocky Mountain J. Math. 9(1979), 631-639.

[117] P. R. Halmos and L. J. Wallen, *Powers of partial isometries*, Indiana Univ. J. Math. 19(1970), 657-663.

[118] R. Hancock and I. Raeburn, *The C^*-algebras of some inverse semigroups*, Bull. Austral. Math. Soc. 42(1990), 335-348.

[119] M. Heller and W. Sasin, *Groupoid approach to noncommutative quantization of gravity*, J. Math. Phys. 38(1997), 5840-5853.

[120] E. Hewitt and K. A. Ross, *Abstract Harmonic Analysis I*, Springer-Verlag, New York, 1963.

[121] E. Hewitt and K. A. Ross, *Abstract Harmonic Analysis II*, Springer-Verlag, New York, 1970.

[122] E. Hewitt and H. S. Zuckerman, *The ℓ_1-algebra of a commutative semigroup*, Trans. Amer. Math. Soc. 83(1956), 70-97.

[123] P. Higgins, *Presentations of groupoids, with applications to groups*, Proc. Camb. Phil. Soc. 60(1964), 7-20.

[124] P. Higgins, *The fundamental groupoid of a graph of groups*, J. London Math. Soc. 13(1976), 145-149.

[125] P. Higgins, *Notes on categories and groupoids*, Vol. 32, Mathematical Studies, Van Nostrand Reinhold Company, London, 1971.

[126] P. J. Higgins and K. C. H. Mackenzie, *Fibrations and quotients of differentiable groupoids*, J. London Math. Soc. 42(1990), 101-110.

[127] P. J. Higgins and K. C. H. Mackenzie, *Algebraic constructions in the category of Lie algebroids*, J. Pure Appl. Algebra 80(1992), 237-272.

[128] P. J. Higgins and K. C. H. Mackenzie, *Duality for base-changing morphisms of vector bundles, modules, Lie algebroids and Poisson structures*, Math. Proc. Cambridge Philos. Soc. 114(1993), 471-488.

[129] N. Higson, *On the K-theory proof of the index theorem*, Contemporary Math. 148(1993), 67-86.

[130] M. Hilsum and G. Skandalis, *Stabilité des C^*-algèbres de feuilletages*, Ann. Inst. Fourier (Grenoble) 33(1983), 201-208.

[131] M. Hilsum and G. Skandalis, *Morphismes K-orientés d'espaces de feuilles et fonctorialité en théorie de Kasparov (d'après une conjecture d'A. Connes)*, Ann. Scient. École Norm. Sup. 20(1987), 325-390.

[132] P. -M. Ho and Y. -S. Wu, *Noncommutative geometry and D-branes*, Physics Letters 398(1987), 52-60.

[133] J. W. Howie, *An introduction to semigroup theory*, Academic Press, London, 1976.

[134] S. Hurder, *The Godbillon measure of amenable foliations*, J. Differential Geometry 23(1986), 347-365.

[135] S. Hurder and A. Katok, *Ergodic theory and Weil measures for foliations*, Annals of Math. 126(1987), 221-275.

[136] M. A. Jaswon and M. A. Rose, *Crystal symmetry: Theory of Colour Crystallography*, Ellis Horwood Limited, Chichester, England, 1983.

[137] B. E. Johnson, *Cohomology of Banach algebras*, Mem. Amer. Math. Soc. 127(1972).

[138] B. E. Johnson, *Approximate diagonals and cohomology of certain annihilator Banach algebras*, Amer. J. Math. 94(1972), 685-698.

[139] P. E. T. Jorgensen, L. M. Schmitt and R. F. Werner, *q-canonical commutation relations and stability of the Cuntz algebra*, Pacific J. Mathematics 165(1994), 131-151.

[140] M. V. Karasev and V. P. Maslov, *Nonlinear Poisson Brackets. Geometry and Quantization*, Translations of Mathematical Monographs, Vol. 119, American Mathematical Society, Providence, 1993.

[141] J. Kellendonk, *The Local Structure of Tilings and their Integer Group of Coinvariants*, Commun. Math. Phys. 187(1997), 155-169.

[142] J. Kellendonk, *Noncommutative geometry of tilings and gap labelling*, Reviews in Mathematical Physics, 7(1995), 1133-1180.

[143] J. Kellendonk, *Topological equivalence of tilings*, J. Math. Phys. 38(1997), 1823-1841.

[144] J. Kellendonk, *Integer groups of Coinvariants associated to Octogonal Tilings*, to appear, Commun. Math. Phys..

[145] J. L. Kelley, *General Topology*, Springer-Verlag, New York, 1985.

[146] A. A. Kirillov, *Elements of the Theory of Representations*, Springer-Verlag, New York, 1976.

[147] S. Kobayashi and K. Nomizu, *Foundations of Differential Geometry*, Vol. 1, Interscience Tracts, No. 15, John Wiley and Sons, New York, 1963.

[148] A. Kumjian, *On localizations and simple C^*-algebras*, Pacific J. Math. 112(1984), 141-192.

[149] A. Kumjian, D. Pask, I. Raeburn and J. Renault, *Graphs, groupoids and Cuntz-Krieger algebras*, J. Funct. Anal. 144(1997), 505-541.

[150] A. Kumjian, P. S. Muhly, J. N. Renault and D. P. Williams, *The Brauer Group of a Locally Compact Groupoid*, preprint, 1998 (52pp.).

[151] G. Landi, *An Introduction to noncommutative spaces and their geometries*, Lecture Notes in Physics, m. 521, Springer-Verlag, New York, 1997.

[152] N. P. Landsman, *Strict deformation quantization of a particle in external gravitational and Yang-Mills fields*, J. Geometry and Physics, 12(1993), 93-132.

[153] N. P. Landsman, *Mathematical Topics Between Classical and Quantum Mechanics*, Springer Monographs in Mathematics, Springer-Verlag, NY, 1998.

[154] S. Lang, *Differential and Riemannian Manifolds*, Graduate Texts in Mathematics, vol. 160, Springer-Verlag, New York, 1995.

[155] H. B. Lawson, *Foliations*, Bull. Amer. Math. Soc. 80(1974), 369-418.

[156] H. B. Lawson, *The qualitative theory of foliations*, CBMS Regional Conference Series in Mathematics, No. 27, American Mathematical Society, Providence, 1977.

[157] M. Lawson, *Inverse semigroups: The theory of partial symmetries*, to appear, World Scientific, 1998.

[158] P. Libermann, *Sur les groupoïdes différentiables et le "presque parallelisme"*, Symposia Math. 10(1972), 59-93.

[159] A. Lichnerowicz, *Déformations d'algèbres associées à une variété symplectique (les $*_\nu$-produits)*. Ann. Inst. Fourier (Grenoble) 32(1982), 157-209.

[160] Z-J. Liu and P. Xu, *Exact Lie bialgebroids and Poisson groupoids*, Geom. and Funct. Anal. 6(1996), 138-145.

[161] Z-J. Liu, A. Weinstein and P. Xu, *Manin triples for Lie bialgebroids*, J. Differential Geometry 54(1997), 547-574.

[162] J-H. Lu and A. Weinstein, *Groupoïdes symplectiques doubles des groupes de Lie-Poisson*, C. R. Acad. Sci. Paris Sér. I Math. 309(1989), 951-954.

[163] D. McDuff and D. Salamon, *Introduction to symplectic topology*, Oxford Mathematical Monographs, Oxford University Press, New York, 1995.

[164] K. C. H. Mackenzie, *Lie Groupoids and Lie algebroids in Differential Geometry*, London Mathematical Society Lecture Note Series, vol. 124, Cambridge University Press, Cambridge, 1987.

[165] K. C. H. Mackenzie, *Integrability obstructions for extensions of Lie algebroids*, Cahiers Topologie Géom. Différentielle Catégoriques 28(1987), 29-52.

[166] K. C. H. Mackenzie, *A note on Lie algebroids which arise from groupoid actions*, Cahiers Topologie Géom. Différentielle Catégoriques 28(1987), 283-302.

[167] K. C. H. Mackenzie, *Classification of principal bundles and Lie groupoids with prescribed gauge group bundle*, J. Pure Appl. Algebra 58(1989), 181-208.

[168] K. C. H. Mackenzie, *Lie algebroids and Lie pseudoalgebras.*, Bull. London Math. Soc. 27(1995), 97-147.

[169] K. C. H. Mackenzie and P. Xu, *Lie bialgebroids and Poisson groupoids*, Duke Math. J. 18(1994), 415-452.

[170] G. W. Mackey, *Ergodic theory and virtual groups*, Math. Ann. 166(1966), 187-207.

[171] K. Maclanahan, *K-theory for partial crossed products by discrete groups*, J. Functional Analysis 130(1995), 118-130.

[172] G. Maltsiniotis, *Groupoïdes quantiques*, C. R. Acad. Sci. Paris 314(1992), 249-252.

[173] M. H. Mann, I. Raeburn and C. E. Sutherland, *Representations of finite groups and Cuntz-Krieger algebras*, Bull. Australian Math. Soc., 46(1992), 225-243.

[174] K. Mikami and A. Weinstein, *Moments and reduction for symplectic groupoids*, Publ. Res. Inst. Math. Sci. 24(1988), 121-140.

[175] I. Moerdijk, *On the weak homotopy type of étale groupoids.*, Integrable systems and foliations (Montpellier, 1995), 147-156, Progress in Mathematics, Vol. 145, Birkhäuser Boston, Boston, MA, 1997.

[176] I. Moerdijk and D. A. Pronk, *Orbifolds, sheaves and groupoids*, K-theory, 12(1997), 3-21.

[177] P. Molino, *Riemannian Foliations*, Progress in Mathematics, Vol. 73, Birkhäuser, Boston, 1988.

[178] C. C. Moore and C. Schochet, *Global Analysis on foliated spaces*, Mathematical Sciences Research Institute Publications, Vol. 9, Springer-Verlag, New York, 1988.

[179] P. S. Muhly, *Coordinates in Operator Algebra*, to appear, CBMS Regional Conference Series in Mathematics, American Mathematical Society, Providence, 180pp..

[180] P. S. Muhly and J. N. Renault, *C*-algebras of multivariable Wiener-Hopf operators*, Trans. Amer. Math. Soc. 274(1982), 1-44.

[181] P. S. Muhly and D. P. Williams, *Transformation group C*-algebras with continuous trace. II.*, J. Operator Theory 11(1984), 109-124.

[182] P. S. Muhly and D. P. Williams, *Continuous trace groupoid C*-algebras*, Math. Scand. 70(1992), 127-145.

[183] P. S. Muhly and D. P. Williams, *Groupoid cohomology and the Dixmier-Douady class.* Proc. London Math. Soc. 71(1995), 109-134.

[184] P. S. Muhly, J. N. Renault and D. P. Williams, *Equivalence and isomorphism for groupoid C*-algebras*, J. Operator Theory, 17(1987), 3-22.

[185] P. S. Muhly, J. N. Renault and D. P. Williams, *Continuous-trace groupoid C*-algebras. III.*, Trans. Amer. Math. Soc. 348(1996), 3621-3641.

[186] G. J. Murphy, *C*-algebras and operator theory*, Academic Press, Boston, 1990.

[187] A. Nica, *C*-algebras generated by isometries and Wiener-Hopf operators*, J. Operator Theory 27(1992), 17-52.

[188] A. Nica, *On a groupoid construction for actions of certain inverse semigroups*, International J. Math. 5(1994), 349-372.

[189] A. Nica, *Inverse semigroups and a Hulanicki-type theorem for partially ordered discrete groups*, preprint, 1993.

[190] A. Nijenhuis, *Theory of the Geometric Object*, Ph. D. Thesis, University of Amsterdam, 1952.

[191] V. Nistor, A. Weinstein and P. Xu, *Pseudodifferential operators on differential groupoids*, preprint, 1997.

[192] A. Ocneanu, *Path algebras*, Lecture Notes, American Mathematical Society meeting, Santa Cruz, 1986.

[193] A. Ocneanu, *Quantized groups, string algebras and Galois theory for algebras*, Operator Algebras and applications, Vol. 2, (ed. D. Evans and M. Takesaki), London Mathematical Society Lecture Notes 136(1988), 119-172.

[194] D. A. Pask and C. E. Sutherland, *Filtered inclusions of path algebras; a combinatorial approach to Doplicher-Roberts duality*, J. Operator Theory 31(1994), 99-121.

[195] A. L. T. Paterson, *Weak containment and Clifford semigroups*, Proc. Roy. Soc. Edinburgh A 81(1978), 223-30.

[196] A. L. T. Paterson, *Amenability*, Mathematical Surveys and Monographs, No. 29, American Mathematical Society, Providence, R. I., 1988.

[197] A. L. T. Paterson, *Inverse semigroups, groupoids and a problem of J. Renault* in *Algebraic methods in Operator Theory*, ed. R. Curto and P. E. T. Jorgensen, Birkhäuser, Boston, 1993, 11 pp..

[198] A. L. T. Paterson, *Smooth groupoids, left Haar systems and the Connes-Higson asymptotic morphism*, talk given at *GroupoidFest 3*, Berkeley, 1997.

[199] A. L. T. Paterson, *The analytic index of pseudodifferential operators on smooth groupoids*, preprint, 1998.

[200] G. Pedersen, *Analysis Now*, Graduate Texts in Mathematics, Vol. 118, Springer-Verlag, New York, 1988.

[201] R. Penrose, *The Role of Aesthetics in Pure and Applied Mathematical Research*, Bull. Inst. Math. and its Appl. 10(1974), 266-71. (pp. 653-655 of *P. J. Steinhardt and S. Ostlund, infra.*)

[202] M. Petrich, *Inverse semigroups*, John Wiley and Sons, New York, 1984.

[203] J. Phillips, *The holonomic imperative and the homotopy groupoid of a foliated manifold*, Rocky Mountain J. Math. 17(1987), 151-165.

[204] J. -P. Pier, *Amenable Locally Compact Groups*, Wiley, New York, 1984.

[205] J. -P. Pier, *Amenable Banach algebras*, Pitman research notes in mathematics, Longman, Harlow, England, 1988.

[206] M. Pimsner, *A class of C^*-algebras generalizing both Cuntz-Krieger algebras and crossed products by* $\mathbf{Z}$, preprint, 1993.

[207] J. F. Plante, *Foliations with measure preserving holonomy*, Annals of Mathematics 102(1975), 327-361.

[208] J. Pradines, *Théorie de Lie pour les groupoïdes différentiables. Relations entre propriétés locales et globales.* C. R. Acad. Sci. Paris Sér. A-B 263(1966), A907-A910.

[209] J. Pradines, *Théorie de Lie pour les groupoïdes différentiables. Calcul différential dans la catégorie des groupoïdes infinitésimaux*, C. R. Acad. Sci. Paris Sér. A-B 264(1967), A 245-A 248.

[210] J. Pradines, *Géométrie différentielle au-dessus d'un groupoïde*, C. R. Acad. Sci. Paris Sér. A-B 266(1968), A 1194-A 1196.

[211] J. Pradines, *Troisième théorème de Lie pour les groupoïdes différentiables.*, C. R. Acad. Sci. Paris Sér. A-B 267(1968), A 21-A 23.

[212] J. Pradines, *Feuilletages: holonomie et graphes locaux.*, C. R. Acad. Sci. Paris Sér. I Math. 298(1984), 297-300.

[213] J. Pradines, *Graph and holonomy of singular foliations*, Differential geometry (Santiago de Compostela, 1984), 215-219, Res. Notes in Math. 131, Pitman, Boston, MA, 1985.

[214] J. Pradines, *Quotients de groupoïdes différentiables.* C. R. Acad. Sci. Paris Sér. I Math. 303(1986), 817-820.

[215] J. Pradines, *Remarque sur le groupoïde cotangent de Weinstein-Dazord*, C. R. Acad. Sci. Paris Sér. I Math. 306(1988), 557-560.

[216] J. Pradines and B. Bigonnet, *Graphe d'un feuilletage singulier*, C. R. Acad. Sci. Paris Sér. I Math. 300(1985), 439-442.

[217] G. B. Preston, *Monogenic inverse semigroups*, J. Australian Math. Soc. 40(1986), 321-342.

[218] J. Quigg and I. Raeburn, *Landstad duality for partial actions*, preprint, 1994.

[219] A. Ramsay, *Virtual groups and group actions*, Advances in Mathematics 6(1971), 253-322.

[220] A. Ramsay, *Topologies for measured groupoids*, J. Functional Analysis, 47(1982), 314-343.

[221] A. Ramsay, *The Mackey-Glimm dichotomy for foliations and other Polish groupoids*, J. Funct. Anal. 94(1990), 358-374.

[222] A. Ramsay, *Lacunary sections for locally compact groupoids*, Ergodic Theory Dynam. Systems 17(1997), 933-940.

[223] A. Ramsay and M. E. Walter, *Fourier-Stieltjes algebras of locally compact groupoids*, Selfadjoint and nonselfadjoint operator algebras and operator theory (Fort Worth, TX, 1990), 143-156. Contemp. Math. 120, Amer. Math. Soc., Providence, RI, 1991.

[224] A. Ramsay and M. E. Walter, *Noncommutative harmonic analysis on groupoids.*, Topics in operator theory, operator algebras and application (1994), 251-264, Rom. Acad., Bucharest, 1995.

[225] A. Ramsay and M. E. Walter, *Fourier-Stieltjes algebras of locally compact groupoids*, J. Funct. Anal. 148(1997), 314-367.

[226] A. Razah Salleh, *Union theorems for groupoids and double groupoids*, University of Wales, Ph. D. Thesis, Bangor, 1976.

[227] A. Razak Salleh and J. Taylor, *On the relation between the fundamental groupoids of the classifying space and the nerve of a local cover*, J. Pure Appl. Alg. 37(1985), 81-93.

[228] G. Reeb, *Sur certaines propriétés topologiques des variétés feuilletés*, Act. Sci. et Ind., Hermann, Paris, 1952.

[229] B. L. Reinhart, *Differential Geometry of Foliations*, Ergebnisse der Mathematik und ihrer Grenzgebiete, Vol. 99, Springer-Verlag, Berlin, 1983.

[230] J. N. Renault, *A groupoid approach to C*-algebras*, Lecture Notes in Mathematics, Vol. 793, Springer-Verlag, New York, 1980.

[231] J. N. Renault, *C*-algebras of groupoids and foliations*, Proc. Sympos. Pure Math., 38(1982), 339-350.

[232] J. N. Renault, *Two applications of the dual groupoid of a C*-algebra*, Lecture Notes in Mathematics, 1132(434-445), Springer-Verlag, New York, 1985.

[233] J. N. Renault, *Répresentation de produits croisés d'algèbres de groupoïdes*, J. Operator Theory, 18(1987), 67-97.

[234] J. N. Renault, *The ideal structure of groupoid crossed product C*-algebras.* (with an appendix by G. Skandalis), J. Operator Theory 25(1991), 3-36.

[235] J. N. Renault, *Multiplicateurs de Fourier et fonctions de Littlewood pour les groupoïdes r-discrets*, C. R. Acad Sci. Paris Sér. I Math. 319(1994), 15-19.

[236] J. N. Renault, *The Fourier algebra of a measured groupoid and its multipliers*, J. Funct. Anal. 145(1997), 455-490.

[237] J. N. Renault and B. Moran, *Ideal structure of groupoid crossed product C*-algebras*, Miniconferences on harmonic analysis and operator algebras (Canberra, 1987), 267-268, Proc. Centre Math. Anal. Austral. Nat. Univ. 16, Austral. Nat. Univ. Canberra, 1988.

[238] M. A. Rieffel, *Induced representations of C*-algebras*, Advances in Mathematics 13(1974), 176-257.

[239] M. A. Rieffel, *Questions on Quantization*, manuscript prepared for the Dartmouth Workshop on E-theory, Quantization and Deformations, September, 1997.

[240] R. M. Robinson, *Comments on the Penrose tiles.* Mimeographed notes, September, 1975.

[241] J. Roe, *Finite propagation speed and Connes's foliation algebra*, Math. Proc. Cambridge Phil. Soc. 102(1987), 459-466.

[242] H. L. Royden, *Real Analysis*, Macmillan Publishing Company, New York, 1988.

[243] W. Rudin, *Functional Analysis*, Tata McGraw-Hill, New Delhi, 1974.

[244] H. N. Salas, *Semigroups of isometries with commuting range projections*, J. Operator Theory, 14(1985), 311-346.

[245] J. Schwartz, *Two finite, non-hyperfinite, non-isomorphic factors*, Comm. Pure Appl. Math. 16(1963), 19-26.

[246] A. K. Seda, *A continuity property of Haar systems of measures*, Ann. Soc. Sci. Bruxelles Sér. I 89(1975), 429-433.

[247] A. K. Seda, *An extension theorem for transformation groupoids*, Proc. Roy. Irish Acad. Sect. A 75(1975), 255-262.

[248] A. K. Seda, *On compact transformation groupoids*, Cahiers Topologie Géom. Différentielle 16(1975), 409-414.

[249] A. K. Seda, *Haar measures for groupoids*, Proc. Roy. Irish Acad. Sect. A 76(1976), 25-36.

[250] A. K. Seda, *Quelques résultats dans la catégorie des groupoïdes d'opérateurs*, C. R. Acad. Sci. Paris Sér. A-B 288(1979), A21-A24.

[251] A. K. Seda, *Transformation groupoids and bundles of Banach spaces*, Cahiers Topologie Géom. Différentielle 19(1978), 131-146.

[252] A. K. Seda, *On the continuity of Haar measure on topological groupoids*, Proc. Amer. Math. Soc. 96(1986), 115-120.

[253] M. Senechal, *Quasicrystals and Geometry*, Cambridge University Press, Cambridge, 1995.

[254] A. J. L. Sheu, *Groupoid and compact quantum groups*, in *Quantum groups and quantum spaces*, Warsaw, 1995, 41-50, Banach Center Publ., 40, 1997.

[255] A. J. L. Sheu, *Compact quantum groups and groupoid C^*-algebras*, J. Funct. Anal. 144(1997), 371-393.

[256] N. Sieben, *C^*-crossed products by partial actions and actions of inverse semigroups*, J. Australian Math. Soc. 63(1997), 32-46.

[257] N. Sieben, *C^*-crossed products by twisted inverse semigroup actions*, J. Operator Theory 39(1998), 361-394.

[258] N. Sieben, *Morita equivalence of C^*-crossed products by inverse semigroup actions*, preprint, 1997.

[259] I. Tamura, *Topology of Foliation: An Introduction*, Translations of Mathematical Monographs, Vol. 97, American Mathematical Society, Providence, 1992.

[260] P. J. Steinhardt and S. Ostlund, *The Physics of Quasicrystals*, World Scientific, New Jersey, 1987.

[261] J. Taylor, *Quotients of groupoids by the action of a group*, Math. Proc. Camb. Phil. Soc. 103(1988), 239-249.

[262] R. Thom, *Généralisation de la Théorie de Morse aux Variétés Feuilletées*, Ann. Inst. Fourier Grenoble 14(1964), 173-190.

[263] I. Vaismann, *Lectures on the Geometry of Poisson Manifolds*, Progress in Mathematics, Vol. 118, Birkhäuser, Boston, 1994.

[264] O. Veblen and J. H. C. Whitehead, *The Foundations of Differential Geometry*, Cambridge, 1932.

[265] A. Weinstein, *A universal phase space for particles in Yang-Mills fields*, Lett. Math. Phys. 2(1978), 417-420.

[266] A. Weinstein, *Symplectic geometry*, Bull. Amer. Math. Soc. 5(1981), 1-13.

[267] A. Weinstein, *Symplectic groupoids and Poisson manifolds*, Bull. Amer. Math. Soc. 16(1987), 101-104.

[268] A. Weinstein, *Coisotropic calculus and Poisson groupoids*, J. Math. Soc. Japan 40(1988), 705-727.

[269] A. Weinstein, *Blowing up realizations of Heisenberg-Poisson manifolds*, Bull. Sci. Math. 113(1989), 381-406.

[270] A. Weinstein, *Noncommutative geometry and geometric quantization*, Symplectic geometry and mathematical physics (Aix-en-Provence, 1990), 446-461, Progr. Math., 99, Birkhäuser Boston, MA, 1991.

[271] A. Weinstein, *Symplectic groupoids, geometric quantization and irrational rotation algebras*, pp. 281-290 of *P. Dazord and A. Weinstein, supra.*

[272] A. Weinstein, *Lagrangian mechanics and groupoids*, Mechanics day (Waterloo, ON, 1992), 207-231, Fields Inst. Commun. 7, American Mathematical Soc. RI, 1996.

[273] A. Weinstein, *Groupoids: Unifying Internal and External Symmetry. A tour through some examples.*, Notices Amer. Math. Soc. 43(1996), 744-752.

[274] A. Weinstein, *Poisson Geometry*, Diff. Geom. Appl., 9 (1998), 213-238.

[275] A. Weinstein, *Tangential deformation quantization and polarized symplectic groupoids*, in *Deformation and symplectic geometry (Ascona, 1996)*, 301-314, Math. Phys. Stud. 20, Kluwer Acad. Publ., Dordrecht, 1997.

[276] A. Weinstein and P. Xu, *Extensions of symplectic groupoids and quantization*, J. Reine Angew. Math. 417(1991), 159-189.

[277] J. J. Westman, *Harmonic analysis on groupoids*, Pacific J. Math. 27(1968), 621-632.

[278] J. J. Westman, *Cohomology for ergodic groupoids*, Trans. Amer. Math. Soc. 146(1969), 465-471.

[279] R. O. Wells, *Differential Analysis on Complex Manifolds*, Graduate Texts in Mathematics, 65, Springer-Verlag, New York, 1986.

[280] H. E. Winkelnkemper, *The graph of a foliation*, Ann. Global Analysis and Geometry 1(1983), 51-75.

[281] J. R. Wordingham, *The left regular *-representation of an inverse semigroup*, Proc. Amer. Math. Soc. 86(1982), 55-58.

[282] P. Xu, *Morita equivalent symplectic groupoids*, pp. 291-311 of *P. Dazord and A. Weinstein, supra.*

[283] P. Xu, *Symplectic groupoids of reduced Poisson spaces*, C. R. Acad. Sci. Paris Sér. I Math. 314(1992), 457-461.

[284] P. Xu, *On Poisson groupoids.*, Internat. J. Math. 6(1995), 101-124.

[285] P. Xu, *Deformation Quantization and Quantum Groupoids*, RIMS, Kyoto University, 1997.

[286] P. Xu, *Flux homomorphism on symplectic groupoids.*, Math. Z. 226(1997), 575-597.

Index of Terms

Index of Symbols